Contesting Earth's History in Transatlantic Literary Culture, 1860–1935

Contesting Earth's History in Transatlantic Literary Culture, 1860–1935

Believers and Visionaries on the Borderlines of Geology and Palaeontology

RICHARD FALLON

Great Clarendon Street, Oxford, OX2 6DP,
United Kingdom

Oxford University Press is a department of the University of Oxford.
It furthers the University's objective of excellence in research, scholarship,
and education by publishing worldwide. Oxford is a registered trade mark of
Oxford University Press in the UK and in certain other countries

© Richard Fallon 2025

The moral rights of the author have been asserted.

All rights reserved. No part of this publication may be reproduced, stored in a retrieval system, transmitted, used for text and data mining, or used for training artificial intelligence, in any form or by any means, without the prior permission in writing of Oxford University Press, or as expressly permitted by law, by licence or under terms agreed with the appropriate reprographics rights organization. Enquiries concerning reproduction outside the scope of the above should be sent to the Rights Department, Oxford University Press, at the address above.

You must not circulate this work in any other form
and you must impose this same condition on any acquirer.

Published in the United States of America by Oxford University Press
198 Madison Avenue, New York, NY 10016, United States of America

British Library Cataloguing in Publication Data

Data available

Library of Congress Control Number: 2025930994

ISBN 9780198926160

DOI: 10.1093/9780198926191.001.0001

Printed and bound by
CPI Group (UK) Ltd, Croydon, CR0 4YY

Links to third party websites are provided by Oxford in good faith and
for information only. Oxford disclaims any responsibility for the materials
contained in any third party website referenced in this work.

The manufacturer's authorised representative in the EU for product safety is
Oxford University Press España S.A. of El Parque Empresarial San Fernando de Henares,
Avenida de Castilla, 2 – 28830 Madrid (www.oup.es/en or
product.safety@oup.com). OUP España S.A. also acts as importer into Spain
of products made by the manufacturer.

Acknowledgements

The researching and writing of this book were made possible thanks to a Leverhulme Trust Early Career Fellowship at the University of Birmingham, with all the generous funding that such a fellowship entails. During my time at Birmingham, I benefited immensely from being part of the scholarly community of its Nineteenth-Century Centre and Lit/Sci Lab. Receiving a subsequent honorary fellowship was also integral in allowing me to complete my revisions in a timely manner. More generally, the book's content has been strengthened through being aired at papers before annual conferences of the British Association for the History of Science, the British Association for Victorian Studies, the British Society for Literature and Science, the Society for the History of Natural History, and the International Research Network for the Study of Science and Belief in Society, as well as in talks to the brilliant 'Popularizing Palaeontology' network, which is organized by Chris Manias at King's College London.

I am grateful for having had the opportunity to discuss my research at greater length following the delivery of longer papers to the Centre for Nineteenth-Century Studies International, the University of Bristol Gothic Reading Group in collaboration with University of Birmingham GOTHICA, the University of Stirling's English seminar, the University of Buckingham's English postgraduate seminar, the Virtual HistSTM group, and the University of Lincoln's Nineteenth-Century Research seminar, and to have had a sample chapter examined during a History of Science research meeting of the Vrije Universiteit Amsterdam. The value of conversations like these cannot be overstated.

I would, moreover, like to thank Margaret Conway, Gowan Dawson, Fern Elsdon-Baker, Christine Ferguson, Diarmid Finnegan, Abraham C. Flipse, Billie Gavurin, Tyler Greenfield, Edward Guimont, Adele Guyton, John Holmes, Ralph O'Connor, Clare Stainthorp, Will Tattersdill, and two anonymous readers for helpful discussions, advice on particular points, and more general support. A Small Grant from the British Society for Literature and Science enabled me to hold a symposium on 'Nonfictional Fiction, 1820–1920' that helped clarify some of my thinking at a late stage in the project, while the Natural History Museum kindly assisted with some of the costs of reproducing images. An earlier version of sections of chapter 2 appeared as 'Seen through Deep Time: Occult Clairvoyance and Palaeoscientific Imagination', *Journal of Victorian Culture*, 28 (2023), 143–62. For the opportunity to revisit it, I thank Oxford University Press.

For access to rare or manuscript material and for help with other archival and bibliographical enquiries, I thank the British Library, Marlene Rothermel of the Andrews University Center for Adventist Research, Jonathan Clatworthy of the Charles Lapworth Archive Collection, Caroline Lam of the Geological Society of London Archives, Alexandra Foulds of Gladstone's Library, Chelsi Cannon of Loma Linda University Libraries, Christine Jankowski of the Lloyd Library and Museum, Adria Seccareccia of McGill University Archives, Justin Gwinnett of the University of Birmingham Main Library, Danielle Spittle of the University of Edinburgh Centre for Research Collections, Thomas Chisholm of the University of Liverpool Library, Delinda Buie and Chad Kamen of the University of Louisville Ekstrom Library, and Taylor Kalloch and Alden Ludlow of Wellesley Historical Society. All material from the Natural History Museum Archives is used here by permission of the Trustees of the Natural History Museum, with thanks to its archivists. Thank you also to Cathy Wilbanks of Edgar Rice Burroughs, Inc., for the permission to reproduce a striking illustration by John Coleman Burroughs. How, moreover, would research like this be possible without access to digitized resources like those available on the Biodiversity Heritage Library, the Internet Archive, and HathiTrust?

As always, I am indebted to the love and patience of my wife, Jennifer. I dedicate this book to our daughter, Imogen, who is too young to object to being the dedicated subject of a monograph about occult time travel, scriptural deluges, and secret underground worlds.

Contents

List of Figures	viii
Introduction	1
1. Envisioning the Days of Genesis	27
2. Deep Clairvoyance	64
3. Hollow Ground	106
4. Submerged in the Public Sphere	146
5. Geohistory, Unimagined	185
Epilogue: To See and To Make Others See	227
Bibliography	240
Index	259

List of Figures

Figure 1.1 John William Grover, *Conversations with Little Geologists on the Six Days of Creation, Illustrated with a Geological Chart* (London: Edward Stanford, 1878). © The British Library Board, 7202.i.1.   28

Figure 1.2 Louis Figuier, *The World before the Deluge*, rev. edn, ed. by Henry W. Bristow (London: Chapman and Hall, 1867), 309. Cadbury Research Library: Special Collections, University of Birmingham.   39

Figure 1.3 (A) John William Dawson, *The Story of the Earth and Man*, 2nd rev. edn (London: Hodder and Stoughton, 1873), 219. (B) John William Dawson, *Nature and the Bible* (New York: Robert Carter and Brothers, 1875), facing 123.   45

Figure 1.4 Samuel Kinns, *Moses and Geology; or, The Harmony of the Bible with Science*, rev. edn (London: Cassell, 1887), facing 281.   49

Figure 1.5 David L. Holbrook, *The Panorama of Creation as Presented in Genesis Considered in Its Relation with the Autographic Record as Deciphered by Scientists* (Philadelphia, PA: Sunday School Times, [1908]), 58.   61

Figure 2.1 [George Winslow Plummer], 'Khei X', *Rosicrucian Fundamentals: An Exposition of the Rosicrucian Synthesis of Religion, Science and Philosophy in Fourteen Complete Instructions* (New York: Flame Press, 1920), 23.   70

Figure 2.2 Photograph of Sherman Foote Denton, n.d., Photograph Collection, Denton Family Papers, Wellesley Historical Society, Wellesley, MA. Courtesy of Wellesley Historical Society, MA.   82

Figure 2.3 (A and B) William Denton, *The Soul of Things; or, Psychometric Researches and Discoveries*, vol. III (Boston: William Denton, 1874), 79, 120. (C) Untitled sketchbook, Sherman F. Denton Papers & Artwork, Box 1, Folder 9, Denton Family Papers.   83

Figure 2.4 (A) Untitled sketchbook, Sherman F. Denton Papers & Artwork, Box 1, Folder 9. Courtesy of Wellesley Historical Society, MA. (B) Benjamin Waterhouse Hawkins's widely imitated design for *Plesiosaurus*. Photograph by Chris Sampson, CC BY 2.0.   86

Figure 2.5 (A) Charles Darwin, *On the Origin of Species by Means of Natural Selection, or, The Preservation of Favoured Races in the Struggle for Life* (London: John Murray, 1859). (B) Curuppumullage Jinarajadasa, *First Principles of Theosophy*, 9th edn (Madras [Chennai]: Theosophical Publishing House, 1951), 116. Reproduced courtesy of the Main Library, University of Birmingham.   97

LIST OF FIGURES   ix

Figure 2.6   Howard V. Brown, front cover of *Astounding Stories*, 17 (1936).   103

Figure 3.1   John Cleves Symmes Jr's signed circular declaring the Earth to be hollow and filled with concentric spheres, distributed across the world in 1818.   107

Figure 3.2   (A) 'Adam Seaborn', *Symzonia: A Voyage of Discovery* (New York: J. Seymour, 1820), frontispiece. (B) William R. Bradshaw, *The Goddess of Atvatabar; Being the History of the Discovery of the Interior World and Conquest of Atvatabar* (New York: J. F. Douthitt, 1892), frontispiece.   113

Figure 3.3   (A) Americus Symmes, *The Symmes Theory of Concentric Spheres, Demonstrating that the Earth Is Hollow, Habitable Within, and Widely Open about the Poles* (Louisville, KY: Bardley & Gilbert, 1878), frontispiece © The British Library Board, 8561.cc.5. (B) William Reed, *The Phantom of the Poles* (New York: Walter S. Rockey, 1906), frontispiece. (C) Marshall B. Gardner, *A Journey to the Earth's Interior; or, Have the Poles Really Been Discovered* (Aurora, IL: Marshall B. Gardner, 1913), frontispiece.   121

Figure 3.4   John Uri Lloyd, c.1915–1920, George Grantham Bain Collection, LC-B2-5190-5, Library of Congress Prints and Photographs Division, Washington, D.C.   132

Figure 3.5   Edgar Rice Burroughs, *Back to the Stone Age* (Tarzana, CA: Edgar Rice Burroughs Incorporated, 1937), facing p. 114. John Coleman Burroughs Pellucidar art copyright © 1937 Edgar Rice Burroughs, Inc. All Rights Reserved. Trademarks Tarzan®, Pellucidar®, At the Earth's Core™, Edgar Rice Burroughs®. Owned by Edgar Rice Burroughs, Inc. Used by Permission.   143

Figure 4.1   Jules Verne, *Vingt mille lieues sous les mers* (Paris: Hetzel, [1872]), 297.   153

Figure 4.2   Undated photograph of a characteristically resolute Henry Hoyle Howorth, wearing ceremonial sword. Letters and Autographs of Zoologists with Biographies and Portraits, vol. 4, Ellen S. Woodward Collection, Blacker Wood Collection, McGill University Archives, Montreal.   161

Figure 4.3   Lewis Spence, *The Problem of Atlantis* (London: William Rider & Son, 1924), facing 198. Copy in author's collection.   170

Figure 4.4   (A) Pertinent newspaper clippings, dated 1925 and 1926, attached to a copy of Spence, *The Problem of Atlantis*. (B) At the end of a chapter on 'The Evidence from Geology', the book's anonymous annotator complains that most of the scientific references are 'very old', pointing to the need for 'absolutely the latest scientific opinion' on this contested topic. Copy in author's collection.   172

Figure 4.5   James Churchward, *The Lost Continent of Mu: The Motherland of Man* (New York: William Edwin Rudge, 1926), 42.   176

List of Figures

Figure 5.1 College of Medical Evangelists Library, c.1912. Loma Linda University Photo Archive, LLU00691, Department of Archives and Special Collections, Loma Linda University, Loma Linda, California. — 187

Figure 5.2 George McCready Price, *The New Geology: A Textbook for Colleges, Normal Schools, and Training Schools; and for the General Reader* (Mountain View, CA: Pacific Press Publishing Association, 1923), 398. — 204

Figure 5.3 Harry Rimmer, *Monkeyshines: Fakes, Fables, Facts Concerning Evolution* (Los Angeles, CA: Research Science Bureau, 1926), 26–27. Copy in author's collection. — 206

Figure 5.4 Price, *The New Geology*, 260. — 210

Figure 5.5 (A) William Buckland, *Geology and Mineralogy Considered with Reference to Natural Theology*, 2 vols (London: William Pickering, 1836), fold-out chart. © The British Library Board, W5/7294.
(B) *Bible Readings for the Home Circle* (Battle Creek, MI: Review and Herald Publishing Company, 1889), 428. — 211

Figure 5.6 Price, *The New Geology*, 694. — 216

Figure 5.7 Price, *God's Two Books; Or Plain Facts about Evolution, Geology, and the Bible* (South Bend, IN: Review and Herald Publishing, 1911), 154–55. — 218

Introduction

How much do we really know about what took place in the prehistoric past? For Charles Gould, writing in his book *Mythical Monsters* (1886), even experts on this subject have cause for humility:

> Does the written history of man, comprising a few thousand years, embrace the whole course of his intelligent existence? or have we in the long mythical eras, extending over hundreds of thousands of years and recorded in the chronologies of Chaldæa and of China, shadowy mementoes of pre-historic man, handed down by tradition, and perhaps transported by a few survivors to existing lands from others which, like the fabled (?) Atlantis of Plato, may have been submerged, or the scene of some great catastrophe which destroyed them with all their civilization?[1]

As indicated by Gould's provocative question mark, suggesting that the famous continent of Atlantis, sunken along with all its once-exalted human inhabitants, was not wholly incredible, even far-fetched myths were to be taken seriously as potential sources of evidence about the prehistoric world. If what those myths contained went against scientific orthodoxy, that was no reason to dismiss them offhand. After all, he argued, were legendary creatures like the dragon, unicorn, and phoenix so much more incredible than palaeontologists' latest discoveries? Gould, an English-born member of the Royal Society of Tasmania and accomplished geological surveyor of that country, advised his readers to expect the unexpected from the results of future research. 'Can we suppose', he wondered, 'that we have at all exhausted the great museum of nature? Have we, in fact, penetrated yet beyond its ante-chambers?'[2]

For Gould's scientific contemporaries, the broad shape of deep history could seem so set in stone that his intimations of radical surprises in store would have been jarring. Geology and palaeontology had only been born in recognizable form during the late eighteenth century and yet, by the mid-nineteenth, Western scientific savants had already sketched a vast and vivid 'geohistorical' outline of Earth's

---

[1] Charles Gould, *Mythical Monsters* (London: W. H. Allen, 1886), 19.
[2] Gould, 19.

changing surface and inhabitants.[3] Its landmarks remain familiar, spanning from the planet's molten origins to the consecutive 'ages' of invertebrates, plants, fish, amphibians, reptiles, mammals, and humans. The reconstruction of Earth's history from fossils and rocks provoked widespread awe at the power of science—a rational power that, as the adage went, put the magic spells of fairy tales to shame, not to mention its ability to reveal the masterful creative handiwork of God Himself. The deep past's gloom seemed to retreat before the illuminating beams of palaeoscience.[4]

By the mid-century decades, moreover, factions of the scientific community were starting to regard evolution as geohistory's principal plot, linking the earliest single-celled organisms to the noble *Homo sapiens*. The result was what has come to be called the 'evolutionary epic': a potent modern myth in which human civilization, an insignificant speck in chronological terms, stands at the climax of an eventful cosmic history. This epic narrative, filled with exciting ingredients like dinosaurs and cavemen, and characterized by biological descent and soaring upward progress, was circulated by diverse practitioners to wide audiences, and in media of all kinds, during the second half of the nineteenth century and beyond.[5]

Notwithstanding the evolutionary epic's mass appeal in the late nineteenth and early twentieth centuries—and the flattering position it gave to humanity as both apex of Earth's history and the sole species able to reconstruct that history—there were many, like Charles Gould, who found it incomplete, or just plain wrong. The harshest critics expressed dissatisfaction with a cosmic narrative that was, they suggested, written with disturbing exclusivity by a cliquish scientific elite, in stark contrast with the more participatory research culture of earlier eras; even worse, critics averred, this elite allowed little room in their narrative for religious meaning, or for the possible veracity of cherished legends like that of Atlantis. Independent-minded thinkers proposed to correct alleged misrepresentations of geohistory through their own works. Most did so from the disreputable fringes of the scientific community, but a few spoke from its top echelons. They are the subject of this book, which combines methods from literary studies and the history of science to explore the form and content of their fascinating claims about Earth's deep history. In so doing, my book brings fresh insights to a rich scholarship on the literature and culture of palaeoscience and on idiosyncratic, heterodox,

---

[3] Martin J. S. Rudwick, *Worlds Before Adam: The Reconstruction of Geohistory in the Age of Reform* (Chicago: University of Chicago Press, 2008). I use 'geohistory' in the sense popularized by Rudwick, referring to the long narrative of geological history and the unfolding of life in deep time.

[4] I model my book's employment of this helpfully concise term on its use in Sumathi Ramaswamy, *The Lost Land of Lemuria: Fabulous Geographies, Catastrophic Histories* (Berkeley: University of California Press, 2004), where it appears as 'paleo-science'.

[5] For example, see Ralph O'Connor, 'From the Epic of Earth History to the Evolutionary Epic in Nineteenth-Century Britain', *Journal of Victorian Culture*, 14 (2009), 207–23; and Bernard Lightman and Bennett Zon, eds., *Evolution and Victorian Culture* (Cambridge: Cambridge University Press, 2014).

or otherwise controversial attempts to contribute to palaeoscientific knowledge. While previous scholars who have taken these attempts seriously have typically discussed subjects like creationism and Atlantis separately, I bring these schools of thought under one roof for the first time, shining new light on their contrasts and cross-pollinations.

Focusing on British and North American science from the 1860s to the interwar decades, *Contesting Earth's History* examines five of the era's most important 'borderline' approaches to palaeoscience, existing on the edges of, or just beyond, scientific respectability: old-earth Christian creationism, occult visionary time travel, hollow-earth theory, sunken-continent catastrophism, and young-earth Christian creationism.[6] To better understand these approaches, I range through varied and often remarkable sources, including a palaeontological epic poem, works of pioneering science fiction, mind-bending hollow-earth novels, monographs investigating Atlantis, self-published palaeoanthropological pamphlets, and a 'Flood Geology' textbook. Some were written by famous names in literature, science, and religion, but many are the work of fascinating and hitherto almost entirely overlooked figures. I also draw upon a range of revealing archival recourses, from 1870s sketchbooks of psychic research housed in Wellesley Historical Society, MA, to letters on young-earth geology sent to the Natural History Museum, London, during the Christian fundamentalist controversies of the 1920s.[7]

By bringing these sources together, we see that their authors were united by the desire to give humans additional agency, whether as more important subjects *in* the geohistorical story or as more resourceful authors and investigators *of* it, than most reputable palaeoscientific researchers allowed. They were not satisfied with inhuman or merely provisional narratives of deep time. As such, they sought firmer records of the prehistoric past than rocks and fossils could provide, gesturing to proofs from sacred texts, myths, and even accounts penned by contemporary visionaries. Some rejected geohistory entirely, showing how obscurantist savants had overcomplicated a subject that could more accurately be deduced by any observer with common sense, whether with reference to basic physics, or the Bible, or both. Across these works, we see strange and fabulous conceptions

---

[6] For old-earth creationism, see James R. Moore, 'Geologists and Interpreters of Genesis in the Nineteenth Century', in *God and Nature: Historical Essays on the Encounter between Christianity and Science*, ed. by David C. Lindberg and Ronald L. Numbers (Berkeley: University of California Press, 1986), 322–50. For young-earth creationism, see Ronald L. Numbers, *The Creationists: From Scientific Creationism to Intelligent Design*, rev. edn (Cambridge, MA: Harvard University Press, 2006). For hollow-earth theory, see Walter Kafton-Minkel, *Subterranean Worlds: 100,000 Years of Dragons, Dwarfs, the Dead, Lost Races & UFOs from Inside the Earth* (Port Townsend, WA: Loompanics Unlimited, 1989). For occult visionary time travel, see Wouter J. Hanegraaff, 'The Theosophical Imagination', *Correspondences*, 5 (2017), 3–39. For sunken-continent catastrophism, see Ramaswamy.

[7] In this book, I follow standard usage in capitalizing the name of our planet, 'the Earth', but I also follow most of the scholars I cite in rendering the terms 'young-earth creationist' and 'hollow-earth theory' in lower-case. Modern fully capitalized and non-hyphenated versions like 'Young Earth Creationist' and 'Hollow Earth Theory' seem rather too official for the phenomena they describe, at least with regard to the nineteenth and early twentieth centuries.

of Earth's deep history take shape. Usually (but not always) rejected by leading naturalists, their cultural impact can hardly be understated. Pronouncements on fossils by famed occultist H. P. Blavatsky, for instance, attracted wide transatlantic, counter-establishment readerships; meanwhile, pulp novelists such as Tarzan's creator, Edgar Rice Burroughs, turned heterodox geology into bestselling lost world romances.

Crucially, my subjects were pushing back against not just the metaphysical direction of palaeoscientific research, but also the way in which it was being written and published: increasingly, that is, in terse accounts of primary research, published in scientific journals aimed at specialists. The figures I discuss took a broader view of what truth claims, and scientific evidence, might look like, but they were not simply isolated cranks. During these febrile decades, it remained far from universally clear what counted as legitimate palaeoscientific research or a legitimate researcher: not just notoriously controverted subjects like the mechanism of evolution or the place of religion, but also matters of appropriate professional training, publishing venue, tone, and literary form, despite decades of scientific standardization, could all still become matters for debate. My subjects saw themselves as contributors to scientific thought and it has hitherto been uncertain how they asserted this fact or how contemporaries determined whether their geohistorical claims were beyond the pale. After all, even notions that appeared heterodox to almost every elite geologist, like the advanced former civilization of Atlantis, were not necessarily absurd to general readers, nor were they conclusively falsifiable.

To help clarify this nebulous situation, I analyse the textual strategies these authors employed. While some writers of borderline palaeoscience courted legitimacy by compromising with the formal standards of technical journals, many subverted prevalent scientific genre conventions or asserted the value of alternative forms like fiction. Their choices and reception reveal a plurality of conceptions of scientific practice, as well as its appropriate textual incarnations and reading audiences. Most importantly, they also represent attempts to shift the balance of palaeoscientific power by creating textual spaces in which modern intellectual hierarchies, dominated by notions of expertise and predicated upon access to extensive scientific resources, could be levelled away. As we shall see, vision was a central theme in these efforts: borderline authors' literary and visual experimentation encouraged readers to see prehistoric events with new eyes, or to unsee what they thought they saw.

As noted above, my focus is situated broadly in a transatlantic anglophone sphere containing Britain, Canada, and the United States. The reason for this focus is organic: many of the figures I am interested in were English speakers who moved between these countries. Creationist George McCready Price, for instance, was born in New Brunswick, Canada, lived most of his life in the United States, and spent much of the 1920s teaching in the British town of Watford. Others maintained constant transatlantic communications, like the

Scottish-educated Canadian geologist John William Dawson, eager to keep up with the geological news in the mother country. At times, I also reach into the British colony of India, following the scientific activities of the occult Theosophical Society from its founding in New York to its second home in London to its long-term headquarters in Adyar, Madras (now Chennai). In science and more broadly, this was the age of growing international connections, one of telegraphic expansion, mass media, and the moment of the greatest reach of most European empires. If a narrowed focus must by necessity be a compromise, the opportunity to undertake fine-grained discussions of the shared scientific practices, literary genres, and religious currents of this geographical triangulation are valuable compensation.[8] This route also acknowledges the variant standards of orthodoxy in different scientific cultures. For example, the contemporaneous career of University of Munich palaeontologist Edgar Dacqué, who espoused views about the existence of humans throughout deep geological time based on mythic intuition, would hardly have been possible in Britain or North America.[9]

The authors I discuss sought to shape and reshape both the story of planetary prehistory and the textual forms appropriate to science. Bringing literary scholarship into entirely new and surprising conversations with research on subjects like Christian fundamentalism and 'pseudoscience', *Contesting Earth's History* shows that the literary borders of palaeoscience in this period required constant negotiation—not just to exclude mavericks on their fringes, but also to makes sense of figures sitting uneasily within their bounds. In the process, I illuminate hitherto ignored strategies used by authors to attest to an enchanted universe in which we humans are the purpose of existence, or older than we seem, or more capable of ranging through time than ever previously imagined. The result is a story of bold, controversial, and often idiosyncratic attempts to write Earth's history, and to see it with unprecedented clarity.

Contesting Geohistory

Palaeoscientific research proliferated during the late nineteenth and early twentieth centuries as the imperial powers carried out fossil excavations, mapping, and

---

[8] For the transatlantic culture of Britain, Canada, and the United States, see Brook Miller, *America and the British Imaginary in Turn-of-the-Twentieth-Century Literature* (Basingstoke: Palgrave Macmillan, 2010); and Diarmid A. Finnegan, *The Voice of Science: British Scientists on the Lecture Circuit in Gilded Age America* (Pittsburgh, PA: University of Pittsburgh Press, 2021).

[9] Nicola Gess, *Primitive Thinking: Figuring Alterity in German Modernity*, trans. by Erik Butler and Susan L. Solomon (Berlin: De Gruyter, 2022), 207–9. For global approaches to the cultural history of geology, see the essays in Alison Bashford, Emily M. Kern, and Adam Bobbette, eds., *New Earth Histories: Geo-Cosmologies and the Making of the Modern World* (Chicago: University of Chicago Press, 2023).

resource extraction across the globe.[10] International savants thereby filled out in greater and greater detail the shape of geohistory and the palaeontological waypoints of the evolutionary epic narrative. For all their general agreement, however, important specifics remained utterly disputed. Among all the consensuses, major areas of dispute—which were ruthlessly targeted by borderline thinkers—can be conceptualized, in part, as clashes between 'historical' and 'causal' explanations. Martin J. S. Rudwick argues that the 'discovery of the Earth's deep history' has been characterized by the former approach, inspired by the study of human history, rather than by the physics-based latter—the historical *what* long predating the causal *how*.[11] Broadly speaking, for thinkers interested in the latter, discussing the *what* alone was a suspect way of doing science. For thinkers concerned with the former, sidelining the *how* had helped the earth sciences to acquire empirical methods and thus to supplant the speculative geotheories promulgated by early modern philosophers.

Three areas of contestation were particularly polarizing. The first was the behaviour of the planet's surface. Did land rise from the sea and sink back into it, in constant oscillation? Was the crust cooling and contracting, as magisterially argued by geologist Eduard Suess in his *Das Antlitz der Erde* (*The Face of the Earth*) (1883–1909)? Were continents static, or, as geophysicist Alfred Wegener explosively proposed in 1912, did they drift? Geoscientific communities could not agree. 'Geology in 1912 was as fragmented, from a theoretical point of view', declares Mott T. Greene, in one of the few detailed general histories of the subject in this period, 'as at any time since its eighteenth-century beginnings'.[12] Clarity on the subject of continents was necessary for understanding and explaining the historical distribution of life on the planet. To explain how similar plants, animals, and even humans could exist in areas separated by seas and oceans, many thinkers theorized networks of sunken former land bridges. Of these, Philip Sclater's quasi-Atlantean continent of 'Lemuria' (a term coined in 1864), linking Madagascar with India, was the most prominent.[13] A unifying theory was called for, but only after the Second World War, and beyond the scope of this book, was an international consensus reached in favour of Wegener's drift (or rather 'plate tectonics').

The second area of contestation, in the wake of the publication of Charles Darwin's *On the Origin of Species* (1859), was evolution. Evolutionary theory was absorbed into the thinking of most naturalists of repute by the late 1870s, but it faced enduring challenges from within the scientific community. In 1866,

---

[10] James A. Secord, 'Global Geology and the Tectonics of Empire', in *Worlds of Natural History*, ed. by H. A. Curry, N. Jardine, J. A. Secord, and E. C. Spary (Cambridge: Cambridge University Press, 2018), 401–17.

[11] Martin J. S. Rudwick, *Earth's Deep History: How It Was Discovered and Why It Matters* (Chicago: University of Chicago Press, 2014), 298.

[12] Mott T. Greene, *Geology in the Nineteenth Century: Changing Views of a Changing World* (Ithaca, NY: Cornell University Press, 1982), 290.

[13] Ramaswamy, 22–23.

the eminent physicist William Thomson (later Lord Kelvin) had argued that the planet was not old enough for evolution by means of Darwin's primary mechanism, natural selection, to have taken effect.[14] He and other vocal physicists maintained this view until the century's end, amid a general backlash, often emerging from Thomson's own evangelical milieu, against 'uniformity'. This term referred, usually in a somewhat caricatured manner, to the views of geologist Charles Lyell, whose book *The Principles of Geology* (1830–33) had memorably discouraged geologists from taking explanatory recourse to rates of geological change inconsistent with those witnessed in the present, and whose work was thus understood as promoting a view of that change as extremely gradual.[15] Natural selection was a product of uniformitarian thinking. Thomson disliked what he, and many other Christians, perceived as its secular implications, but natural selection was also being criticized as inadequate by biologists and palaeontologists, who supplemented it with mechanisms like neo-Lamarckism, orthogenesis, and saltationism, most of which posited a less gradual, or more teleologically directed, version of evolution than Darwin had envisaged.[16] To conservative publics at the dawn of the twentieth century, this pluralism indicated a collapse of evolutionary theory itself.[17] Evolution's death was greatly exaggerated, but, again, a consensus on mechanism (or the *how*) emerged only after the period under discussion.

Resistance to evolution was connected to a third issue on which naturalists developed divergent interpretations: human prehistory. Although the extinct Neanderthals became widely recognized as human ancestors from the 1860s on, human antiquity was the hardest aspect of evolution to reconstruct, and to swallow. In 1870, naturalist Ernst Haeckel hypothesized that humans had originated on Sclater's Indian continent of Lemuria, and an Asian origin for humanity, seemingly indicated by the Bible, was a common assumption.[18] Ancestors were sporadically unearthed—Eugène Dubois's 'Java Man' in the 1890s; the British 'Piltdown Man' in 1912; the South African 'Taung Child' in the 1920s—each one fuelling controversy. It did not help matters that, as anti-evolutionists insistently pointed out, prehistoric human remains were almost invariably found in scattered fragments. Nationalism and racism shaped interpretations of these meagre remains, while eminent naturalists could wield their status to promote peculiar interpretations. Henry Fairfield Osborn, President of the American Museum of

---

[14] Rudwick, *Earth's Deep History*, 232–33.
[15] For modern summaries of Lyell's ideas, see Adelene Buckland, *Novel Science: Fiction and the Invention of Nineteenth-Century Geology* (Chicago: University of Chicago Press, 2013), chapter 3; and Rudwick, *Earth's Deep History*, 164–74.
[16] Peter J. Bowler, *Life's Splendid Drama: Evolutionary Biology and the Reconstruction of Life's Ancestry, 1860–1940* (Chicago: University of Chicago Press, 1996).
[17] Numbers, *The Creationists*, 51–52.
[18] Ramaswamy, 26.

Natural History in New York from 1908 to 1933, for example, endorsed a 'polyphyletic' genealogy that distanced humans from apes, and human races from each other.[19] Many thinkers far further from the corridors of palaeoanthropological power shared Osborn's desire to dignify, rather than bestialize, humanity's origins.

These matters are major determinants of the chronology of this book, which begins around the time of the publication of Darwin's evolutionary ideas and ends in the interwar period, during and after which time cultural, scientific, and religious shifts—discussed briefly in my epilogue—turned palaeoscience in new directions. In the decades under discussion, the lack of consensus about these important palaeoscientific topics—which also included the nature of the planet's unseen poles and interior—opened up rich borderlands of thought. My main areas of discussion will not be such respectable controversies as the mobility of continents, but rather the considerably more 'unorthodox' hypotheses that flourished in this exciting scientific climate, such as the possibility of encountering mammoths on the inverse side of Earth's crust. Considering the plurality of opinion described above, however, how do we determine what was 'orthodox', and to whom?

Back in the early nineteenth century, demarcating scientific orthodoxy on the nature of the planet's history had been a murky affair. For wide demographics in Britain and the United States, the 6,000-year age of Earth remained, despite challenges, a commonsensical assumption, and, as Noah Heringman has recently argued, geology held no monopoly on the notion of deep timescales in anglophone culture.[20] Elaborate accounts of planetary formation over aeons, penned by learned naturalists, were not taken as gospel. The Romantic poet Charlotte Smith, in a footnote to her geological poem 'Beachy Head' (1807), briskly declared that she had 'never read any of the late theories of the earth', nor was she 'satisfied' with what she had heard of attempts to explain fossils.[21] Two years later, Washington Irving's satirical *History of New York* (1809) irreverently ran through the speculative geotheories of natural philosophers like the late Georges-Louis Leclerc, Comte de Buffon, before leaving 'readers at full liberty to choose among' these 'mighty soap-bubbles'.[22] Despite the emergence of specialist organizations like the Geological Society of London (founded in 1807), the early geological community remained heterogeneous in this period, although one of its first moves was to marginalize the geotheories derided by Irving and disregarded by Smith. Speaking

---

[19] Ronald Rainger, *An Agenda for Antiquity: Henry Fairfield Osborn & Vertebrate Paleontology at the American Museum of Natural History, 1890–1935* (Tuscaloosa: University of Alabama Press, 1991), 231.

[20] Noah Heringman, *Deep Time: A Literary History* (Princeton, NJ: Princeton University Press, 2023).

[21] Charlotte Smith, *Beachy Head with Other Poems* (London: J. Johnson, 1807), 159.

[22] 'Diedrich Knickerbocker' [Washington Irving], *A History of New York, from the Beginning of the World to the End of the Dutch Dynasty*, 2 vols (New York: Inskeep & Bradford, 1809), I, 21–22.

of geology in 1830s London, Rudwick has described a 'social topography' divided into 'graduated zones' of 'ascribed competence', at the top of which sat 'elite geologists' like the gentlemanly Lyell.[23]

These subtle gradations were not self-evident to many figures attempting to contribute to palaeoscientific thought. Formalized hierarchies of expertise were even more limited in the antebellum United States, where, immersed in Jacksonian democracy, individuals routinely asserted their right to pronounce upon scientific matters.[24] This represented the survival of an eighteenth-century notion that all rational actors might contribute to knowledge creation in the public sphere. That ideal persisted in Britain, as elite geologists learnt when the journalist Robert Chambers anonymously published his wide-ranging pre-Darwinian argument in favour of evolution, *Vestiges of the Natural History of Creation* (1844).[25] Richard Yeo calls this important book, the product of Chambers' extensive scientific reading, a 'borderline case'.[26] Its excited reception outside gentlemanly geological communities revealed the limitations of relying on 'ascribed competence' to demarcate expertise: the sophisticated *Vestiges* left many readers with a powerful impression of the anonymous author's cross-disciplinary proficiency, but readers with specialist knowledge detected serious errors. Of course, even among undisputed experts, nineteenth-century palaeoscientific research could rarely be delegated into disciplinary specialisms, instead ranging promiscuously across not just geology and palaeontology but also archaeology, anatomy, ethnology, geography, zoology, and history.

Embarrassments such as the *Vestiges* affair helped motivate the scientific community more formally to police its own boundaries and standards. Reforms were espoused vociferously in the second half of the century by the group known as scientific naturalists, led by British savants like Darwin's agnostic ally Thomas Henry Huxley. Their bid to separate scientific from religious spheres contributed to what Ronald L. Numbers has called the 'privatization' of religion's role in scientific practice.[27] In both the United States and the United Kingdom, geohistory had formerly been interpreted by most practitioners, from the liberal Lyell and Chambers to theologically conservative evangelicals like Scottish geologist-journalist Hugh Miller and Yale College grandee Benjamin Silliman, as part of a theistic worldview. By the second half of the century, even pious geologists

---

[23] Martin J. S. Rudwick, 'Charles Darwin in London: The Integration of Public and Private Science', *Isis*, 73 (1982), 186–206 (190).

[24] Daniel Patrick Thurs, *Science Talk: Changing Notions of Science in American Popular Culture* (New Brunswick, NJ: Rutgers University Press, 2007), 30–31, 33–41.

[25] James A. Secord, *Victorian Sensation: The Extraordinary Publication, Reception, and Secret Authorship of* Vestiges of the Natural History of Creation (Chicago: University of Chicago Press, 2000).

[26] Richard Yeo, 'Science and Intellectual Authority in Mid-Nineteenth Century Britain: Robert Chambers and *Vestiges of the Natural History of Creation*', *Victorian Studies*, 28 (1984), 5–31 (11).

[27] Ronald L. Numbers, 'Science, Secularization, and Privatization', in *Eminent Lives in Twentieth-Century Science and Religion*, ed. by Nicolaas A. Rupke, 2nd edn (Frankfurt: Peter Lang, 2009), 349–62.

were beginning to 'privatize' their beliefs, omitting mention of Christian frameworks in their publications, while most scientific journals sidelined religion.[28] Christian geology maintained its prestige in North America for longer, dominated by evangelicals like John William Dawson, but, by the early twentieth century, it was rare for a work of scholarly palaeoscientific research explicitly to discuss God.

Although their attacks on religion were hotly contested by fellow savants, the scientific naturalists were joined by all manner of reformers in the establishment of new institutions and paid positions in science. An age of museum building culminated in the 1881 opening of the Natural History Museum in London and the founding of American museums bankrolled by wealthy tycoons, such as what became Osborn's domain, the American Museum of Natural History, opening in 1877.[29] Meanwhile, the number of salaried palaeoscientific jobs increased, thanks to new and often science-focused universities like Mason College, Birmingham, established in 1875 and given university status in 1900. Science was becoming something that took place more exclusively in well-equipped spaces like these, but the reconfiguration of expertise went deeper. In an approach typified by the commercial science journal *Nature* (founded in 1869), acceptable science was coming to be conceived as primarily first-hand specialist research, published chiefly in technical articles by dedicated contributors, among whom the litterateurs, popularizers, or politicians who had contributed to science earlier in the century were viewed with suspicion.[30] In *Nature* and periodicals like the *Proceedings of the Royal Society*, moreover, scientific papers were becoming more compressed and jargon-heavy.[31] What it meant to be a scientific practitioner, and who could access scientific research, was changing.

These shifting practices are of interest to the scholar of science's borderlines due to their extremely uneven unfolding. Many of the later nineteenth century's leading savants, including Darwin, demonstrated little interest in paid research posts, nor in technical silos. Great prestige was still attached to projects that drew from many disciplines and resulted in distinguished books, sometimes still addressed to educated general audiences. The nineteenth century, as Pratik Chakrabarti

---

[28] Nicolaas A. Rupke, 'Down to Earth: Untangling the Secular from the Sacred in Late-Modern Geology', in *Science without God: Rethinking the History of Scientific Naturalism*, ed. by Peter Harrison and Jon H. Roberts (Oxford: Oxford University Press, 2019), 182–96.

[29] Lukas Rieppel, *Assembling the Dinosaur: Fossil Hunters, Tycoons, and the Making of a Spectacle* (Cambridge, MA: Harvard University Press, 2019).

[30] Melinda Baldwin, *Making* Nature: *The History of a Scientific Journal* (Chicago: University of Chicago Press, 2015), 77, 83. See also Martin J. S. Rudwick, *Bursting the Limits of Time: The Reconstruction of Geohistory in the Age of Revolution* (Chicago: University of Chicago Press, 2005), 220, for earlier shifts away from the proverbial armchair.

[31] Julie McDougall-Waters and Aileen Fyfe, 'The Rise of the *Proceedings*, 1890–1920s', in *A History of Scientific Journals: Publishing at the Royal Society, 1665–2015*, ed. by Aileen Fyfe, Noah Moxham, Julie McDougall-Waters, and Camilla Mørk Røstvik (London: UCL Press, 2022), 363–402 (394). See also Alex Csiszar, *The Scientific Journal: Authorship and the Politics of Knowledge in the Nineteenth Century* (Chicago: University of Chicago Press, 2018), 31, 37, 211.

observes, 'was the period when the greatest synthesis of ideas' existed between geology and other fields.[32] During the early twentieth, titans like Osborn pushed back against 'extreme modern specialization' and sought to emulate the rounded careers of the older generation of naturalists.[33] James A. Secord points to an era of '[g]rand theories of the Earth', running from the mid-nineteenth century into the 1930s, during which time geologists and palaeontologists contributed to a 'traffic between natural historical, philosophical and literary perspectives', all 'contributing to a common understanding' of geohistory.[34] In fiction, this non-disciplinary view of palaeoscience was bombastically represented by Professor Challenger, the polymath co-protagonist of Arthur Conan Doyle's romance *The Lost World (1912)*.

If even the elites refused fully to embrace the specialization and professionalization ongoing around them, it is unsurprising that the methods, activities, and career paths connected to palaeoscience remained somewhat disparate. One of geology's founding myths, after all, had stressed the utility of diverse practitioners in collecting observations and specimens across the world, whether in provincial England or distant colonies.[35] This participatory ideology remained attractive. The gatekeeping role of bastions of modern science like *Nature* should, moreover, not be exaggerated, as the exigencies of commercial publishing were not always compatible with scientific austerity.[36] Even religion did not entirely disappear from the technical writings of eminent palaeoscientific researchers. As late as 1932, the belligerent but respected Scottish-born palaeontologist Robert Broom concluded his decisive monograph on *The Mammal-Like Reptiles of South Africa and the Origin of Mammals* with speculations on the 'intelligent controlling power' that had 'guided' evolution towards 'man'.[37]

Outside specialist scientific circles, faith often motivated more subversive attitudes towards the scientific establishment. A fascination with science, and antipathy to many of its most vaunted practitioners, characterized major religious movements like the occult Theosophical Society (founded in 1875) and Christian fundamentalism (which emerged in the 1910s). It is in this complicated climate—a time when the borders of palaeoscience were being policed like never before, but also vigorously strained against from inside and out—that the action of *Contesting Earth's History* unfolds.

---

[32] Pratik Chakrabarti, *Inscriptions of Nature: Geology and the Naturalization of Antiquity* (Baltimore, MD: Johns Hopkins University Press, 2020), 22.
[33] Quoted in Rainger, 142.
[34] Secord, 'Global Geology and the Tectonics of Empire', 405.
[35] Rudwick, *Bursting the Limits of Time*, 467.
[36] Melinda Baldwin, 'The Business of Being an Editor: Norman Lockyer, Macmillan and Company, and the Editorship of *Nature*, 1869–1919', *Centaurus*, 62 (2020), 1–14.
[37] Quoted in Jesse Richmond, 'Design and Dissent: Religion, Authority, and the Scientific Spirit of Robert Broom', *Isis*, 100 (2009), 485–504 (498).

Borderline Palaeoscience

In discussing subjects like creationism, hollow-earth theory, and Atlantis, I am entering into fields frequented by many previous scholars. Despite all this attention, much of what it meant for these subjects to be scientifically heterodox, and to whom, has gone uninterrogated. Studies of moments when geology and palaeontology's legitimate borderlines appeared decidedly hazy to historical actors have, with the exception of the field that was, in the twentieth century, dubbed 'cryptozoology', been surprisingly few.[38] This sparseness and lack of comparison makes for a striking contrast with the sustained attention paid to the contemporaneous explorations of savants like Alfred Russel Wallace on the psychic fringes of the human and physical sciences.[39] My subject necessitates a different approach to the work of scholars investigating this area, who usually focus on nuances of experimentation and observation in spaces like the Spiritualist séance room and laboratory.[40] The most controversial matters in palaeoscience were events never seen by human eyes, and thus harder to measure than the surreptitious movements of a spirit medium's finger. Names of comparable stature to that of Wallace were embroiled in these controversies. Given that he was an acclaimed geologist and principal of McGill University, Montreal, John William Dawson's claim that geohistory was entirely compatible with the Book of Genesis was hard to dismiss. Other figures, like British politician Henry Hoyle Howorth, could marshal social and intellectual resources to see their surprising claims taken seriously, and, if they did not necessarily convince, they were rarely refuted. For people on the religious extremes who attributed their knowledge of prehistory to revelation or clairvoyance, the problems of proof and disproof were drawn most markedly.

From old-earth creationists like Dawson to Theosophists, Christian fundamentalists, and hollow-earth-propounding businessmen living in the American Midwest, these heterogeneous groups, some utterly transatlantic and others locally constituted, shared features of belief. Opposing the notion that humans were a circumstantial evolution from the apes, they framed us either as the teleological climax of the universe, as psychically free to escape the limitations of time, as extending back in a civilized form far deeper into geohistory than savants claimed, or as some combination of all these factors. Most rejected what Rudwick considers the defining feature of the geohistorical narrative: the 'contingent' nature of

---

[38] For cryptozoology, see Daniel Loxton and Donald R. Prothero, *Abominable Science! Origins of the Yeti, Nessie, and Other Famous Cryptids* (New York: Columbia University Press, 2013).

[39] For example, note the absence of palaeoscience in important collections and overviews like Lara Karpenko and Shalyn Claggett, eds., *Strange Science: Investigating the Limits of Knowledge in the Victorian Age* (Ann Arbor: University of Michigan Press, 2017); and Christine Ferguson and Efram Sera-Shriar, 'Spiritualism and Science Studies for the Twenty-First Century', *Aries*, 22 (2022), 1–11.

[40] Efram Sera-Shriar, *Psychic Investigators: Anthropology, Modern Spiritualism, and Credible Witnessing in the Late Victorian Age* (Pittsburgh, PA: Pittsburgh University Press, 2022).

the Earth's development.[41] Sacred texts showed that plan lurked beyond all apparent contingencies, but, even without recourse to religion, scientific rebels were able to argue that the shape of Earth's history could be deduced from rational thinking and simple experiments alone, in the confident worldbuilding vein of early modern geotheory. The hyper-anthropocentrism of my subjects accords with what Vybarr Cregan-Reid, in one of very few literary approaches to this subject in a late nineteenth-century context, calls a 'resistance to accepting the strangeness and darkness of deep time'.[42] I differ from Vybarr Cregan-Reid, however, in examining this subject entirely through palaeoscientific writings, rather than his, to my mind, more circuitous focus on diverse literary and artistic sources that represent a vaguer 'ideological shift in late-Victorian culture'.[43]

It is when he does focus on geology that Cregan-Reid raises another incisive point that I have found productive for the study of borderline palaeoscience. In the 1870s, archaeologist George Smith translated what was later called the *Epic of Gilgamesh* from Neo-Assyrian tablets, provided a corroborating source for flood legends like the biblical deluge. As Cregan-Reid points out, the first volume of Suess's *Das Antlitz der Erde*, one of the century's most important geological works, took *Gilgamesh*'s deluge story seriously.[44] Such respectful treatment of the descriptions of natural phenomena in ancient texts had long been unfashionable. As Rhoda Rappaport argues, the proto-geological philosophers of the late seventeenth and early eighteenth centuries had, like Suess, valued the ancient testimonies of 'human witnesses to natural events'.[45] She sees the beginning of the end of this tradition in the work of Buffon, whose *Histoire naturelle* (*Natural History*) (1749–1804) framed ancient texts as collateral evidence for ancient events more reliably reconstructed from physical evidence, and whose *Les époques de la nature* (*The Epochs of Nature*) (1778) was one of the first major works to describe ages in which no humans had existed.[46] In Suess's day, however, *Gilgamesh* was not even the only exciting evidence that ancient textual sources reflected, in distorted manner, useful descriptions of ancient geographies, beings, and phenomena. British geologists in India insisted that the mythic beings of Hindu antiquity had natural, albeit prosaic, origins, while Heinrich Schliemann was lauded for proving that Homer's Troy was not pure fiction.[47]

Gould's *Mythical Monsters*, with which I began, and many of the authors I will discuss in subsequent chapters, merely took this argument to the margins

---

[41] Rudwick, *Earth's Deep History*, 180.
[42] Vybarr Cregan-Reid, *Discovering Gilgamesh: Geology, Narrative and the Historical Sublime in Victorian Culture* (Manchester, 2013), 203.
[43] Cregan-Reid, 191.
[44] Cregan-Reid, 198.
[45] Rhoda Rappaport, *When Geologists Were Historians 1665–1750* (Ithaca, NY: Cornell University Press, 1997), 261.
[46] Rappaport, 261.
[47] Chakrabarti, e.g. 113.

of propriety, insisting on the substantial value of literary and artistic documents, including the Bible and the classics, for shedding light on the deep past. They held human documents in higher evidentiary esteem than most palaeoscientific practitioners, in part because most believed in the articulateness and divine inspiration of these documents, but also because many were unconvinced that what was inexactly called Earth's prehistory had truly lacked human witnesses. Daniel Lord Smail has argued that anxious late nineteenth-century historians alleviated the 'vertigo' induced by deep time by restricting history proper to the 'short chronology' during which humans have produced written 'documents', relegating the rest to the new concept of prehistory.[48] The subjects of *Contesting Earth's History* blithely recolonized this prehistory, insisting on the power of written documents to reveal the truth about super-ancient human civilizations and even, through visionary power, prehuman eras.

These borderline approaches to palaeoscience were characterized by what scholars call *enchantment*. In using this term, which is not conventionally linked to palaeontology and geology, I draw upon research interested in the historical belief that modern Western life has become progressively secularized and thus 'disenchanted'. This notion, associated with the early twentieth-century thinker Max Weber, is now discredited as sociological analysis, but historical actors unquestionably wrestled with what they felt was the unnatural dethroning of religious meaning from a naturalized world.[49] Cautiously framing borderline palaeoscience within this self-conscious reaction against a perceived disenchantment is profitable. After all, even mainstream palaeoscience sits uneasily within Weber's paradigm, given its persistent links with Christianity and the efforts of popularizers to prove that its discoveries were even more enchanting than myths and fairy tales.[50] Advocates of enchanted palaeoscience capitalized on these associations, taking them further and more literally than those geologists and palaeontologists who had 'privatized' their faith or committed to more cautious methodologies. Christian evangelicals and fundamentalists opposed the godless implications of scientific naturalism, while other thinkers, like the Theosophists, applied what Egil Apsrem calls an 'open-ended' naturalism that allowed researchers of paranormal and theistic phenomena to insist upon their use of empirical methodologies, at least to their own satisfaction.[51] The aforementioned Howorth chided geologists for failing to recognize the facts contained in sacred and mythic texts, all while denying that he had any religious axes to grind.

---

[48] Daniel Lord Smail, *On Deep History and the Brain* (Berkeley: University of California Press, 2008), 49–50.

[49] Egil Asprem, *The Problem of Disenchantment: Scientific Naturalism and Esoteric Discourse, 1900–1939* (Leiden: Brill, 2014), 1–2.

[50] Ralph O'Connor, 'Introduction: Varieties of Romance in Victorian Science', in *Science as Romance*, vol. VII of *Victorian Science and Literature*, gen. eds Gowan Dawson and Bernard Lightman (London: Pickering & Chatto, 2012), xi–xxxvi; Melanie Keene, *Science in Wonderland: The Scientific Fairy Tales of Victorian Britain* (Oxford: Oxford University Press, 2015), chapter 1.

[51] Asprem, 10.

While a minority of these figures communicated work in leading scientific periodicals, or with respected scientific book publishers, the majority were excluded from these outlets, whether voluntarily or involuntarily. By consequence, they addressed their claims either to interest groups, like the readerships of denominational periodicals, or to the remnants of the public sphere of knowledge so successfully addressed by Chambers in his *Vestiges*. This realm of debate and discussion, conceptualized by philosopher Jürgen Habermas as having formed in the eighteenth century, had a tenuous existence at the end of the nineteenth: Thomas Broman argues that science's escalating cultural 'authority' and the growth of its professional structures rendered the public sphere—in an appropriate metaphor— 'a hollow shell', 'public in principle but recondite in practice'.[52]

One result of the survival of direct scientific democracy's superficial 'shell' was that countless authors took little heed of the hierarchies of 'recondite' expertise. The dustjacket of Eugen Georg's *The Adventure of Mankind* (1931), a Theosophically-inflected occult treatise touching on humanity's Mesozoic prehistory, for instance, declared perusal 'THE DUTY OF ALL ENLIGHTENED AMERICANS'.[53] While strange contributions like this were only occasionally noticed in the corridors of intellectual power, as when a reader of *Science* wondered why the 'reputable' firm E. P. Dutton had published Georg's dissident book, they could arouse the interest of more widely conceived communities of consumers.[54] The science fiction novelist Stanton A. Coblentz, writing in the ubiquitous *New York Times*, found that Georg's description of humans from the 'era of the dinosaur', while not entirely convincing, depicted 'that which might have been', and in so doing displayed 'a breadth of historical outlook that is seldom attempted'.[55] Utterly heterodox palaeoscience could, in other words, provide even urbane New Yorkers with thought-provoking counterfactual reading material.

Other writers addressing the public sphere paid mere lip service to expertise. The Pune-based Sanskrit scholar Narayan Bhavanrao Pavgee, in the preface to *The Vedic Fathers of Geology* (1912), placed 'before the *Public*' his argument, intended to be 'reviewed by the *Savants* of the East and the West', that ancient Hindu texts, the Vedas, exhibited astonishing knowledge of palaeoscience.[56] In a backhanded compliment, Pavgee praised 'Western Geologists' for their 'commendable perseverance' as latecomers in a scientific field begun by Vedic fathers thousands of years ago.[57] The preface's authorial alternation between humility and confidence soon disproportionately favoured the latter as the book's argument progressed. Even had

---

[52] Thomas Broman, 'The Habermasian Public Sphere and "Science *in* the Enlightenment"', *History of Science*, 34 (1998), 123–49 (142–43).
[53] Eugen Georg, *The Adventure of Mankind*, trans. by Robert Bek-Gran (New York: E. P. Dutton, 1931).
[54] Charles Stuart Gager, 'At the Top Is Magic', *Science*, 74 (1931), 569–70 (569).
[55] Stanton A. Coblentz, 'Man Through the Ages', *New York Times*, 17 January 1932, BR20.
[56] Narayan Bhavanrao Pavgee, *The Vedic Fathers of Geology* (Poona [Pune]: Arya-Bhushan Press, 1912), x. See Chakrabarti, 100–101.
[57] Pavgee, 24.

they been offered more flattery, few 'Western Geologists' in the Anglosphere were likely to have respectfully considered Pavgee's nationalistic evidence for the antiquity and insights of Hindu science, at least on his terms, although claims of this kind were the bread and butter of Orientalist organizations like the Theosophical Society.

Framing maverick work as popularization was another subterfuge. In *Beginnings, or Glimpses of Vanished Civilizations* (1911), Irish litterateur Marion McMurrough Mulhall professed simply to communicate established science. Soon after a deferential preface, however, she denounced the 'shadowy' evidence 'which sufficed to convince some of the great anti-Christian scientists of the length of man's appearance on the earth', arguing instead for a syncretic interpretation of Genesis in which Noah's grandson 'Votan' reached America via Atlantis.[58] The Catholic weekly *America* observed that Mulhall's 'profession of writing only for "her young friends, boys and girls", is merely a modest way of presenting to the general public' a work of 'serious' intent.[59] In a sense, Mulhall's book simply substantiates modern scholars' conclusions that popularizers reshape scientific knowledge in important ways, or even that they participate in constructing scientific knowledge in a manner that problematizes the very term 'popularizer'.[60] Although I am indebted to both ideas, the term itself, whether accepted or understood merely as an actors' category, is not critical to my argument. The majority of my subjects bypassed even Mulhall's half-hearted pretence of 'popularizing' knowledge already created, instead asserting their right to contribute to science.

Clearly, some early twentieth-century writers still operated in a democratic mode of palaeoscientific practice that would not have been out of place a century prior. *Contesting Earth's History* reveals many previously untold stories from this fascinating world. My contribution extends much further, however, than simply expanding our historical knowledge of controversial thinkers. In previous critical examinations of heterodox fields like young-earth creationism, strictly literary matters like the style, genre, and mode of publications have rarely been the centre of attention and are usually left undescribed. As such while we may have a sophisticated understanding of many heterodox theories and theorists, we have, hitherto, had almost no conception of their aesthetics. It is this book's contention that, by analysing the literary modes and genres adopted by borderline palaeoscientific

---

[58] Marion McMurrough Mulhall, *Beginnings, or Glimpses of Vanished Civilizations* (London: Longmans, Green, 1911), 2, 88.

[59] Untitled review of *Beginning, or Glimpses of Vanished Civilizations*, by Marion McMurrough Mulhall, *America*, 5 (1911), 284.

[60] For reshaping knowledge, see Bernard Lightman, *Victorian Popularizers of Science: Designing Nature for New Audiences* (Chicago: University of Chicago Press, 2009). For contributing to knowledge, see James A. Secord, 'Knowledge in Transit', *Isis*, 95 (2004), 654–72; and Jonathan R. Topham, 'Rethinking the History of Science Popularization/Popular Science', in *Popularizing Science and Technology in the European Periphery 1800–2000*, ed. by Faidra Papanelopoulou et al. (Basingstoke: Ashgate, 2009), 1–20.

researchers, we can better understand the variant notions of scientific practice and method on display. Such a project entails bringing the tools of literary studies to the thorny topic of 'pseudoscience'.

On the Textual Fringe

Distinguishing merely controversial interpretations of geohistory from ones beyond the pale of consideration was a more organized task at the end of the nineteenth century than it had been at the century's beginning, but it was still no simple matter, even for experts. This is hardly surprising, given that Michael D. Gordin's compelling research on the demarcation of science from pseudoscience shows that no all-purpose, ahistorical philosophical criteria can be applied to the demarcation problem.[61] Reconstructions of geohistorical events, in particular, are not straightforwardly testable or falsifiable, two oft-cited determinants of scientific method. Thus the method-obsessed young-earth creationist George McCready Price could, in 1917, dismiss all such reconstructions as 'pseudo-science'.[62] I have been intrigued by Gordin's observation that this term, which came into regular use on both sides of the Atlantic in the late nineteenth century, relies upon the ancient 'literary' concept of 'mimesis': that is to say, opponents dismiss as pseudoscientific those practitioners they see as mimicking merely the 'trappings' of science.[63] This can be a credible charge, given that, as Gordin notes, 'counterestablishment' researchers in the modern world replicate 'professionalizing markers' of 'establishment science', such as scholarly journals, falsifiable methods, and specialist organizations.[64]

This reference to 'mimesis' tantalizingly implies that, in the absence of a universal criterion of demarcation, attention to literary form has helped authors to present fringe work as scientific, and helped readers to determine whether content is pseudoscientific. While Gordin attends to the role of publishers in mediating controversies, he does not pursue these matters into the more fine-grained textual territory of literary analysis. The academic field of literature and science has so far had little to say to his work on pseudoscience, despite the potentially privileged role Gordin here appears to accord to literary methods. *Contesting Earth's History* takes up the baton, showing how borderline thinkers signalled their conceptions of valid scientific practice through the extent of their participation in the dominant

---

[61] Michael D. Gordin, *The Pseudoscience Wars: Immanuel Velikovsky and the Birth of the Modern Fringe* (Chicago: University of Chicago Press, 2012).
[62] George McCready Price, *Q. E. D., or New Light on the Doctrine of Creation* (New York: Fleming H. Revell, 1917), 142.
[63] Gordin, *The Pseudoscience Wars*, 232.
[64] Michael D. Gordin, *On the Fringe: Where Science Meets Pseudoscience* (Oxford: Oxford University Press, 2021), 43.

genres of scientific writing. In fact, mimetic imitation of a drily professionalized style, while common in the mid-to-late twentieth century, Gordin's domain, was only one of various ways these writers engaged with the forms of science. Rejection of a technical register in favour of more dazzling and creative stylings was a popular alternative, as was hypercorrection: appearing more scientific than the scientists (to use the word that was gradually replacing the older term 'savant' as a self-identification in the early twentieth century). By refusing to comply fully, or sometimes at all, with the generic conventions expected of contributions to scientific knowledge, these borderline thinkers sought terrain more advantageous to themselves.

My approach is not intended to imply that negotiating scientific authority was purely a disembodied literary affair. In addition to the developing hierarchies of expertise noted above, modern gradations of Rudwickian 'ascribed competence' continued to communicate tacit signals; these will typically, in the case of Pavgee, have included prejudices regarding race, and gender in the case of Mulhall (and religion in the case of both). Rather than relegating science to pure text, as some borderline thinkers themselves attempted, thinking about the ways in which their writings were literary constructions as well as truth claims allows us to speak with new nuance about how contemporaries conceptualized scientific method. This factor reflects upon their material conditions. After all, for thinkers who lacked access to specimens, laboratories, journals, or the aid of learned societies, authority had to be constructed, and method justified, textually. As such, these thinkers asserted that a meaningful contribution to palaeoscience could be submitted without a primary basis in conventional first-hand research and without the necessity of specialized qualifications. As noted above, this was the kind of approach being phased out of the most prestigious scientific journals, although wider intellectual publics by no means considered it discredited. It was also an extension of these writers' frequent insistence that clues about geohistory contained in literary and artistic sources, sacred or otherwise, had just as much evidentiary value as fossils and rocks.

This insistence on the validity of 'literary' approaches to palaeoscience, both in terms of methods and evidence, was the latest evolution in these sciences' complex relationship with textuality. Writers on palaeoscientific subjects were predisposed to take literary concerns seriously. As scholars of literature and science, especially Ralph O'Connor and Adelene Buckland, have shown, many British geological and palaeontological writers in the early and mid-nineteenth century were renowned for their belletristic craft. While this could be valuable for these fields' prestige and popularity, it was far from unproblematic. 'Doing geology', Buckland observes, 'meant writing it too', and 'writing was a useful—albeit often a dangerous—tool'.[65]

---

[65] Buckland, *Novel Science*, 13, 22.

As she demonstrates, early and mid-nineteenth-century geologists like Lyell chastised writers whose incaution led them textually to arrange neutral rocky data into elegant, writerly 'plots', whether geotheoretical, geohistorical, or evolutionary.[66] For Lyell and his colleagues, the *Vestiges*' success crystallized the menace of skilful writing and compelling plot. During the later period that is my focus, some of these complications were undoubtedly being ironed out. Geohistorical and evolutionary 'plots' were far more widely accepted, and the tendency towards utilitarian prose in scientific writing not addressed to general readers signalled that, as O'Connor observes, 'aesthetic' qualities were to be seen as 'popular' and unrelated to the nitty-gritty of science, leading to reduced generic ambiguity.[67]

It is my contention that, as a result of these changes, the writing process became *more* rather than *less* important for thinkers on the borders of palaeoscientific respectability, as their literary choices stood out in ever-starker relief. Buckland provocatively declares that 'literature was science' for nineteenth-century geologists, because they were 'constantly alert to the importance of choosing the best literary forms with which to write geological science into existence'.[68] This could not be truer for the subjects of my chapters, who have been selected to evidence the spectrum of textual strategies adopted by borderline practitioners in their quest to recruit readers to their interpretations of geohistory. While John William Dawson, for instance, roughly bifurcated his career, publishing work that addressed scientific colleagues in technical journals and producing elegant old-earth creationist content in accessible genres for evangelical audiences, he also found opportunities for cross-contamination, airing serious new palaeontological claims in avowedly popular books. His less esteemed fellow travellers, too, sought generic models that bypassed the scrutiny, and agnosticism, of technical journals, from re-enactments of prophetic visions to one schoolmaster's attempt to collect testimonials to the validity of his data, if not to his creationist conclusions (chapter 1). Occultists embraced the high-stakes charismatic authority of experiential narratives and oracular declarations based on minimal material evidence (chapter 2), while maverick hollow-earthers rejected unified models of authorship and firm divisions between fact and fiction, favouring a more open-minded approach both to scientific investigation and to writing science (chapter 3).

If the authors discussed in these first three chapters embraced a range of aesthetically attractive options, those in my remaining chapters were more circumspect, exploiting the scepticism towards imagination and textuality usually seen as part of the rhetorical armoury of elite science. Figures hoping to be taken seriously by the palaeoscientific establishment, despite the controversial nature of their claims about ancient floods and sunken continents, stressed the scholarly nature of their

---

[66] Buckland, *Novel Science*, 13–22.
[67] Ralph O'Connor, *The Earth on Show: Fossils and the Poetics of Popular Science, 1802–1856* (Chicago: University of Chicago Press, 2007), 244.
[68] Buckland, *Novel Science*, 15.

methods and presentation (chapter 4), albeit while also pushing back on the idea that scientific research could not be conducted entirely in the comfort of libraries. Finally, those young-earth creationists most combative towards the establishment turned the tables on it by dismissing the textual and visual conventions of their scientific foes as mere literary and artistic constructs (chapter 5). Overall, these explorations of geohistory, both radical and conservative, uniformly challenged the notion that writing up was simply an after-thought in the scientific process.

Understanding these texts requires substantial literary analysis of scientific non-fiction. The task is complicated by our limited understanding of the genres under discussion. O'Connor argues that literary scholars analysing scientific non-fiction tend to perform 'fine-grained' close readings of choice passages, ignoring matters of genre and thereby leaving readers 'to conclude that the only large-scale form embodied in a work of scientific non-fiction is that of an argument' or 'explanation'.[69] The result is that the genres of scientific non-fiction remain 'mostly unmapped beyond vague labels' that have little to say about literary form.[70] Attending to genre, rather than simply to argument, requires a stable definition of the former term. In John Frow's lucid theorization, 'genre' is a series of meaningful structural constraints shared by authors and readers. These constraints, which generate meaning at a 'deeper' level than the 'explicit 'content' of a text', include *'formal features'*, *'thematic structure'*, and *'frame'*.[71] 'Modes' share some of these attributes, but, unlike genres, they are 'adjectival' rather than 'structural'.[72] George McCready Price, to use a clear example, sometimes dips into a satiric mode while employing the reputable genre of scientific textbook in his young-earth masterpiece, *The New Geology* (1923). The 'mapping' of non-fictional genres cannot be hard-edged, but, as O'Connor observes, greater precision is necessary when discussing literary genre alongside scientific content.

Contesting Earth's History takes the literary nature of scientific non-fiction seriously. I move between fine-grained and larger-scale approaches, attempting to avoid, when possible, that easy slippage between form and argument. By zooming in, I provide close readings of passages from writings by important figures in the history of borderline palaeoscience, whose works may, at first appear to resist this kind of analysis. By zooming out, I demonstrate their engagement with the wider genres of scientific and imaginative literature. Although I work towards clustering some non-fictional texts within genres, the systematic mapper must more zealously shun the desire to zoom in, and embrace the urge to coin generic terms,

---

[69] Ralph O'Connor, 'The Meanings of "Literature" and the Place of Modern Scientific Nonfiction in Literature and Science', *Journal of Literature and Science*, 10 (2017), 37–45 (41).

[70] O'Connor, 'The Meanings of "Literature"', 41. For a more detailed discussion by the same author, see 'Science for the General Reader', in *The Routledge Research Companion to Nineteenth-Century British Literature and Science*, ed. by John Holmes and Sharon Ruston (London: Routledge, 2017), pp. 155–71.

[71] John Frow, *Genre* (London: Routledge, 2005), 9–10, 19.

[72] Frow, 65.

more often than I do. The sustained close reader, too, must stay longer with one text at a time than I find it useful to do in my pursuit of generalizations. In avoiding either extreme, however, I make space for cultural and biographical details that contextualize the lives of borderline practitioners. In this way, I trace what Frow calls the 'symbolic action' of literary conventions, the way 'generic organisation of language' and 'images ... makes things happen'.[73] To again use Price as an example, *The New Geology* adheres copiously if superficially to the generic conventions of the geological textbook, but, on closer inspection, his book turns out to be a Trojan horse that, in a conspicuously 'symbolic action', demolishes far more than it instructs.

Borderline palaeoscience could be so compellingly written that even non-believers took note, drawing inspiration from it in the construction of ambitious works of fiction. In examining this subject, I speak to Michael Saler's research on verisimilitude in the fantastic fiction of the period. Saler argues that, appealing to increasingly secular readerships enamoured with science, popular authors like H. P. Lovecraft depicted 'logically cohesive worlds intended to reconcile reason and enchantment' while exhibiting a certain 'self-reflexive irony'.[74] They did so by balancing their absorbing worldbuilding with more knowing touches, playfully drawing occasional attention to the fictionality of their content. As I discuss in chapter 2, the thrillingly weird concepts and verisimilar, matter-of-factual language of occult palaeoscience scratched the itch of simulated enchantment for non-believers such as Lovecraft. However, I also show that the most overtly 'ironic' literary techniques could also, counter-intuitively, be employed in the service of enchanted science and the rejection of secularism. Indeed, I demonstrate this by delving deep into the work of an intriguing author who, though cited briefly by Saler, was discontented with secular worldviews: Cincinnati pharmacist John Uri Lloyd.[75] Lloyd's self-reflexive hollow earth romance, *Etidorhpa* (1895), challenged his readers' conceptions of scientific method, and of the boundaries between fiction and non-fiction, and irony and earnestness, promoting a radically empirical approach to knowledge. Here, as in many of the texts I examine, non-utilitarian literary form was part of the author's scientific argumentation.

Literary Technologies of Total Visibility

Despite the insights generated by treating these texts as more than pure 'argument', it must be recognized that their authors crafted them persuasively to stake claims about what could be known about the deep past. That might be almost

---

[73] Frow, 2.
[74] Michael Saler, *As If: Modern Enchantment and the Literary Prehistory of Virtual Reality* (Oxford: Oxford University Press, 2012), 15.
[75] Saler, 76–77.

everything; alternatively, it might be nothing outside what was stated in the Bible. In either case, borderline thinkers wished their readers to see geohistory differently. As chapter after chapter of this book shows, often this 'seeing' was quite literal: my time-transcending visionaries include the author of Genesis (as featured in evangelical hypotheses), 'psychometer' Elizabeth Denton, Theosophical clairvoyant Charles Webster Leadbeater, and Seventh-day Adventist prophet Ellen White. While certain visionaries were concerned with helping others to peer more perceptively into prehistory, vision could be a zero-sum game. White's followers demonstrated that, as she had personally verified that the world was created in six literal days, in the manner described in Genesis, other attempts to envision Earth's history were nefarious. In this case, the scaffolds built up to assist the wider public imaginatively to see the prehistoric past, a triumph of nineteenth-century scientific communication, had to be dismantled by White's geological acolyte, Price.

To understand clashes over visual access to long lost aeons, I employ a now-classic concept from the history of science. In *Leviathan and the Air-Pump* (1985), Steven Shapin and Simon Schaffer, speaking of early modern experimental science, argued that 'three *technologies*' were developed in the attempt to establish 'matters of fact': 'a *material technology*', the scientific instrument; 'a *social technology*', the relationships between natural philosophers; and 'a *literary technology*'.[76] I am concerned with this final 'technology', which, in Shapin and Schaffer account, consisted of verisimilar accounts of experiments, published in scientific journals or the like, designed to make a reader into a 'virtual witness' to that experiment.[77] Claims about geohistory were not so easily reduced to experiments, virtual or otherwise. Nonetheless, palaeoscience's own rhetorical tools have been explored by Rudwick, whose influential 1976 article on geology's 'visual language' showed how the development of genres like the stratigraphic section allowed for the communication of considerable quantities of theoretical content in the early nineteenth century.[78] Rudwick went on explicitly to argue that lifelike illustrated 'scenes from deep time'—such as paintings of duelling dinosaurs—were a way of recruiting wider audiences as Shapin-and-Schaffer-style 'virtual witnesses' to the otherwise unseeable events of prehistory.[79] The notion of prehistoric scenes as means of recruiting 'witnesses' has been developed by Victoria E. M. Cain, who examines how the paintings on display in the American Museum of Natural History in the

---

[76] Steven Shapin and Simon Schaffer, *Leviathan and the Air-Pump: Hobbes, Boyle, and the Experimental Life*, new edn (Princeton, NJ: Princeton University Press, 2011), 25.
[77] Shapin and Schaffer, 61.
[78] Martin J. S. Rudwick, 'The Emergence of a Visual Language for Geological Science 1760–1840', *History of Science*, 14 (1976), 149–95.
[79] Martin J. S. Rudwick, *Scenes from Deep Time: Early Pictorial Representations of the Prehistoric World* (Chicago: University of Chicago Press, 1992), 1. See also Rudwick, *Bursting the Limits of Time*, 74–75.

twentieth century were used to instil Osborn's controversial views about palaeoanthropology into visitors (although Cain also shows that serious complications diluted the instilling process).[80]

Indeed, the illusionism of palaeontological art, models, and mounted skeletons has been a major subject of scholarly analysis.[81] On paper, successfully recruiting witnesses for unseeable events required literary technologies of immense power, presented by authors of overwhelming authoritativeness. As scholars have noted, the language of many of the most eminent palaeoscientific writers of the past several centuries was strikingly pictorial or rich in visual metaphor. For titans like Buffon, Horace Bénédict de Saussure, Alexander von Humboldt, Lyell, Darwin, John Lubbock, and Eduard Suess, reconstructing geohistory was figuratively akin to taking a panoramic view only available to the most clear-sighted of savants.[82] Adelene Buckland observes that these men uniquely claimed 'the power to range across seemingly unbridgeable chasms of time and space', making time-transcending vision fundamental to geologists' self-fashioning, as well as providing a justification for colonial supremacy.[83] While they assembled the epic fragments of Earth's history, a far wider assortment of figures helped to develop the literary and visual technologies with which virtual witnesses would be appealed to. Thus eloquent geologists and disparate writers of all kinds, artists, lecturers, modellers, and museum-builders contributed to the project of making prehistory visible both in the eye and in the mind's eye. These were the tools of the 'fantasy of total visibility' that, in O'Connor's words, has driven the presentation of prehistory.[84]

As I noted at the start, this book begins when the epic narrative of Earth's history—in the eyes of some, an evolutionary epic narrative—was already firmly established in European and North American science and widely popular with general audiences. The attempt to recruit virtual witnesses to this narrative had been extremely effective: O'Connor's *The Earth on Show* (2007) and Rudwick's *Worlds Before Adam* (2008), two major studies of the intellectual and cultural emergence of geology and palaeontology, end between the 1840s and the 1850s, with the geohistorical narrative, and the literary and visual technologies through which it was conceived and communicated, substantially constructed. I show what happened to these technologies next in the little-understood publications of borderline researchers, where practitioners, recognizing that they were writing against

[80] Victoria E. M. Cain, '"The Direct Medium of the Vision": Visual Education, Virtual Witnessing and the Prehistoric Past at the American Museum of Natural History, 1890–1923', *Journal of Visual Culture*, 9 (2010), 284–303.
[81] Rieppel, chapters 5 and 6.
[82] Heringman, 97, 197 (Buffon and Lubbock); Rudwick, *Bursting the Limits of Time*, 17, 233 (Saussure); Secord, 'Global Geology', 401–5 (Humboldt and Suess).
[83] Adelene Buckland, '"Inhabitants of the Same World": The Colonial History of Geological Time', *Philological Quarterly*, 97 (2018), 219–40 (224).
[84] O'Connor, *Earth on Show*, 432.

the grain, were attentive to the workings of the literary technologies of virtual witnessing.

Price's project was to dismantle these technologies, dismissing the likes of stratigraphic sections, geological columns, and evolutionary displays as unreal. However, the enhanced anthropocentrism that I have argued motivated borderline researchers, and the possibility of transcending time that fascinated them, was frequently inspired less by opposition to mainstream palaeoscience than by an outsized enthusiasm. While their distance from respectable scientific practice was, in these cases, a matter of degree rather than kind, the degree could be very large indeed. In chapter 2, I examine the transcripts and sketches provided by the 'psychometric' Denton family in the 1860s and 1870s—evidence, they claimed, of mental journeys back to prehistoric times. This might be considered the ultimate expression of a participatory scientific practice rendering scientific resources and facilities superfluous, were it not for the fact that these mental journeys were activated by holding fossils and rocks collected by the family's geologist patriarch, William Denton. While the psychometers' claims were extreme, their ways of seeing can regularly be traced back to the geological writing conventions and scenes from deep time in which the Dentons' psychic imaginations were steeped. Engagement with the more schematized literary technologies of palaeoscience could, alternatively, be foregrounded. In the same chapter we shall see the Theosophist Curuppumullage Jinarajadasa commissioning redesigns of famous scientific diagrams, with added occult dimensions. Unlike Price, Jinarajadasa engaged with this visual language not to refute scientists' understanding of deep time, but rather to show how Theosophy's understanding of the universe went one step further.

Between these extremes, it was possible more cautiously to simulate witnessing. The 'day-age' creationists I discuss in chapter 1 argued that the creation story in Genesis, when broadly interpreted, was the result of the author's divinely revealed glimpse of past geological periods. To prove their argument, they placed readers in the position of the Genesis seer, credibly explaining how his experiences had resulted in the brief biblical text by experimenting with different literary genres, from the humble children's 'conversation' to the epic poem, English literature's most prestigious form. These day-agers counted on the persuasive power of virtual witnessing, but theorists who kept claims about lost civilizations on the margins of respectability felt that only small doses were healthy. Thus, when politician Ignatius Donnelly's book, *Atlantis: The Antediluvian World* (1882), a major subject of chapter 4, brought new feasibility to the story of that lost city, the closest readers came to virtually seeing Atlantis was through second-hand allusion to a recent *voyage extraordinare* by Jules Verne. All other sights, as appropriate for a work of scientific induction, were indirect—a quotation from Plato here, an engraving of an Aztec glyph here. The urge to provide more virtual viewing material than could strictly be accounted for was a strong one. Scottish scholar-poet Lewis Spence closed *The Problem of Atlantis* (1924) by admitting that the intuition of his mind's eye played a stronger role in proving the existence of Atlantis than any of the concrete evidence he had striven to provide.

* * *

The world of borderline palaeoscience was a vast one. The five fields of enquiry I have chosen to evidence my arguments, while far from comprehensive, exhibit a representative range of authors, contrasting genres, literary technologies, and target demographics. Plenteous threads of connection will recur between these five fields, including shared obsessions with contesting the claims of Charles Lyell, shared doubts about humans' genealogical connections with other animals, and a shared refusal to believe that mammoths had lived in cold climates. Chapter 1 discusses the nineteenth-century heyday of the theory that the six *days* of creation in Genesis represented six prehistoric *ages* witnessed by the inspired author. 'Day-age' vision theory put human presence into deep time, balancing scientism with religious experientialism and spectacle. I argue that evangelical proponents of the theory circumvented the sceptical attitudes of intellectual journals by appealing to alternate literary forms of persuasion. The principle of vision is taken further in chapter 2, which examines occult researchers who allegedly saw through time. I argue that visionaries like the Denton psychometers and Theosophical clairvoyants literalized palaeontologists' figurative claims to have revivified the past, motivated by Romantic faith in the truthfulness of imagination. I also show how their works influenced two twentieth-century pioneers of science fiction: H. P. Lovecraft and E. T. Bell.

In chapter 3, I turn to the legacy of American officer John Cleves Symmes Jr, who declared in 1818 that Earth is hollow, with vast holes at each pole. His geotheory was revived in the 1870s, inspiring a spectrum of publications ranging from utopian fiction to non-fiction monographs. I argue that these hollow-earth theorists took a ludic approach to truth claims, the ambiguous framing of their works embodying their authors' challenges to the exclusivity and naturalism of elite science. The final section of the chapter considers the mainstream appropriation of the hollow earth in Edgar Rice Burroughs's Pellucidar stories. Lauded by pulp magazine editors for their restless and manly ethos, even these stories were not without their own playful approach to authorial framing and unexpected forays into the occult.

Keeping in the realm of lost worlds, chapter 4 examines research on diluvial catastrophes that wiped out ancient populations and sank continents. Books by men thinking along these lines were often taken seriously—whether as convincing, intriguing, or dangerous—in mainstream periodicals and scientific journals. Beginning with Ignatius Donnelly, I argue that these catastrophists developed the 'suppositional synthesis': a genre in which empirical aesthetics worked in productive tension with sensational claims. These thinkers, I contend, also tapped into a widespread cultural sympathy for independent-minded polymaths working on a proudly 'literary' model of scientific method. Chapter 5 returns to creationism. Whereas earlier scholars have focused on the social and theological aspects of the early twentieth-century revival of young-earth creationism, I show how its textual strategies were founded on a rejection not just of mainstream geology and palaeontology but also of their visual languages and literary technologies. Finally, in an epilogue, I explore the changing fortunes of borderline palaeoscience in the decades after the Second World War.

As these chapters cumulatively demonstrate, *Contesting Earth's History* takes us through deep and shallow time alike, ranging from the bottom of the ocean to the centre of the planet and diving with equal enthusiasm into the pages of heavy textbooks and those of thrilling adventure novels. We will begin with a group of thinkers captivated by the notion that the story of creation outlined at the start of Genesis betrayed evidence of the author's optical encounter with the prehistoric world.

1
Envisioning the Days of Genesis

On the left of the chart, a yellow sun pronounces God's creation of the heavens and the Earth (Figure 1.1). Curved lines project rightwards, subdivided into six creative days—and the seventh, on which God rested. Each day is accompanied by the appropriate quotation from Genesis. On the right, however, the quotations are followed by scientific information about different eras of geohistory, above corresponding illustrations of dinosaurs, pterodactyls, and mammoths. The message is clear: the Genesis days—known collectively as the 'Hexaemeron'—were not twenty-four-hour periods, but rather long geological ages. John William Grover, a respected London civil engineer, designed this elegant lithograph for his *Conversations with Little Geologists on the Six Days of Creation, Illustrated with a Geological Chart* (1878). 'Familiar' conversations, a genre of scripts in which adults inform children about nature, had once been the premiere model of popular science.[1] Grover's chart displayed the Bible's compatibility with palaeoscience, with the script adding further elaborations (including a refutation of the 'foolish' theory of evolution).[2] This compatibility, as the conversation's earnest young characters discover, will bring new religio-scientific awe into daily life. George, for example, recognizes that the Portland stone of which St Paul's Cathedral is constructed came from the Jurassic period—part of the fifth 'day-age' of Genesis. Now, 'when I look at St. Paul's', he says, 'I shall think about the fifth day, and the great lizards which lived in those times'.[3]

The conversation genre was an unfashionable model of science popularization by the 1870s, when Grover's chart was published. One might think that his day-age interpretation of Genesis was equally passé. Despite its Christian proponents' claims that palaeoscience presented startling circumstantial evidence for the divine inspiration and accuracy of the Bible's creation story, the day-age position enjoyed support from few leading British naturalists during the late nineteenth century. Significantly, Andrew J. Brown's authoritative account of day-age theory ends around 1860, when the theory 'passed its peak in the mainstream scientific world' due not least to robust scholarly challenges to the historicity

---

[1] Melanie Keene, 'Familiar Science in Nineteenth-Century Britain', *History of Science*, 52 (2014), 53–71.
[2] John William Grover, *Conversations with Little Geologists on the Six Days of Creation Illustrated with a Geological Chart* (London: Edward Stanford, 1878), 10.
[3] Grover, 18.

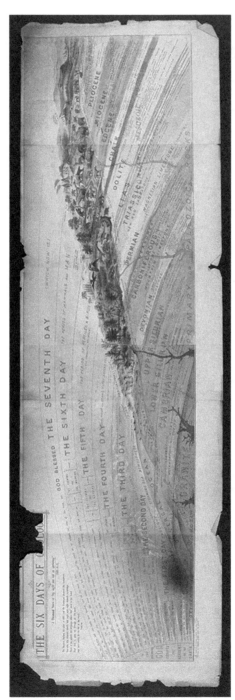

Figure 1.1 John William Grover, *Conversations with Little Geologists on the Six Days of Creation, Illustrated with a Geological Chart* (London: Edward Stanford, 1878). Grover's lithographic chart demonstrates the compatibility between the six-day creation account in Genesis and the ages of geology. © The British Library Board, 7202.i.1.

of the events described in the Bible.[4] It was not, however, ejected from scholarly circles. As Brown admits, the theory retained currency in North America, especially among Calvinist savants, and its promotion into the 1890s by the Canadian geologist John William Dawson has often been discussed by scholars.[5] In the British context, a mid-1880s clash between pious politician W. E. Gladstone and scientific naturalist T. H. Huxley is usually seen as the theory's last public gasp.[6] Until now, little concerted attention has been paid to its other late-century manifestations.

I contend that the importance of day-age theory in this period has been underrated. This fact does not reflect lack of attention to the strategies of reconciling Genesis with geology, which have been discussed almost to exhaustion by scholars of science and religion.[7] Rather, it is the *literary technologies* of day-age theory—to maintain Simon Schaffer and Steven Shapin's influential term, discussed in my introduction—that have rarely been considered, and never with regard to the late nineteenth century.[8] This oversight is significant, because I argue that it is the literary aspect of the theory that is of most interest in this period. Discussing in new detail famous proponents like Dawson and Gladstone, as well as revealing the significance of many lesser-known figures, this chapter shows that day-age theory provided scope for the exercise of literary skill in works ranging from obscure labours of love to texts hotly debated across the periodical press. In a trend that will recur throughout *Contesting Earth's History*, proponents attempted to circumvent the standards of evaluation required in the specialized scientific press and in liberal intellectual periodicals, in this case adopting carefully chosen literary genres and strategic publication routes. Perhaps surprisingly, many proponents shared with opponents the desire to portray day-age theory—which, given its shared reliance on geology, philology, theology, and what today's literary scholars would call 'close reading' skills, could not be pronounced upon with sole authority by any discipline—as a subject uniquely available for evaluation by laypeople. Friends and foes alike thus deemed the theory's persuasiveness a fit topic for the public sphere, rather than one best restricted to a scholarly cloister where the experts knew best (even though various titans

---

[4] Andrew J. Brown, *The Days of Creation: A History of Christian Interpretation of Genesis 1:1–2:3* (Leiden: Brill, 2012), 220. My account of the early history of day-age theory is indebted to Brown.

[5] Susan Sheets-Pyenson, *John William Dawson: Faith, Hope, and Science* (Montreal: McGill-Queen's University Press, 1996); Nanna Katrine Lüders Kaalund, 'Of Rocks and "Men": The Cosmogony of John William Dawson', in *Historicizing Humans: Deep Time, Evolution, and Race in Nineteenth-Century British Science*, ed. by Efram Sera-Shriar (Pittsburgh, PA: University of Pittsburgh Press, 2018), 44–67.

[6] David Bebbington, *The Mind of Gladstone: Religion, Homer, and Politics* (Oxford: Oxford University Press, 2004), chapter 8.

[7] For a classic overview, see James R. Moore, 'Geologists and Interpreters of Genesis in the Nineteenth Century', in *God and Nature: Historical Essays on the Encounter between Christianity and Science*, ed. by David C. Lindberg and Ronald L. Numbers (Berkeley: University of California Press, 1986), 322–50.

[8] For the only sustained literary analysis of day-age writings, see Ralph O'Connor, *The Earth on Show: Fossils and the Poetics of Popular Science, 1802–1856* (Chicago: University of Chicago Press, 2007), chapter 10.

of nineteenth-century geology and theology weighed in alongside more humble voices).

In addition to the levelling aspect of its non-specialist accessibility, what made day-age theory so enduring was its direct appeal to another persistent theme central to this book: vision. The fact that the geological accuracy of the Hexaemeron was usually explained as the result of a miraculous, time-transcending vision undergone by the author, whether Moses or an earlier patriarchal seer (often called the 'Mosaist'), gave day-age theory a distinctly visual, experiential quality. The theory became a subject in which authors' convincing, verisimilar textual presentation of how the seer had *seen* became, in effect, evidence for the theory's validity: each reader was invited to become a virtual Mosaist. Tellingly, the exegetical and scientific arguments rehearsed by proponents of the theory between the 1850s and the 1900s were fairly similar. It was, rather, their ability to breathe new literary life into these hackneyed claims—their success in recruiting virtual witnesses— that really counted. This imaginative aspect of day-age theory led to speculative reconstructions of the seer's experience in which authors navigated the Bible's use of poetry and history, literalism and metaphor, realism and artistic convention. Although these required imaginative exertion, they spoke to a desire to go beyond imagination: the desire to believe that *somebody* had seen the prehistoric past, and God's work, first-hand. For sceptics, however, these supposed visions were too obviously mere literary constructions.

Pictorial Representation of the Creation

A robust vein of ancient Christian tradition legitimized the belief that Genesis did not refer to literal, twenty-four-hour days, but the interpretation of the Hexaemeral days as geological periods developed alongside the birth of geology itself in eighteenth-century Francophone science.[9] The notion received support in Britain during the early nineteenth century, when geologists demonstrated that an age of sea monsters had been followed by one of land mammals—not unlike the trajectory laid out in the fifth and sixth days of Genesis's creation story, in which sea and land creatures, respectively, are formed. That a happy concordance between geological ages and Hexaemeron (flexibly interpreted) had been discovered by accident became a refrain, just as apologists had long found in pagan myths 'collateral evidences' supporting Christianity.[10] Proponents often arranged these concordances in columnar form, assigning the lines of Genesis to relevant

---

[9] For day-age theory's older precedents, see Brown, *Days of Creation*, esp. chapter 2. For early geological versions, see Martin J. S. Rudwick, *Bursting the Limits of Time: The Reconstruction of Geohistory in the Age of Revolution* (Chicago: University of Chicago Press, 2005), 148, 328–29, 447.

[10] See James Parkinson, *Organic Remains of a Former World: An Examination of the Mineralized Remains of the Vegetables and Animals of the Antediluvian World; Generally Termed Extraneous Fossils*, 3 vols (London: Sherwood, Neely, and Jones, 1804–11), III, 451–52. For 'collateral evidences', see Colin Kidd, *The World of Mr Casaubon: Britain's Wars of Mythography, 1700–1870* (Cambridge: Cambridge University Press, 2016), 39.

geological ages.[11] It was eagerly stressed that the Hebrew word used for 'day' in Genesis, '*yom*', was employed figuratively elsewhere in the Bible.

Day-age theory initially struggled against other ways of explaining geology on Christian terms, whether straightforward young-earth creationism (discussed in chapter 5) or the mediatory 'gap theory', which placed the geological ages in a period before God created our present world in literal days. By the 1840s, however, Christian geologists found these two alternatives increasingly hard to swallow. The German 'higher criticism', with its mercilessly historicized interpretations of biblical authorship, and challenge to the veracity of much of the Bible's factual content, was slow to establish itself in Britain, but progressives increasingly adopted the position of Anglican mathematician Baden Powell, who admitted that discrepancies and inaccuracies in scripture could not be argued away.[12] If liberal Christians therefore accepted a necessary non-concordance between Genesis and geology, conservatives sought watertight evidence of the Bible's revealed wisdom. Day-age theory thus became a valuable counterargument both to critics of the Bible's divine inspiration and to attempts to erect boundaries between religious and scientific knowledge. Among its early British commentators can be numbered the chemist Humphry Davy and the novelist Anne Brontë.[13]

Interpretations of Genesis were hotly debated in colleges and theological seminaries on the other side of the Atlantic too.[14] A major reason for day-age theory's eminence in the United States was the arrival, in 1848, of Swiss geologist-geographer Arnold Guyot, who soon took up a position at the College of New Jersey (Princeton). In the winter of 1840, day-age theory had 'flashed upon my mind', Guyot recalled, like a Pauline conversion.[15] He converted James Dwight Dana, geology professor at Yale College, to day-age theory at a memorable meeting in August 1850.[16] Meanwhile, in Canada, the up-and-coming young geologist John William Dawson was hinting to his famous British colleague, Charles Lyell, that Christian geologists needed to take measures to refute liberal theology. In November 1846 he told Lyell (a deist who disliked reconciliation schemes) that geologists had to undertake 'a thorough and scientific enquiry' in order to combat the 'shallow' theories of 'English critics' like 'Prof Powell'.[17] Over the 1850s, Dawson aired his own day-age scheme in lectures and drafted a rebuttal to the 'flimsy'

---

[11] Brown, *The Days of Creation*, 229.

[12] For higher criticism, see Michael C. Carhart, *The Science of Culture in Enlightenment Germany* (Cambridge, MA: Harvard University Press, 2007), chapter 5. For Baden Powell, see Pietro Corsi, *Science and Religion: Baden Powell and the Anglican Debate, 1800–1860* (Cambridge: Cambridge University Press, 1988), 134.

[13] Clare Flaherty, 'A Recently Rediscovered Unpublished Manuscript: The Influence of Sir Humphry Davy on Anne Brontë', *Brontë Studies*, 38 (2013), 30–41.

[14] Rodney L. Stiling, 'Scriptural Geology in America', in *Evangelicals and Science in Historical Perspective*, ed. by David N. Livingstone, D. G. Hart, and Mark A. Noll (Oxford: Oxford University Press, 1999), 177–92.

[15] Arnold Guyot, *Creation; or, The Biblical Cosmogony in the Light of Modern Science* (New York: Charles Scribner, 1884), vii.

[16] James Dwight Dana, 'Creation; or, the Biblical Cosmogony in the Light of Modern Science', *Bibliotheca Sacra*, 42 (1885), 201–24 (221).

[17] John William Dawson to Charles Lyell, 14 November 1846, Papers of Sir John William Dawson (1820–1899), GB237 Coll-192, Heritage Collection, University of Edinburgh Library.

permissive theology of 'Powell and others'.[18] The Congregationalist Dana and Presbyterians Guyot and Dawson all belonged, notably, to the Reformed branch of Protestantism, with its historic predilection for rational and biblically affirming scientific enquiry.[19] This triumvirate, further energized by their evangelical inheritance, gave day-age theory scientific respectability in North America until the century's end.

Although many attributed the theory's emergence to independent intuition, as in Guyot's epiphany, its subsequent theorization was shaped by the artistic metaphors of German Protestant theologians. Pioneering biblical critic Johann Gottfried Eichhorn had argued that the Mosaist's account was 'more pictures and poems than pure history'.[20] Eichhorn was a notoriously liberal interpreter intent on historicizing scripture, but the idea of the creation story as pictorial-poetic did not necessarily imply scepticism about its facticity. Conservative Pietist Georg Christian Knapp argued, in a work translated into English in 1831, subsequently a classic in American seminaries, that the creation story was a '*picture*' intended to be intelligible to 'common men'.[21] Elaborating, Knapp compared the Hexaemeron to 'six separate *pictures*' based on 'natural truth' but not painted with 'mathematical or scientific accuracy', and as such the Bible's 'pictorial representation of the creation' was not 'literally and exactly true'.[22] More meticulously, the Lutheran Johann Heinrich Kurtz theorized that the story consisted of six 'prophetico-historical tableaux', 'scenes' of a creation 'drama' revealed to the inner eye of a witness who described it '*as* he had seen it.'[23] This talk of intelligibility, theatricality, and artistic licence would intrigue subsequent thinkers.

Neither Knapp nor Kurtz subscribed to day-age theory, but the notion of Genesis as a vision became wedded with this interpretation in the anglophone sphere. In a series of essays and lectures, compiled in his posthumously published book *The Testimony of the Rocks* (1857), Scottish Free Church (Presbyterian) geologist Hugh Miller simulated how the Mosaist's visionary experience of prehistory might have resulted in Genesis's terse, six-day framework. Miller's 1854 lecture in London's Exeter Hall, for instance, recounted a primitive sociological experiment: he claimed that labourers visiting the British Museum's fossil galleries were, in his experience, most struck by the very same specimens that the Mosaist had wished to inform his similarly intelligent but unlettered contemporaries about.[24]

---

[18] Dawson to Lyell, received 26 November 1859, Papers of Sir John William Dawson.

[19] George M. Marsden, *Fundamentalism and American Culture*, 2nd edn (Oxford: Oxford University Press, 2006), 14–17; David N. Livingstone, *Dealing with Darwin: Place, Politics, and Rhetoric in Religious Engagements with Evolution* (Baltimore, MD: Johns Hopkins University Press, 2014), 49.

[20] Quoted in Carhart, 169.

[21] George [sic] Christian Knapp, *Lectures on Christian Theology*, 2 vols (New York: G. & C. Carvill, 1831–33), I, 356; Andrew Kloes, *The German Awakening: Protestant Renewal After the Enlightenment, 1815–1848* (Oxford: Oxford University Press, 2019), 137.

[22] Knapp, I, 364.

[23] John Henry Kurtz, *The Bible and Astronomy; An Exposition of the Biblical Cosmology, and Its Relations to Natural Science*, 3rd rev. German edn, trans. by T. D. Simonton (Philadelphia, PA: Lindsay & Blakiston, 1857), 110.

[24] Hugh Miller, *The Testimony of the Rocks; or, Geology in Its Bearing on the Two Theologies, Natural and Revealed* (Edinburgh: Thomas Constable, 1857), 144–46.

This explained why the Genesis story lacked attention to so many geologically important but visually uninteresting details.

Elsewhere, he hinted that evidence from imaginative literature aided interpretation. John Milton's revered epic poem *Paradise Lost* (1667; 1674), Miller pointed out, contains two relevant depictions of the logistics of revelation: verbal, as in archangel Raphael's relation of the creation story to Adam, and visual, as in archangel Michael's relation of the future. Here, Michael removes the 'film' from Adam's eyes with a herbal mixture, piercing 'to the inmost seat of mental sight' and causing Adam's 'spirits' to become 'entranced'.[25] Citing Kurtz, Miller judged that the Genesis account bore the hallmarks of Michael's visual communication. Employing mellifluous prose, Miller developed this idea by fleshing out the metaphors of visual spectacle established by German predecessors, imagining Genesis as a geological series of 'prophetico-historical tableaux' presented to the Mosaist—envisioned as resembling the operations of a modern theatrical device, the diorama.[26] Finally, he called for an audacious successor to Milton who would pen the scientific *Paradise Lost* of the nineteenth century (and, as Ralph O'Connor observes, implicitly demonstrated his own fitness for the task).[27]

Miller's death by suicide in 1856 cut this project short, but his writing cemented day-age theory as the most prominent Genesis-geology reconciliation scheme, worthy of high literary endeavour—a visionary theory calling for visionary representation. Reviewing his argument, the Protestant *Dublin University Magazine* declared that 'nothing could be more convincing'.[28] This being said, day-age theory remained extremely controversial. It was subject to heavy criticism, Brown notes, from both the most conservative and the most liberal biblical interpreters.[29] The progressive *Westminster Review*, for example, crushingly characterized Miller's theory as the 'cooking of a case', the fraudulent concoction of which had 'racked his brain to suicidal insanity', arguing that it presented especial danger when appearing to enjoy official Free Church endorsement.[30] Similar objections, albeit expressed in calmer terms, were common, while most Christian critics considered the case attractive but likely unprovable.

During the 1860s, the equipoise around reconciliation schemes crumbled in the British intellectual sphere, not least because the publication of the so-called *Essays and Reviews* (1860) by a group of liberal churchmen finally brought German-style historicized interpretations of the Bible into the mainstream of religious discussion.[31] Palaeoscience was shedding its Christian scaffolds, and the

---

[25] John Milton, *Paradise Lost*, ed. by Stephen Orgel and Jonathan Goldberg (Oxford: Oxford University Press, 2004), XI.412–20.
[26] O'Connor, *Earth on Show*, chapter 10, esp. 411–15.
[27] O'Connor, *Earth on Show*, 414.
[28] 'Hugh Miller and Geology', *Dublin University Magazine*, 50 (1857), 596–610 (608).
[29] Brown, *The Days of Creation*, 277.
[30] Untitled review of *The Testimony of the Rocks*, by Hugh Miller, *Westminster Review*, 68 (1857), 176–85 (181, 184).
[31] Brown, *Days of Creation*, 269–75.

geological elite's fairly rapid conversion to belief in evolution and human antiquity, facilitated by the efforts of scientific naturalists like T. H. Huxley, further weakened the case for exact concordance.[32] Questions of how the Hexaemeron might correspond with geological epochs became off-limits topics in most highly regarded scientific journals and technical monographs. The technical *Geological Magazine* (founded 1864), despite its otherwise open-minded approach to contributors' idiosyncrasies, had little time for explicitly exegetical interpretations of geology, and *Nature* (founded 1869) was consistently hostile. Conservatives opposed to overly creative interpretation were merciless too: German theologian Franz Delitzsch pointed out the utter absence of textual evidence in the Bible for his fellow Lutheran Kurtz's claims about prophetico-historical tableaux.[33] Day-age theory was losing respectable publication venues in which to be taken seriously.

In North America, however, the theory's sun had barely risen. In 1863, it received its most significant reification in Dana's authoritative *Manual of Geology*.[34] Nonetheless, even in this more amenable climate, day-agers had to move with the times. Dawson's first treatise on the subject, *Archaia* (drafted by 1855 and published in 1860), was, Nanna Katrine Lüders Kaalund notes, a 'commercial failure'.[35] Over the following decade, Dawson, who had become Principal of McGill College, Montreal, in 1855, was disturbed by the influence of the higher criticism and scientific naturalism on the other side of the Atlantic, but he admitted that reconciliation schemes might be premature. Regarding human antiquity, geologists, he told Lyell, must 'necessarily have a longer chronology than that of the Bible', although such 'apparently' different views may one day be reconciled'.[36] This willingness to accept short-term discordance manifested in Dawson's unusual concordance scheme, which placed all known geological periods in the fifth and sixth days of Genesis. As a result, the biblical creation of plants on the third day was as yet unreconcilable with science. This was bold, as prehistoric plants were Dawson's speciality, but, conveniently, plants were also a thorn in the side of evolutionists.[37] Elsewhere, promising evidence emerged in support of the Bible's accuracy. In May 1861, Dawson expressed pleasure at the profusion of 'materials for a new chapter of *Archaia*', including the fossil discoveries of palaeontologist Hugh Falconer, whose 'Mediterranean river with hippopotami', Dawson believed,

---

[32] Nicolaas Rupke, 'Down to Earth: Untangling the Secular from the Sacred in Late-Modern Geology', in *Science without God: Rethinking the History of Scientific Naturalism*, ed. by Peter Harrison and Jon H. Roberts (Oxford: Oxford University Press, 2019), 182–96.

[33] Carl Friedrich Keil and Franz Delitzsch, *Biblical Commentary on the Old Testament*, 4 vols, trans. by James Martin (Edinburgh: T. & T. Clark, 1872), I, 44–46.

[34] James Dwight Dana, *Manual of Geology: Treating of the Principles of the Science with Special Reference to American Geological History, for the Use of Colleges, Academies, and Schools of Science* (Philadelphia, PA: Theodore Bliss, 1863), 741–46.

[35] Kaalund, 44.

[36] Dawson to Lyell, 11 March 1863, Papers of Sir John William Dawson.

[37] See Dawson to Lyell, 26 February 1862, Papers of Sir John William Dawson.

was potentially the mysterious Edenic river Gihon.[38] In the winter of 1866, he and his wife, Margaret, invited students into their home to hold evening discussions on 'the Natural History and Ethnology of Genesis'.[39]

Behind the Veil

The following decades would see the theory take on new literary forms. At Christmas 1871, English naturalist Cuthbert Collingwood sent copies of his new day-age epic poem, *A Vision of Creation*, published by Longmans and Green, to potentially receptive figures across the literary and scientific world. One, addressed to anti-evolutionary palaeontologist Louis Agassiz of Harvard, is the copy digitized today on Google Books; another went to Lord Tennyson, presumably intended to offer additional scientific grounding for the hard-won faith expressed in the Poet Laureate's lyric cycle *In Memoriam* (1850).[40] This was the most ambitious attempt to promote day-age theory to the elites of literature and science since Miller's heyday.

The most substantial study of Collingwood's life, an article written by Nora Fisher McMillan, focuses on his career as a travelling naturalist and Fellow of the Linnean Society, including his participation in the earliest debates over Charles Darwin's interpretation of evolution, of which he was a lifelong opponent.[41] Collingwood's poem received little attention from McMillan and his increasingly marginal religious beliefs have, until now, remained effectively unexplored. Medically trained, Collingwood spent the period from 1858 to 1866 in Liverpool, where he held various posts, including Lecturer in Botany at Royal Infirmary Medical School. After an unsatisfactory span as a ship's surgeon-naturalist between 1866 and 1867, Collingwood moved to London. During this period, he struggled to find scientific work, unsuccessfully lobbying the famous botanist Joseph Dalton Hooker to support his application for Sherardian Professor of Botany at Oxford.[42] In these years he was becoming alienated from the scientific community: as he wrote to the curator of the Liverpool Museum in October 1870, Collingwood had been disappointed not to receive more invitations from friends to attend the recent meeting of the British Association in Liverpool. His travels, he suspected, had

---

[38] Dawson to Lyell, 15 May 1861, Papers of Sir John William Dawson.
[39] Enclosed in Dawson to Lyell, 12 December 1866, Papers of Sir John William Dawson.
[40] Dennis R. Dean, 'Tennyson and Creation', *Tennyson Research Bulletin*, 9 (2007), 22–41. For geology and *In Memoriam*, see Michelle Geric, *Tennyson and Geology: Poetry and Poetics* (Cham: Palgrave Macmillan, 2017), chapter 4.
[41] Nora Fisher McMillan, 'Picture Quiz: Cuthbert Collingwood (1826–1908)', *Linnean*, 17 (2001), 9–20.
[42] Cuthbert Collingwood to Joseph Dalton Hooker, 4 June 1868, modern handwritten copy, Collingwood, Dr. Cuthbert; with Research Papers of Mrs Nora McMillan, MBE, MSc, MRIA (1853)–[1990s], D859/9/11, quoted by courtesy of the University of Liverpool Library.

rendered him 'out of sight' and 'out of mind', and as a result he chose not to attend the event.[43]

A Vision of Creation was an opportunity to contribute to science in a different manner. Collingwood's introductory matter explained that the poem was intended to put day-age theory 'in a pictorial form and in a vivid light', constituting a irenic 'step' towards reconciling the 'unscientific Christian' and 'the free-thinking man of science'.[44] Citing Knapp and Kurtz (xv–vi), he opposed the liberal view, promulgated in *Essays and Reviews*, that the creation story was purely symbolic, insisting that 'we are seeking for something *besides* symbolism' (xvi). Text and image, prose and poetry were united in Collingwood's multimedia characterization of the creation story as 'a *poetic* picture' but also a 'true transcript' of a vision (xvii). Offering the day-age theorist's conventional claim to originality, in this case stating that he had read Miller's *Testimony* only after having begun his research (vii), Collingwood nonetheless used as his epigraph Miller's call for a geologically up-to-date Victorian Milton, unequivocally implying that he, Collingwood, was the man for the job.

A Vision of Creation is an epic blank verse poem in sixteen short books (each with prose synopsis). It begins, in a 'Proem', with the epic poet's conventional call for inspiration, asking God Himself to 'instruct/My muse aright' (3). What follows is the story of a seer from 'a primitive and patriarchal age' who finds himself '[w]rapp'd in an awful trance' (7):

> Pervaded with a godlike consciousness
> Of Nature, and of secrets buried deep
> In Nature's womb. Oh that I could but look
> Behind the veil! Oh that I might conceive
> The mystery of Creation! (8)

Most readers would have recognized a verbal echo of Canto LVI of *In Memoriam* here—not least Tennyson himself, if he read the poem he was sent—when Tennyson's speaker is at his most desperate. Agonizing over the dubiety of life after death in a world where God has allowed so many organisms, such as the prehistoric '[d]ragons of the prime', to fall extinct, the speaker longs for 'answer or redress/Behind the veil, behind the veil!'[45] Unlike Tennyson's, Collingwood's speaker's hunger for gnostic knowledge is immediately satisfied when an archangel appears on behalf of God, who wishes that 'His acts/Should not lack witness' (54). Transporting the seer to a mountain-top, the archangel presents, in now-familiar terms, 'a series of living pictures . . . that shall occupy six day-like periods'

---

[43] Collingwood to T. J. Moore, 6 October 1870, Collingwood, Dr. Cuthbert; with Research Papers of Mrs Nora McMillan, D859/11/2.

[44] Cuthbert Collingwood, *A Vision of Creation: A Poem* (London: Longmans, Green, 1872), vi, xiii. Subsequent references included in text.

[45] [Alfred Tennyson], *In Memoriam* (London: Edward Moxon, 1850), 81.

(70). Collingwood's evidence is presented twice: first, in prose argument, and second, from the seer's perspective, in the poem. Scientific details of the latter are clarified in conversation with the archangel, or by paratextual glosses indicating, for example, that the 'creatures terrible*/And dire to look on' in the poem's fifth 'day-like' period are the dinosaurs '*Iguanodon, Megalosaurus,* &c'. (151).

Collingwood's choice to relate the action in the voice of a character, the seer, might appear to lower the poem to the humbler genre of dramatic monologue. After all, his model, *Paradise Lost*, and most other extant epic poems, told their stories using a mediatory narrator, like that briefly appearing in Collingwood's 'Proem'. Rather than moderating his ambition, the decision in fact swells it: the lack of clear distinction between the 'I' of the inspired poet of the 'Proem' and the 'I' of the seer gives the impression that Collingwood himself is a 'patriarchal' visionary. Indeed, this ambitious work shows that the geologico-literary project that O'Connor dubs 'Upstaging Milton', which he sees as reaching its inventive apex in Miller's final works in the 1850s, did not end with Miller's death.[46] For Collingwood, of course, upstaging Milton also entailed upstaging Miller (and, as we have just seen, Tennyson).

By drawing on ancient notions of epic inspiration, Collingwood claimed the highest cultural authority. In so doing, he opted chiefly for thematic rather than formal innovation, borrowing archaic vocabulary and syntactical constructions from *Paradise Lost* and other early modern epic poems. His expository aims, however, demanded a simpler structure. Unlike Milton, who employed the complex Homeric technique of nested storytelling, Collingwood's *Vision* lacks surrounding narrative entirely. As a result, it is less a story than a didactic epic fragment. The model for this fragment, however, does come from *Paradise Lost*: following Miller, Collingwood reconfigured the prophetic narrative of Christ-ward progress communicated by Milton's archangel Michael. Herbert Tucker, in his vast study of nineteenth-century epic, does not discuss Collingwood, although his description of the epics of the century's later decades—eclectic, historicist, scientific vehicles for a 'hyper-myth of progress'—rings true in Collingwood's case.[47] In *Paradise Lost*, Michael's revelation is a chastening tale of the fallen human frailty that will culminate in Christ's redemption. The archangel of *A Vision of Creation*, in contrast, offers up a march of progress towards Collingwood's steam-age present. He admits that humanity will degrade itself with 'ignoble violence' (195), but this comes across as something of an afterthought. While both authors' archangels conclude in a prediction of 'New Heavens, and a New Earth' (208), Collingwood's triumphant millennium, in contrast with the distant salvation prophesized by Michael, seems close at hand.[48]

---

[46] O'Connor, *Earth on Show*, 411.
[47] Herbert F. Tucker, *Epic: Britain's Heroic Muse 1790–1910* (Oxford: Oxford University Press, 2008), 466.
[48] Milton, XII.549 ('New heavens, new earth').

Although the seer's visions—'living pictures' (70)—are explicitly *actual*, rather than symbolic, they are, in the tradition of Knapp, mediated by divine artistry. The unfortunate animals of the Triassic period, for instance, are not revealed by visions, as they lack 'the elements of a picture' (xxviii). This language also hints at Collingwood's familiarity with palaeo-artistry. In 1865, Louis Figuier's book *La Terre avant le déluge* (*The World before the Deluge*) (1863), with its sensational series of engravings of past geological periods by Édouard Riou, had been translated into English. Martin Rudwick sees Figuier's book as conventionalizing the 'scenes from deep time' genre, which, he notes, evolved from the early modern tradition of illustrated biblical scenes, especially those depicting the days of Genesis.[49] The gilt cover of Collingwood's book depicts extinct animals based on Riou's restorations (Grover, too, would purloin designs from Figuier's handy volume). A distinctive reference to the 'restoration' of the Carboniferous plant '*Lepidodendron Sternbergii*' by 'M. Eugène Deslongchamps' (128), moreover, provides further evidence that Collingwood was using Figuier as a source.[50] Looking closer, we find that some of Collingwood's descriptions even appear to be ekphrastic responses to Riou's plates. For instance, compare the seer's description of the sixth day-age with Riou's 'Ideal landscape of the Miocene Period' (Figure 1.2). The 'unwieldy' *Mastodon* and *Deinotherium* (164), Collingwood's 'tuskéd beasts', frequent the 'marshy sward' and 'miry pools' of 'the river's banks' in 'luxurious sunshine' (165). Moving rightwards on Riou's plate keeps us in sync with Collingwood's seer, who observes 'wooded dells' where 'creeping stems'—identified as those '*Bambusinites* and *Smilacites*' that, in Figuier's translated prose, 'interlaced' the Miocene forests—create 'tangled mazes' inhabited by the ape *Dryopithecus*, with 'its foul semblance to humanity' (166).[51]

We have already heard Collingwood characterize his poem as 'pictorial' (vi) in much the same way the Genesis creation story was, he said, both '*poetic* picture' and 'true transcript' of prehistoric events (xvii). Where did poetry end and transcript begin? Understanding *A Vision of Creation* as, in part, a reading of Figuier's book clarifies Collingwood's mixed metaphors. Riou's illustrations, like most scenes from deep time, were labelled as 'ideal scenes': they followed natural history conventions that combined realistic anatomical representation with an ecologically stylized assemblage of species. In the same way, for Collingwood, the Mosaist saw an idealized version of the prehistoric past in which characteristic fauna and flora were clustered: a rearrangement of scenes that truly happened, compositionally tweaked for didactic clarity. Fittingly, Collingwood's ekphrastic response to Riou re-enacted the experience of the biblical seer, who had gropingly attempted

---

[49] Martin J. S. Rudwick, *Scenes from Deep Time: Early Pictorial Representations of the Prehistoric World* (Chicago: University of Chicago Press, 1992), 24, 173.

[50] See Louis Figuier, *The World before the Deluge*, trans. by W. S. O. (London: Chapman and Hall, 1865), 120.

[51] Also compare Figuier, plates XXI and XV, with Collingwood, *A Vision of Creation*, 151–52 and 149, respectively.

to describe the fabulous prehistoric scenes presented to his eyes. Of course, the product of this attempt had been the terse Genesis account, rather different in form to Collingwood's loquacious, revisionist epic. The relationship between biblical and palaeoscientific illustration was coming full circle: rather than reading geohistory into biblical scenes, Collingwood produced biblical scenes from geohistorical ones. The persistently Christian tone of Figuier's text facilitated this reinterpretation.

Nonetheless, Collingwood was not satisfied with his poem's reception—or his theory. Although the second edition excerpted many positive reviews, these were mostly from provincial British newspapers rather than the metropolitan scientific and literary authorities he had courted.[52] Perhaps his Miltonic pastiche seemed worn alongside recent epic experiments like Robert Browning's *The Ring and the Book* (1868). In any case, Collingwood abandoned poetry and, like his inspired seer, began to peer behind the veil more directly. At the exact time *A Vision of Creation* was first being printed, Cuthbert's brother Samuel had persuaded him to join the New Church, a Dissenting group propounding the esoteric Christianity

Figure 1.2 Louis Figuier, *The World before the Deluge*, rev. edn, ed. by Henry W. Bristow (London: Chapman and Hall, 1867), 309. Cuthbert Collingwood's poetic description of Miocene life in *A Vision of Creation* (1872) simulates a left-to-right reading of this 'Ideal landscape of the Miocene Period', illustrated by Édouard Riou. Cadbury Research Library: Special Collections, University of Birmingham.

---

[52] Cuthbert Collingwood, *A Vision of Creation: A Poem*, 2nd edn (Edinburgh: William Paterson, 1875), unpaginated front matter.

of the Swedish mystic Emanuel Swedenborg.[53] This group, always fairly limited in membership in Britain, had become even more marginal since its heyday in Collingwood's youth.[54] In his eyes, though, it was more scientifically respectable than other heterodox religious moments. Writing at the end of his life, he dubbed his fellow naturalist Alfred Russel Wallace, who had become a Spiritualist in the 1860s, 'a "crank" of the first water'.[55] Collingwood's newfound Swedenborgianism shifted his thinking on day-age theory. Sending a copy of *A Vision of Creation* to Dawson in Canada in April 1878, he lamented its lack of 'publicity' but added that he had now '*advanced* a step', darkly alluding to a 'truth which (not self-deceived)' he hoped 'someday' to share with the world, although 'the time' had 'not yet come'.[56] He tested the waters of controversy with an anonymous volume of *New Studies in Christian Theology* (1883), which one reader, Gladstone, branded 'heretical' for its views on the Trinity.[57]

The 'time' came in 1886, when Collingwood's *The Bible and the Age* was published. His scope now was wider than merely the interpretation of Genesis: he aimed to lay out a 'Mystical Interpretation of Scripture' that would establish a proper understanding of the relationship between matter and spirit.[58] Establishing that relationship was important to Collingwood's metaphysics, because, if no such relationship exists, he felt, then perception is an illusion, and the world might as well '[d]issolve' like 'the baseless fabric of a vision'.[59] This allusion to the fairy masque scene in Shakespeare's *The Tempest* cast a troubling retrospective light on his epic poem, in which the seer's 'unsubstantial phantasy' had 'dissolved/Into mere nothingness' at each vision's close (133). Indeed, as Collingwood now admitted, the scientific accuracy of Genesis was 'intangible and illusory'.[60] Rejecting the idea that the Bible was intended to teach science, albeit without entirely renouncing reconciliation, Collingwood now argued that the esoteric concept of 'correspondences' between material and spiritual planes, of the kind described by Swedenborg, were far more important.[61] Among other things, Collingwood's 'key' of correspondences explained the spiritual 'higher signification' of Genesis's

---

[53] 'Obituary', *New-Church Magazine* (1908), cutting, 575–76, Collingwood, Dr. Cuthbert; with Research Papers of Mrs Nora McMillan, D859/1.

[54] Ian Sellers, 'The Swedenborgian Church in England', in *Reinventing Christianity: Nineteenth-Century Contexts*, ed. by Linda Woodhead (Ashgate: Aldershot, 2001), 97–104 (97).

[55] Cuthbert Collingwood, 'Personal Reminiscences', typescript (1907), page 71, Collingwood, Dr. Cuthbert; with Research Papers of Mrs Nora McMillan, D859/14.

[56] Collingwood to Dawson, 4 April 1878, Dawson-Harrington Families Fonds, 1022-2-1-124-0003, McGill University Archives, Montreal.

[57] W. E. Gladstone, annotation in [Cuthbert Collingwood] 'A Graduate of Oxford', *New Studies in Christian Theology: Being Thirty-Three Lectures on the Life and Teaching of Our Lord* (London: Elliot Stock, 1883), Glynne-Gladstone Archive, GLA WEG/E 10 /COL, Gladstone's Library, Hawarden, UK.

[58] Cuthbert Collingwood, *The Bible and the Age; or, An Elucidation of the Principles of A Consistent and Verifiable Interpretation of Scripture* (New York: James Pott, 1887), 9.

[59] Collingwood, *The Bible and the Age*, 36.

[60] Collingwood, *The Bible and the Age*, 78,

[61] James F. Lawrence, '*Correspondentia*: A Neologism by Aquinas Attains its Zenith in Swedenborg', *Correspondences*, 5 (2017), 41–63.

description of the material world.[62] Each creation day, for instance, signified a 'stage of terrestrial development' in its lower, natural sense, but corresponded, in its higher sense' with a 'stage ... of spiritual progress'.[63] Collingwood had previously hinted, to Dawson, at his increasing scorn for those obsessed with the lower, material world, dismissing as misguided all pantheistic 'Nature-worshippers', like the scientific naturalist John Tyndall.[64]

Perhaps jaded by the unprepossessing reception of *A Vision of Creation*, Collingwood admitted to Gladstone, then in the process of his own Hexaemeral research, that his occult and Casaubonic '*key to all the mythologies*' was best communicated not in text but in 'conversation', when the author could 'answer difficulties as they may arise'.[65] Even the most persuasive literary technologies, in other words, left something wanting. A disapproving critic in the *Wesleyan-Methodist Magazine*, joining the choir of the universally negative reception of *The Bible and the Age*, scorned Collingwood's 'Christian theosophy'.[66] Indeed, the *London Quarterly Review* wondered why 'no mention of Swedenborg's name occurs in the book' given its 'plain' basis in the Swedish mystic's 'doctrine of "Correspondence"'.[67] Graduating from the poetic authority of epic inspiration to the direct access to spiritual truth supplied by Christian esotericism, Collingwood had again attempted—with even less public success—to vindicate Genesis on terms transcending mere evidentiary argumentation.

A Broad, General Sketch

Despite his Linnean Society membership and record of uncontroversial zoological and botanical publications, Collingwood lacked the literary and scientific sway to convert eminent thinkers to day-age theory, let alone to his esoteric symbology. Dawson, by this point a savant of international reputation, was in a stronger position. Suggestively, however, his major phase of day-age writing was catalysed by an awkward misunderstanding about distinctions of genre. In 1870, Dawson delivered the Royal Society of London's prestigious Bakerian Lecture, but he was subsequently mortified by the Society's unusual decision not to publish it in its famous *Philosophical Transactions*. As Susan Sheets-Pyenson explains, it was not just the Canadian's anti-evolutionary and 'Amerocentrist' approach to

---

[62] Collingwood, *The Bible and the Age*, 83.
[63] Collingwood, *The Bible and the Age*, 84.
[64] Collingwood to Dawson, 4 April 1878, Dawson-Harrington Families Fonds, 1022-2-1-124-0003.
[65] Collingwood to Gladstone, 7 January 1886, Glynne-Gladstone Archive. GLA/GGA/2/7/2/15/130.
[66] Untitled review of *The Bible and the Age*, by Cuthbert Collingwood, *Wesleyan-Methodist Magazine*, 11 (1887), 558.
[67] Untitled review of *The Bible and the Age*, by Cuthbert Collingwood, *London Quarterly Review*, 68 (1887), 159–60 (160).

palaeobotany that vexed British savants, but also his delivery of a 'popularization' lecture that employed 'loose' terminology, rather than communicating specialized original research.[68] After this embarrassment, Dawson, Sheets-Pyenson argues, 'determined to make his mark as a popularizer and generalist'.[69] Nonetheless, he would continue to blur the boundary between contributions to knowledge and popularization, although in future he did so with more strategic intent.

Among the first fruits of his turn to popularization were literary 'Sketches of the Geological Periods as They Appear in 1871' in the evangelical Religious Tract Society's monthly *Leisure Hour*. The editor, James Macaulay, told Dawson that it was 'a high merit to attract popular attention without lowering scientific tone', an achievement of rare individuals like Miller, and promised payment 'at the highest rate that our periodical gives'.[70] Dawson ultimately received £120 for his 'Sketches', and the Religious Tract Society's decision not to compile them into a book allowed him to pursue a favourable contract with another popular Christian publisher, Hodder and Stoughton, who kept the resulting work, *The Story of the Earth and Man* (1873), and its sequels, in print into the twentieth century.[71] Popular science was potentially lucrative, and, as Dawson confessed in the book's preface, provided a venue for promoting research unacceptable in technical journals. The *Leisure Hour* series had evaded 'the prejudices of specialists', meaning that geologist readers would 'find here and there facts which may be new to them, as well as some original suggestions'.[72] These suggestions, 'though stated in familiar terms', were not 'advanced without due consideration' (ix). Among his clandestine contributions were searing criticisms of evolution and palaeoanthropology. Dawson's generic sleight of hand, taking fresh research to wider audiences, did not go unnoticed. The *Spectator*'s critic observed that his 'somewhat controversial' stance 'may be conjectured from the medium through which his work was given to the world'.[73]

In line with the *Leisure Hour*'s subdued approach to evangelical messaging, the 'Sketches of the Geological Periods' did not primarily constitute an argument for Dawson's day-age theory.[74] Instead, he wove his assumptions, often parenthetically, into the fabric of a popular genre, the prose 'epic of Earth history'.[75] Related to epic poems like Collingwood's in content rather than form, this was the same

---

[68] Sheets-Pyenson, *John William Dawson*, 113.
[69] Sheets-Pyenson, *John William Dawson*, 118.
[70] James Macaulay to John William Dawson, 17 January 1871, Dawson-Harrington Families Fonds, 1022-2-1-062-0021.
[71] Macaulay to Dawson, 1 January 1872, Dawson-Harrington Families Fonds, 1022-2-1-076-0001. See Sheets-Pyenson, *John William Dawson*, 123.
[72] John William Dawson, *The Story of the Earth and Man*, 2nd rev. edn (London: Hodder and Stoughton, 1873), vii, ix. Subsequent references included in text.
[73] Untitled review of *The Story of the Earth and Man*, by John William Dawson, *Spectator*, 46 (1873), 1314–15 (1314).
[74] Aileen Fyfe, *Science and Salvation: Evangelical Popular Science Publishing in Victorian Britain* (Chicago: University of Chicago Press, 2004), 265, 270.
[75] Ralph O'Connor, 'From the Epic of Earth History to the Evolutionary Epic in Nineteenth-Century Britain', *Journal of Victorian Culture*, 14 (2009), 107–23.

expository model Figuier had employed. By calling his articles tentative 'Sketches', Dawson may have intended to distinguish them from the epic of Earth history's ever-more-fashionable descendant, the 'evolutionary epic', which told a connected story of biological affinities.[76] The sketchy, discontinuous aspect of the articles, with its subtle creationist implication, was somewhat compromised when Dawson was asked to retitle them for book publication. In pursuit of saleability, the sketches because a '*Story*', and Dawson's proposed subtitle, clarifying that the story consisted of 'a series of sketches of the geological periods', was omitted by Hodder and Stoughton—although, in the text, the 'sketch' remained the organizing principle (e.g. 22, 199, 307).[77] The notion of a bravura but necessarily impressionistic sketch alluded to Dawson's belief in the humble state of geohistorical knowledge (as indicated by the as-yet undiscovered Eozoic vegetation of the third day-age, alluded to above), but it also connected his writing to the style of Genesis.[78] A few years later, he would explain the Mosaist's remarkable combination of perspicuity and terseness by comparing the first verses of Genesis to 'a broad, general sketch from the pen of a historian'.[79] In 1877, when Dawson's revised *Archaia* was published as *The Origin of the World, According to Revelation and Science*, this language persisted. The Genesis story was a 'bold sketch'.[80]

The language of 'sketching' balanced humility with ambition, and, in this case, divinity. It was also in keeping with the pictorial traditions of day-age writing. Discussing the Carboniferous tree-lizard *Dendrerpeton* with Lyell, Dawson gleefully hoped that fellow naturalists would provide a 'glimpse into the life of these old forests'.[81] When discussing the Palaeozoic ecosystems with which he was most familiar, Dawson, whose writerly ambitions have received little previous critical attention, was liable to don Miller's imaginative mantle.[82] Woven into his otherwise expository prose, Dawson's authorial voice, with the help of the first-personal plural, sometimes took readers on fantastic voyages through space and time, 'casting our dredge and tow-net into the Primordial sea' (39) and approaching the Devonian land masses as a 'voyager through the Silurian' (82). His longest stretch of this mode was a discussion of Carboniferous life, framed as an exploration of the steaming forests (119). This 'imaginary excursion' allowed Dawson to describe Carboniferous plants like *Sigillaria* as if on the spot, as when he suggested 'we approach one of these trees' (121) and even 'cut into its stem' (125). Deictic immersion, previously used to great effect in Miller's lectures, is a subject I will

---

[76] For non-evolutionary storytelling, see Adelene Buckland, *Novel Science: Fiction and the Invention of Nineteenth-Century Geology* (Chicago: University of Chicago Press, 2013), 204–5.

[77] Sheets-Pyenson, *John William Dawson*, 129, 248.

[78] For the literary sketch, see Alison Byerly, 'Effortless Art: The Sketch in Nineteenth-Century Painting and Literature', *Criticism*, 41 (1999), 349–64.

[79] John William Dawson, *Nature and the Bible* (New York: Robert Carter and Brothers, 1875), 120.

[80] John William Dawson, *The Origin of the World, According to Revelation and Science* (New York: Harper & Brothers, 1877), 49.

[81] Dawson to Lyell, 23 November 1852, Papers of Sir John William Dawson.

[82] For a brief discussion of Miller's literary influence on Dawson, see Hugh Miller, *The Old Red Sandstone; or, New Walks in an Old Field, Edited with a Critical Study and Notes*, 2 vols, ed. by Ralph O'Connor and Michael A. Taylor (Edinburgh: National Museums Scotland Publishing, 2023), I, 162.

return to in the next chapter. After dozens of pages, Dawson conceded that 'the reader may be wearied with our long sojourn in the pestilential atmosphere of the coal swamps' (152).

In a departure from Miller's technique, which boldly relied on word-paintings alone, Dawson provided illustrated ideal scenes. His wording indicates that they were designed under his instructions, possibly drawn by himself or his daughter Anna, and they were engraved by the firm Butterworth & Heath. Multimedia effort was required to communicate the truly alien nature of ecosystems like the Devonian seascape: 'we must try', he urged, 'with the help of our illustration, to paint these old inhabitants of the waters as distinctly as we can' (96). Nonetheless, the virtual witnessing encouraged by the conjunction of prose and visual imagery could be an underwhelming experience, as Dawson suggested in his explanation of an ideal scene of Miocene life:

> In the illustration, I have grouped some of the characteristic Mammalian forms of the Miocene, as we can restore them from their scattered bones, more or less conjecturally; but could we have seen them march before us in all their majesty, like the Edenic animals before Adam, I feel persuaded that our impressions of this wonderful age would have far exceeded anything that we can derive either from words or illustrations. (255)

Here Dawson invoked unmediated, even prelapsarian visions of prehistoric fauna for which prose and illustration were but unsatisfactory surrogates. Unsurprisingly, as he explained in *The Origin of the World*, Dawson agreed with Miller that the creation story probably represented a vision experienced by Moses or by a seer whose account Moses edited.[83] Whatever the representative limitations of his illustrations and word-paintings in the 'Sketches' and *The Story of the Earth and Man*, they were, implicitly, glimpses into Genesis. That Dawson saw them as such was confirmed when he repurposed them for a lecture series at New York's Union Theological Seminary, published as *Nature and the Bible* (1875). In the earlier work, Dawson had suggested, as did most day-agers, that the 'great tanninim' (whales or sea monsters) of Genesis 1:21, created by God on the fifth day, were the aquatic reptiles of the 'Mesozoic age' (151). To make the point clearer, *Nature and the Bible* reproduced his earlier Mesozoic scene in edited form, and with a new title: 'Tanninim of the Fifth Day' (Figure 1.3). Another caption turned a Cambrian scene of trilobites and other invertebrates (40) into a depiction of the 'Early Sheretzim of the Waters' [sic], establishing that these extinct invertebrates constituted the 'creeping things' mentioned in Genesis 8:19 and elsewhere.[84]

As the cases of Grover, Collingwood, and Dawson indicate, scenes from deep time were being tied consistently to day-age theory in the 1870s. During the next decade, scenes painted by British artist Benjamin Waterhouse Hawkins, whose

---
[83] Dawson, *Origin of the World*, 26, 49–50, 66.
[84] Dawson, *Nature and the Bible*, 118.

Figure 1.3 (a) John William Dawson, *The Story of the Earth and Man*, 2nd rev. edn (London: Hodder and Stoughton, 1873), 219. (b) John William Dawson, *Nature and the Bible* (New York: Robert Carter and Brothers, 1875), facing 123. Repurposing a restoration of Mesozoic reptiles from his popular geology writing in a theological work (with hastily added dinosaur), Dawson indicates that both images depict Genesis's fifth 'day-age'.

work was cited as the inspiration for some of Dawson's illustrations (281), accompanied *Creation* (1884), the final major publication of Princeton geologist Arnold Guyot—who, over three decades prior, had cemented day-age theory's prestige in the United States. A slim, elegant statement of his concordance scheme, *Creation* was illustrated by reproductions of the oil paintings that Hawkins, a staunch anti-evolutionist, had recently produced for Princeton's museum.[85] Thus this scientific-artistic genre, which Rudwick sees as achieving a kind of autonomy by this period, was, in fact, being repeatedly reabsorbed back to its intellectual roots: the illustration of Genesis.[86]

Dawson continued to pen works on biblical geology up until his death in 1899. The most theologically involved were usually addressed to transatlantic evangelical audiences by publishers like Fleming H. Revell, while Harper (in the United States) and Hodder and Stoughton (in Britain) covered books with less strident Christian appeal. Meanwhile, Dawson could share his apologetics with sympathetic thinkers at the respectable but, for the most part, scientifically marginal Victoria Institute in London, established in 1865 as a venue for evangelical scientific research.[87] He remained a respected scientific administrator, involved in both the American Association for the Advancement of Science and the British Association for the Advancement of Science. Technical scientific journals mostly ignored his theistic interpretations, but still provided a platform for Dawson's assimilable scientific opinions. In fact, these latter were often more controversial than his scriptural geology, as idiosyncrasies on a naturalistic plane were less easily demarcated than religious ones. His most fiercely contested contribution to science concerned the true nature of the object he called *Eozoön canadense*, either the earliest known animal fossil (as Dawson claimed) or an inorganic piece of rock (as most other geologists contended).[88]

A Vision of the Divine Work

The busy principal of one of Canada's leading universities, Dawson would remain on the sidelines during the fierce debates over day-age theory that broke out in the British public sphere during the 1880s. The first of these began, in effect, on 23 June 1880, when Samuel Kinns, Principal of 'The College' at Highbury New Park,

---

[85] See Guyot, facing 114; Valerie Bramwell and Robert M. Peck, *All in the Bones: A Biography of Benjamin Waterhouse Hawkins* (Philadelphia, PA: The Academy of Natural Sciences of Philadelphia, 2008), 86–89.
[86] Rudwick, *Scenes from Deep Time*, 24, 173.
[87] Stuart Mathieson, *Evangelicals and the Philosophy of Science: The Victoria Institute, 1865–1939* (London: Routledge, 2021), 94–97.
[88] Juliana Adelman, 'Eozoön: Debunking the Dawn Animal', *Endeavour*, 31 (2007), 94–98.

Islington, was put forward for Fellowship of the Geological Society of London.[89] His name had been vouched for by comparative anatomist Richard Owen, Superintendent of Natural History at the British Museum, which was then in the process of moving its scientific collections to South Kensington, and William Carruthers, the Museum's Keeper of Botany. On 17 November, before the ballot could take place, Kinns seemingly withdrew his name in response to an anonymous objection.[90] Kinns appears never to have discovered what the objector said, although, as we shall see, he suspected that Carruthers, who had doubts about Kinns's suitability almost from the start, knew more than he claimed. Several years later, returning to an affair that 'kept me awake for many long hours nights after nights', Kinns explained to the botanist that he had simply desired access to the Society's library to carry out research for his forthcoming book.[91]

That book was *Moses and Geology* (1882), published internationally by Cassell and in some early editions called by its subtitle, *The Harmony of the Bible with Science*. Kinns defended day-age vision theory on probabilistic grounds: the odds that an uninspired Moses could have arranged geohistorical events in such an accurate order, Kinns claimed, were statistically impossible. He also rejected evolution and provided scientific justifications for various miracles, such as the sun standing still in the Old Testament Book of Joshua. The controversy that would subsequently dog Kinns, foreshadowed in the Geological Society affair, has received very occasional scholarly attention, but, thanks to hitherto unconsulted archival sources, there is far more to say about this important event in the history of concordance debates.[92] Reaching around 12,000 copies printed by the late 1880s and reissued in 1895, *Moses and Geology* sold quite well for what was—as its title indicated—a work of conservative scriptural geology and an expensive giftbook, selling at $3 in the United States, or, in the United Kingdom, 10s.6d. (with an extra two shillings for gilt edges). Although, as usual, critics were divided on the persuasiveness of day-age theory, the book reviewed surprisingly favourably in high places, including a sympathetic piece in *The Times* calling it 'a vision of the Divine work' that would leave 'a new mark on the memory, and on the heart as well'.[93] Outwardly structured in chapters based around successive quotations

---

[89] For a brief biography, see R. S. Simpson, 'Kinns, Samuel (1826–1903)', in *Oxford Dictionary of National Biography*, online edn, September 2004 (Oxford: Oxford University Press, 2004) <https://doi.org/10.1093/ref:odnb/34333> [accessed 30 October 2022].

[90] Application GSL/F/1/10, Geological Society of London Archives.

[91] Samuel Kinns to William Carruthers, 18 May 1883, Natural History Museum Archives DF BOT/404/1/12/3, Department of Botany, Carruthers, William: 'Moses and Geology' by Samuel Kinns: Correspondence, Notes and Newscuttings. By permission of the Trustees of The Natural History Museum. Subsequently NHM.

[92] Richard England, 'Aubrey Moore and the Anglo-Catholic Assimilation of Science in Oxford', unpublished PhD thesis, University of Toronto (1997), 147–65; Robert W. Smith, 'The "Great Plan of the Visible Universe": William Huggins, Evolutionary Naturalism and the Nature of the Nebulae', in *The Age of Scientific Naturalism: Tyndall and His Contemporaries*, ed. by Bernard Lightman and Michael S. Reidy (London: Pickering and Chatto, 2014), 113–36 (129).

[93] 'Moses and Geology', *The Times*, 23 August 1882, 4.

from Genesis, a closer look reveals more heterogeneous contents: Kinns's day-age scheme was a Christmas tree on which to hang copious curious and sentimental factoids. As the contents page indicates, a chapter on human antiquity also deals with 'Phrenology—Power of a mother's love—Example of Queen Victoria', while a discussion of Eden segues from Agassiz to Milton to 'Flowers—Perfumes—Shakespeare's description of Cleopatra'.[94] The book's perambulatory nature may well hint as its origins in lectures given to schoolchildren at the British Museum. Miscellaneous contents held together by Kinns's familiar didactic exposition about Genesis, *Moses and Geology* represents yet another literary shape for day-age argumentation.[95]

Kinns did not intend narrative immersion, but, at key moments, he did contribute to the tradition of turning readers into virtual Mosaists. Here, he shifts mid-passage from the past-tense history of Moses's experience to his present-tense vision:

> Casting his eyes upwards he saw the cloud which surrounded him moving, and soon a pitch-black sky was above him; whilst wondering what this might mean, he heard the same voice of God saying
> *'Let there be light,'*
> and immediately the whole of the expanse was filled with luminous æthereal masses of every variety of form. He gazes with admiration, and as he does so these nebulæ condense, and suns and stars blaze forth throughout the whole canopy of Heaven. (20)

Commenting on Moses's supposed description of Mesozoic saurians like *Plesiosaurus* as 'sea-monsters', Kinns had 'no doubt' that the prophet 'actually saw them sporting in the deep' during the fifth day-age. To assist readers in so doing, he asked them, in a somewhat counterproductive instruction, perhaps surviving from his original lectures, to 'close their eyes for a few minutes and imagine they are accompanying me to the shore of one of those ancient scenes' (278). Kinns's following prose description of a clash between *Ichthyosaurus* and *Plesiosaurus* was keyed to an adjacent plate, reproducing Riou's classic illustration of the same 'Imaginary Fight' (Figure 1.4). This borrowing was lubricated by Cassell's ownership of new editions of Figuier's *The World before the Deluge*. Kinns being of limited capacity as a word-painter, Riou's image would, presumably, make readers' imaginative efforts to see as Moses had done considerably easier. An adjacent

---

[94] Samuel Kinns, *The Harmony of the Bible with Science; or, Moses and Geology*, 2nd edn (New York: Cassell, Petter, Galpin, 1882), xxii–ii. Subsequent references included in text.

[95] For the miscellany, see Jonathan R. Topham, 'John Limbird, Thomas Byerley, and the Production of Cheap Periodicals in the 1820s', *Book History*, 8 (2005), 75–106. For familiar didactic exposition, see Ralph O'Connor, 'Introduction: Varieties of Romance in Victorian Science', in *Science as Romance*, vol. VII of *Victorian Science and Literature*, gen. eds Gowan Dawson and Bernard Lightman (London: Pickering & Chatto, 2012), xi–xxxvi (xvi).

XI.—Imaginary Fight between an Ichthyosaurus and a Plesiosaurus.

Figure 1.4 Samuel Kinns, *Moses and Geology; or, The Harmony of the Bible with Science*, rev. edn (London: Cassell, 1887), facing 281. Having suggested, like others before him, that Moses saw Mesozoic reptiles in a vision of the fifth day-age, Kinns helps readers to do the same by reproducing Riou's illustration from *The World before the Deluge*.

claim that the ichthyosaur's blood was 'not so red as that of our whales' but rather cold and 'slimy' (282) also hints at acquaintance with Collingwood's poem, which described saurian blood as not 'warm and crimson' like that of 'noble brutes,— but black and chill/Like some envenomed stream' (152). These were surprisingly late airings of the view that Mesozoic saurians were not simply primitive, but also positively malefic in nature.[96]

Some of the book's literary debts were too strong even for sympathetic readers. The Catholic *Dublin Review* suggested that 'imaginative appeals' like Kinns's 'ideal sketches of battle royal [*sic*] between palæolithic monsters' were becoming 'a little too familiar to most readers'.[97] The critic was likely thinking not just of Kinns's debts to Figuier's 'ideal sketches' but also to another book illustrated by Riou, Jules Verne's *Voyage au centre de la Terre (Journey to the Centre of the Earth)* (1864). Translated into English in the 1870s, Verne's scientific romance featured a memorable duel between *Ichthyosaurus* and *Plesiosaurus*. The *Review* also suggested that the influence of Hugh Miller was 'a little too patent throughout the work', a point taken up by the non-denominational *Christian World*, in which a critic

---

[96] For Satanic saurians, see O'Connor, *Earth on Show*, 417.
[97] Untitled review of *Moses and Geology*, by Samuel Kinns, *Dublin Review*, 9 (1883), 239–41 (240).

sarcastically dubbed Kinns's depiction of revelation a mere 'magic-lantern exhibition'.[98] Miller had memorably characterized the six day-age visions as resembling the shifting scenes of a diorama in the 1850s, but comparing God's miraculous work with mundane optical technologies risked bathos in less skilled hands.

By the 1880s, the value of literary borrowings for the defence of scriptural inerrancy was more dubious than ever. In *Moses and Geology*, however, paratextual strategies were as important as textual ones. Intending his book as the ideal Sunday School prize, Kinns front-loaded it with conservative religious and political clout. The first edition began with a seven-page list of subscribers, including earls, MPs, and Anglican bishops (vii–xiii), while subsequent editions added numerous obsequious dedications citing any feasible connection between the author, the aristocracy, and royalty.

At the head of his subscribers was Anthony Ashley-Cooper, the Earl of Shaftesbury, the great evangelical philanthropist and Kinns's most powerful ally. Shaftesbury had reason to support Kinns's efforts: four years prior, he had conspicuously left the Society for the Promotion of Christian Knowledge for its publication of liberal interpretations of Genesis in a book by geologist-clergyman Thomas George Bonney.[99] When Bonney complained about Shaftesbury's public criticism, the latter quoted Acts 17:11, insisting that 'the modern laity ... have our equal right and duty with the Laity of old, who "searched the Scriptures daily, whether these things were so"'.[100] Deeming his tense correspondence with Bonney worthy of publication in the conservative evangelical *Record*, Shaftesbury insisted upon his right to evaluate scientific claims, citing the lack of 'harmony' among savants as justification for members of 'the common herd' to 'judge' scientific theories.[101] Matters like this had to be put to the Christian public sphere, and Shaftesbury's influence would help Kinns to do just that.

Pour Encourager Les Autres

In Canterbury Cathedral Library on 3 January 1884, *The Times* reported, Kinns delivered the first of an ambitious series of travelling lectures. These lectures, organized by a Shaftesbury-led committee, were intended to combat religious scepticism using the arguments set forth in *Moses and Geology* two years earlier. *The Times* flatteringly remarked upon the lecture's 'large and influential audience', including the Dean of Canterbury (Richard Payne Smith), adding fatefully that

---

[98] '*Moses and Geology*', *Dublin Review*, 240; 'The New Genesis by Dr. Kinns', *Christian World*, 7 February 1884, clipping, NHM DF BOT/404/1/12/52.

[99] Geoffrey B. A. M. Finlayson, *The Seventh Earl of Shaftesbury 1801–1885* (Vancouver: Regent College Publishing, 1981), 555.

[100] Lord Shaftesbury to Thomas George Bonney, 3 January 1878, LDGSL/776, Geological Society of London Archives.

[101] Shaftesbury to Bonney, 10 January 1878, LDGSL/776.

'eminent men upon the various staffs of the Royal Observatory, the Geological Survey, and the British Museum had testified to the accuracy' of Kinns's 'facts'.[102] These words set off a controversy in which, thanks to Kinns's polarizing approach, very different understandings of the hierarchies of scientific and religious authority, and of the literary technologies capable of delivering persuasive scientific truth claims, came into conflict.

Things turned ugly the day after *The Times* item was published. Several staff from the recently opened Natural History Museum—the botanist William Carruthers, who had signed Kinns's ill-fated application for Fellowship of the Geological Society, and the Keeper of Geology, Henry Woodward—wrote to deny that they had testified to the accuracy of Kinn's facts.[103] In the same letter and several following, Carruthers demolished Kinns's credentials, bringing up the schoolmaster's failure to join the Geological Society. Kinns contacted *The Times* on 7 January to defend himself, naming the scientific men who endorsed his lecture content. He pointed out that they were the men who had allowed Kinns 'to insert their names' in *Moses and Geology*'s third edition as having approved his proof-sheets.[104] Most had been initial subscribers to the book, including Henry William Bristow, Director of the Geological Survey (and editor of Cassell's revised edition of Figuier's *The World before the Deluge*). In subsequent editions, Kinns printed letters of endorsement from the savants themselves. These letters praised his pursuit of facts, if not necessarily their use as day-age ingredients, and Kinns would adamantly contend that only his interpretation, not his ingredients, was up for debate. The limitations of this pragmatic delegation of authority were highlighted by the necessity of later inserting a postscript to Bristow's endorsement, explaining that the geologist 'scored under the word "facts"' and thus implicitly distanced himself from Kinns's arguments.[105]

Liberal opinion on the lecture series was scathing. The West Country *Western Morning News*, in an article Kinns would repeatedly call libellous, sneered that his stance on Genesis had left 'titled ladies in raptures'.[106] After several weeks of fierce exchanges in *The Times* between Carruthers, Kinns, and many of the latter's scientific endorsers, the newspaper refused further letters. The controversy continued elsewhere, as Kinns's combative personality was exacerbated by the threat accusations of incompetence and ungentlemanliness posed to his school's pupil intake. Richard England, the only scholar to discuss the controversy in detail, focuses on unsuccessful behind-the-scenes efforts by powerful figures like Henry Acland, Regius Professor of Medicine at Oxford, to persuade Kinns's allies to

---

[102] 'The Scientific Accuracy of the Bible', *The Times*, 4 January 1884, 10.
[103] 'The Scientific Accuracy of the Bible', *The Times*, 5 January 1884, 7.
[104] 'The Scientific Accuracy of the Bible', *The Times*, 8 January 1884, 7.
[105] Samuel Kinns, *Moses and Geology; or, The Harmony of the Bible with Science*, 8th thousand (London: Cassell, 1885), xx.
[106] Untitled, *The Western Morning News*, 7 January 1884, clipping, NHM DF BOT/404/1/12/16.

drop their patronage or risk associating Christianity, especially Anglicanism, with outdated science. As Charles Pritchard, clergyman and Oxford's Savilian Professor of Astronomy, told Carruthers, Kinns's book was 'pernicious nonsense' and 'pour encourager les autres it would be well to explode it thoroughly'.[107] Here Pritchard referred to a famous quip from Voltaire's *Candide* (1759), which satirically depicted the British navy as occasionally executing failed admirals 'so as to encourage the others'.[108]

Although England notes that only 'a minority of Anglican clergy' supported Kinns, they were powerful and undiscouraged advocates.[109] When calls to stop the lecture series became deafening, the Kinns Committee, including Shaftesbury and Dean Payne Smith, refused to back down.[110] Payne Smith wrote to Kinns on 25 January, decrying Pritchard and the naturalists in *The Times* for their 'unscientific intolerance and fanaticism'.[111] In May, Shaftesbury investigated Carruthers' claim that Kinns would 'deservedly' have been 'blackballed' from the Geological Society had his application not been withdrawn, and found nothing incriminating Kinns.[112] The lectures continued, with Kinns speaking at Exeter Hall and the Bishop's Palace in Bath, as well as in Wales, Yorkshire, and Merseyside in the following months. Even allowing for his exaggerated self-promotion, Kinns's subscription list and various loving reports of his lectures evinced the serious second-hand support day-age theory continued to enjoy in the Church of England (and evangelical culture more widely).

This success gave the thin-skinned Kinns no peace. The fact that his bitter argument with Carruthers often bordered on legal action is a likely explanation for the latter's compilation of relevant clippings and letters as evidence in a vast scrapbook (still held by the Natural History Museum). On 6 March 1884, Kinns's solicitors requested that Carruthers make a 'satisfactory public apology'.[113] Legal threats hung over their tense correspondence, in which Kinns attempted to convince the palaeobotanist to retract his criticisms. The embarrassment they were causing to his lecture series and school led Kinns to claim that 'if I had not been a Christian man I should certainly have committed suicide'.[114] Although he regretted these consequences, Carruthers insisted to an intermediary that he 'had a public duty' to save 'both religion and science from serious wrong' resulting from Kinns's

---

[107] Charles Pritchard to Carruthers, 6 January 1884, NHM DF BOT/404/1/12/14.
[108] Voltaire, *Candide and Other Stories*, ed. by Roger Pearson (Oxford: Oxford University Press, 2006), 65.
[109] England, 164.
[110] Kinns, *Moses and Geology*, 8th edn, xix.
[111] Richard Payne Smith to Kinns (copy), 25 January 1884, NHM DF BOT/404/1/12/48.
[112] 'The Scientific Accuracy of the Bible', *The Times*, 9 January 1884, 7. Carruthers learnt about Shaftesbury's investigation of his claims only in early 1889. See Charles E. De Rance to Carruthers, 30 April 1889, NHM DF BOT/404/1/12/85.
[113] Langton and Son to Carruthers, 6 March 1884, NHM DF BOT/404/1/12/56.
[114] Kinns to Carruthers, 23 April 1884, NHM DF BOT/404/1/12/66.

'public mission'.[115] As an evangelical Presbyterian and editor of the *Messenger for the Children of the Presbyterian Church of England*, a periodical filled with missionary content, Carruthers would have been particularly disturbed by the way Kinns's lecture series was repeatedly referred to as a domestic 'mission'.[116] Dissenting voices of all kinds agreed that the Anglican hierarchy who formed the majority of Kinns's backers had abused their power in giving him a platform. The Unitarian *Inquirer* mockingly considered it 'a matter of public importance' that 'the bishops and deans' supporting Kinns clarify if they wished 'the people to believe that he has discovered and explained the right way of reconciling Genesis and Geology?'[117]

Crucial to the nature of this dispute was the fact that Carruthers, unlike his liberal co-combatants, but like Kinns, was an evolutionary sceptic. Declaring in an 1876 address that palaeobotany provided no evidence of evolution, Carruthers' expertise had posed a serious challenge to Darwin.[118] Although he never seems to have said as such, Carruthers was surely concerned by Kinns's public association of evangelical anti-evolutionism with scientific amateurism. Aside from mockery of Kinns's optical discussion of the sun standing still for Joshua, his specific criticisms tended to highlight the schoolmaster's relatively minor mistakes of palaeobotanical nomenclature. Carruthers never engaged specifically with Kinns's day-age framework, leaving it to be implied that fastidious reconciliation was misguided. That Presbyterians like Carruthers and his fellow palaeobotanist, Dawson, should have such opposing interpretations of the Hexaemeron's relationship with science is unsurprising, given the attested variability of the denomination's responses to evolutionary theory.[119]

A likely explanation for Carruthers' stance is his formative relationship with Free Church minister John Fleming, who taught Natural Science at New College, Edinburgh, while Carruthers studied there in the mid-1850s. It was Fleming who encouraged him to pursue science, and the elderly mentor's preference for, in Mark Harris's words, letting science 'speak for itself without introducing unnecessary theological or philosophical glosses' likely shaped the student's thinking.[120] Answering Kinns's complaints in detail, Carruthers let slip a rare admission of his attitude: Kinns's 'errors are not verbal—they belong to the essence of the book'.[121] His stance on the schoolmaster's theory echoed that of other prominent Dissenting voices. The *Nonconformist* kept specific criticisms light, as 'our objection . . .

---

[115] Carruthers to John Thain Davidson (copy), 26 April 1884, NHM DF BOT/404/1/12/67.

[116] Kinns inaccurately denied that he or his colleagues used this term. For Kinns's work promoted as a 'mission', see Thomas Richardson to Jabez Hogg, 14 March 1884, NHM DF BOT/404/1/12/58.

[117] 'Dr. Kinns and His Critics', *Inquirer*, 4 October 1884, NHM DF BOT/404/1/12/75.

[118] Richard J. A. Buggs, 'The Origin of Darwin's "Abominable Mystery"', *American Journal of Botany*, 108 (2021), 22–36 (31–32).

[119] Livingstone, *Dealing with Darwin*, passim.

[120] 'Eminent Living Geologists: William Carruthers', *Geological Magazine*, 9 (1912), 193–99 (193); Mark Harris, 'Natural Selection at New College: The Evolution of Science and Theology at a Scottish Presbyterian Seminar', *Zygon*, 57 (2022), 525–44 (532).

[121] Kinns to Carruthers, annotated by the latter, 23 April 1884, NHM DF BOT/404/1/12/65.

lies deeper by far', while the astronomer William Huggins, a so-called 'Dissenter of Dissenters' also implicated in the Kinns affair, objected to 'the tone and spirit of the book as a whole'.[122]

Carruthers' letters in *The Times* received a flurry of heterogeneous support, although condemnation of Kinns and his indecorous arguments did not always mean condemnation of day-age theory. James Macaulay, who had printed Dawson's creationist 'Sketches' in the *Leisure Hour*, lamented that Shaftesbury had sent 'such a man going out as a champion of science and religion'.[123] Others denounced Kinns but asked Carruthers to evaluate their own reconciliation methods: one Rector of Alresford outlined the day-age scheme he had 'been accustomed to propound' to parishioners, asking Carruthers 'if you think I have dealt honestly and *not unscientifically* with the text'.[124] Even less welcome were appeals from radicals. When Charles Voysey, a heretical theist, sent Carruthers the pungent sermon he had preached on Kinns, the palaeobotanist firmly told Voysey that his own views were 'diametrically opposed to yours as to the Word of God'.[125] On the other end of the spectrum, Carruthers turned down an offer of page 'space' to spread the 'truth' about Kinns via the '5000 monthly' circulation of the belligerent *Protestant Times*.[126] Yet others punctured Kinns's pretensions, hinting at financial and social chicanery: 'The so-called *College*', scoffed a particularly vociferous correspondent, 'is one room at the back of his house, where *he has a dozen little boys!*'[127]

With these non-sectarian forces rallied against him, Kinns recognized that the authority of his book and lectures relied on their paratextual chainmail of endorsements. As such, he reinforced *Moses and Geology* with every new edition and vigorously circulated letters from his defenders in local newspapers and periodicals. He had, at first, not understood why reprinting testimonials from aristocrats, or from ageing geologists like Bristow, did not strengthen his hand. As the affair continued, however, he recognized that he needed specialized endorsements. In August 1885 he wrote to Dawson, now knighted, in pursuit of a 'a little note' for his eighth edition, explaining that 'it is of the utmost importance that *specialists*, in each department of knowledge treated of by me, should testify to the accuracy' of the book's facts.[128] A Presbyterian palaeobotanist, Dawson was an ideal foil against Carruthers. Wary of being drawn into an issue with more baggage than scriptural geology alone, however, Dawson reluctantly allowed Kinns merely to claim that

---

[122] 'Science and the Bible.—Dr. Kinns', *Nonconformist and Independent*, 10 January 1884, clipping, NHM, DF BOT/404/1/12/22; 'The Scientific Accuracy of the Bible', *The Times*, 19 January 1884, 8. For Huggins's religious beliefs, see Smith, 'The "Great Plan of the Visible Universe"', 128–29.
[123] James Macaulay to Carruthers, 17 January 1884, NHM DF BOT/404/1/12/35.
[124] W. O. Newnham to Carruthers, 5 January 1884, NHM DF BOT/404/1/12/13.
[125] Carruthers to Charles Voysey (copy), 22 January 1884, NHM DF BOT/404/1/12/44.
[126] J. L. Scott to Carruthers, 7 October 1885, NHM DF BOT/404/1/12/81.
[127] James Robertson Reid to Carruthers, 11 January 1884, NHM DF BOT/404/1/12/26.
[128] Kinns to Dawson, 14 August 1885, Dawson-Harrington Families Fonds, 1022-2-1-210-0008.

'the proof-sheets of my chapter upon fossil botany have been examined and generally approved by Sir William Dawson'.[129] With characteristic tenacity, Kinns hoped that, for the next edition, 'you may perhaps be able to allow me to drop out the word "generally"'.[130]

By this point in late 1885, the affair flared down. Kinns (whose father had been a Congregationalist minister) was ordained a deacon of the Church of England that year, helping him to find a respectable living following the ignominious collapse of his Highbury school. In 1886, he became a priest. Only in 1889, when Kinns was settled as Vicar Designate of Holy Trinity, Minories, in the City of London, did tensions resurface. This time, Carruthers was alerted to evidence that Kinns had been privately defaming his name back in 1884 and had, in fact, never stopped doing so.[131] Although Carruthers arguably held the legal upper hand by this point, having been sent potentially compromising materials about (and by) Kinns over the years, his preference for ending the matter quietly seems to have won out.

The People of Average Opinions

Kinns insisted that a string of scientific facts, when endorsed by scholars and authority figures, could legitimately be interpreted by a generalist like himself. Ultimately, despite some success at the middlebrow level, his divisive character helped to prevent this strategy from being taken seriously in most of the heavyweight periodicals and scholarly journals. Nonetheless, the idea that day-age theory was open to evaluation by an informed non-specialist public was spreading. This was not a notion solely propagated by those on the fringes: James Dwight Dana, one of the most esteemed geologists in the United States, took this view. Writing in the *Bibliotheca Sacra*, the prestigious theological journal of Oberlin College in Ohio, Dana noted in April 1885 that neither scientific nor linguistic interpreters could exclusively claim an authoritative reading of Genesis. Proposing 'a higher style of interpretation' based on 'a broad view of all well-established knowledge', Dana provided 'a brief review' of relevant geology, to help the 'reader ... to make himself a judge of the scientific facts fundamental to the interpretations'.[132] This judicial register was also used by opponents. Writing in the *Agnostic*, the freethinker Constance Naden condemned day-agers for misleading the jury. Thanks to their metaphors, she argued, the 'British juryman' had absorbed a 'clear, pictorial, and dramatic' conception of geological periods as having a day-like 'beginning, middle,

---

[129] Kinns, *Moses and Geology*, 8th edn (xvii).
[130] Kinns to Dawson, 7 October 1885, Dawson-Harrington Families Fonds, 1022-2-1-212-0007.
[131] See Carruthers to Kinns (copy), 2 March 1889, NHM DF BOT/404/1/12/90.
[132] Dana, 'Creation', 203, 222.

and end'.[133] Asking the jurors to listen to her 'cross-examination', Naden stressed that geological periods, while superficially day-like, were a convenient 'fiction' subdividing swathes of unbroken deep time.[134]

The liberal intellectual monthly magazines, with their openness to controversial disputes, were ideal courtrooms for this case. In November 1885, Gladstone, briefly out of prime ministerial office, wrote to the *Nineteenth Century* in defence of day-age theory, responding to an article doubting the inspiration of Genesis. In December he was opposed by the eminent scientific naturalist T. H. Huxley, leading to a back and forth over the following months that is the only discussion of day-age theory in late Victorian intellectual culture well known to scholars.[135] Huxley was already famous for belligerence towards Genesis, having, in a memorable New York lecture of September 1876, excoriated its cosmogony under the guise of analysing *Paradise Lost*.[136] Both men had belonged to James Knowles's Metaphysical Society, a club for debates over science and religion, the elite membership of which regularly aired their opinions in the *Nineteenth Century* during Knowles's subsequent editorship.[137] This clubbish segment of the public sphere was, of course, far from a democracy. In January 1886, writing to Gladstone to thank him for defending Genesis, Cuthbert Collingwood lamented that he found 'publishers, & even reviews, most chary of offering to print' his own contributions to the subject. Indeed, having sent his esoteric theory of biblical symbolism to the *Nineteenth Century* and waited 'two months', the manuscript was 'returned *apparently unopened*'.[138]

Unlike the marginalized Swedenborgian naturalist, however, Gladstone received effectively unrestricted access to exclusive periodicals and to friends in the highest scientific echelons. Preparing his scientific evidence for the geological accuracy of Genesis, he turned to Owen and Henry Acland, both minor players in the fading Kinns controversy, who assisted the politician while tactfully attempting to dissuade him. Gladstone's argument, like others before it, hinged upon the similarity between the order in which animal groups had appeared in Genesis and the order in which they had appeared in geohistory. His opponent Huxley vehemently denied any similarity. Predictably, Kinns himself wasted no time in sending Gladstone the latest edition of *Moses and Geology* and was soon pleased to learn

---

[133] Constance Naden, 'Geological Epochs', *Agnostic*, 1 (1885), 304–8 (305).

[134] Naden, 306, 308.

[135] Bebbington, *The Mind of Gladstone*, 238–41; Adrian Desmond, *Huxley: Evolution's High Priest* (London: Michael Joseph, 1997), 162–65.

[136] T. H. Huxley, *American Addresses, with a Lecture on the Study of Biology* (New York: Appleton, 1877), 7–29.

[137] Bernard Lightman, 'Science at the Metaphysical Society: Defining Knowledge in the 1870s', in *The Age of Scientific Naturalism: Tyndall and His Contemporaries*, ed. by Bernard Lightman and Michael S. Reidy (Pittsburgh, PA: University of Pittsburgh Press, 2016), 188–206.

[138] Collingwood to Gladstone, 7 January 1886, Glynne-Gladstone Archive, GLA/GGA/2/7/2/15/130.

that the statesman had been 'examining my book' (which was 'backed up by the highest specialists').[139]

Gladstone, however, had an even shakier grasp on modern ways of ascribing authority and expertise than Kinns. Having originally cited Georges Cuvier, who died in 1832, on the reconcilability of geology with the Hexaemeron, despite learning from Owen that Cuvier's word now stood for little, Gladstone insisted in January on the long-dead savant's cosmogonical 'authority', along with that of several other 'venerable names'.[140] In a perhaps ingenuous instance of the originality topos regularly used by day-age writers, he demonstrated little familiarity with prior scholarship. If he did, indeed, read Kinns's gift copy, he did not adopt the schoolmaster's positions, and even proposed that the animals referred to in Genesis were not extinct ones, robbing the verses of insight into prehistory. For the curate-botanist George Henslow, Gladstone had inexplicably discarded day-age theory's smoking gun: if the *tanninim* of the fifth day represented the 'gigantic' reptiles of the Mesozoic, Henslow explained to the statesman, then Genesis '*does bear a strong resemblance to the state of things*'.[141]

Despite Gladstone's paltry research, his response to Huxley in January 1886 echoed the familiar argument that Genesis was accurate but that its poetic language rendered it immune to pedantic scrutiny. Rather than offering the 'scientific precision' of 'a lecture', it was a 'popular' statement or 'sermon'.[142] Huxley was uncooperative with these attempts to shield the Bible from empirical criticism. In February, he replied that Gladstone was implying that a lecture 'may be taken seriously, as meaning exactly what it says, while a sermon may not'.[143] Gladstone's reluctance to break down emotive religious controversies to falsifiable matters of fact, David Bebbington points out, made him uncomfortable with judicial analogies like those that were enveloping these day-age debates.[144] Nonetheless the printed subtitle of his response to Huxley, 'A Plea for a Fair Trial', gave his interlocutor an angle of attack. Embracing the courtroom atmosphere, Huxley summarized the case for 'my clients—the people of average opinions'.[145] His bald arguments allowed even a master of rhetoric like Gladstone no quarter. For the politically Unionist Huxley and his scientific naturalists, it was not just Gladstone's insistence

---

[139] Kinns to Gladstone, 7 November 1885, Glynne-Gladstone Archive, GLA/GGA/2/7/2/49; and 14 December 1885, Glynne-Gladstone Archive, GLA/GGA/2/7/2/49.

[140] W. E. Gladstone, 'Proem to Genesis: A Plea for a Fair Trial', *Nineteenth Century*, 19 (1886), 1–21 (4).

[141] George Henslow to Gladstone, 16 January 1886, GLA/GGA/2/7/3/14/1/2, Glynne-Gladstone Archive.

[142] Gladstone, 'Proem to Genesis', 5.

[143] T. H. Huxley and Henry Drummond, 'Mr. Gladstone and Genesis', *Nineteenth Century*, 19 (1886), 191–214 (198).

[144] Bebbington, *The Mind of Gladstone*, 230–32.

[145] Huxley, 'Mr. Gladstone and Genesis', 198.

on revelation that had to be publicly discredited, but with it his endorsement of Irish Home Rule.[146]

Huxley's appeal to his 'clients' represents the terms upon which day-age theory would be debated throughout the remainder of the decade. Oxford's Regius Professor of Hebrew, Samuel Rolles Driver, was particularly concerned with levelling the popular authority of North American day-ager geologists. Writing in the Massachusetts-based *Andover Review*, a liberal theological periodical, in December 1887, Driver stressed that Dana and Dawson were authorities on geology, not Genesis. Explaining the significance of the original Hebrew terms and citing work by savants opposed to reconciliation, Driver (despite the periodical's implied readership of American theologians) commended the evidence to the 'British juryman', whose 'general education allows him to discriminate between the arguments addressed to him by opposing advocates'.[147] Dana (before the *Nineteenth Century* debates), Huxley (during them), and Driver (subsequent to them) all professed to give independent readers the information with which to judge the case.

In April, one anonymous contributor to the *Bibliotheca Sacra*, pleased that Driver had empowered the 'ordinary British juryman', took up his offer.[148] Characterizing Genesis as a popular scientific description of a vision, the 'juryman' argued that Dana, who combined scientific skill with literary perceptiveness, had successfully translated this popular language through attention to 'rhetorical' techniques like the clustering of 'salient features' and the use of words 'in a pregnant sense'.[149] The inability of Huxley to recognize these techniques, the contributor concluded, was due to 'an excess of physical and a lack of literary study'.[150] Close reading Genesis like this bypassed the literalistic interpretations of the most conservative and most liberal theological scholars alike, removing it safely from their specialist authority.

This accessibly unscientific approach to the language of scripture was also adopted by Gladstone, in his final stratagem. Unable to win a concession from Huxley in the *Nineteenth Century*, the statesman took a compromise to the more sympathetically pious *Good Words* in 1890. The Genesis days, he now contended, were 'CHAPTERS IN THE HISTORY OF THE CREATION' and thus, as in a textbook, these thematic chapters 'overlap' chronologically.[151] Gladstone gestured, this time, to his own specialist expertise. Refusing to admit the 'Hebraist' and 'Scientist' as the sole 'authoritative witnesses', he pointed out that his own 'specialism', namely 'the study' of communicating 'to the mass of men', was necessary to understand

---

[146] Desmond, *Huxley* (1997), 164, 212–13.
[147] Samuel Rolles Driver, 'The Cosmogony of Genesis: A Defense and a Critique', *Andover Review*, 8 (1887), 639–49 (641).
[148] 'The Cosmogony of Genesis: Professor Driver's Critique of Professor Dana', *Bibliotheca Sacra*, 45 (1888), 356–65 (356)
[149] 'Cosmogony of Genesis', *Bibliotheca Sacra*, 357, 359, 361.
[150] 'Cosmogony of Genesis', *Bibliotheca Sacra*, 365.
[151] W. E. Gladstone, 'The Creation Story', *Good Words*, 31 (1890), 300–11 (304).

the Mosaist's intentions.[152] The confusing metaphor of chapters, however, strayed far from the narrative clarity day-age theory usually provided. The jury was still out.

Broadly Conceptual

By the *fin de siècle*, compelling new presentations of day-age theory were scarce. In 1887, when a former student wrote to Dawson to request the latest research, he was surprised to find that 'there would seem to be practically little difference between the state of the question now & at the time "Archaia" was written' back in the 1850s.[153] In a well-worn convention, Dawson's late work *The Meeting-Place of Geology and History* (1894), presented 'a series of word-pictures' of geohistory, only to observe a few pages later that 'our six pictures are in some degree parallel with the "days" of creation', disingenuously adding that this was 'not an intentional reconciliation'.[154] He was the only respected geologist to express such confidence in the theory. The following year, Dana retracted it from his final edition of the *Manual of Geology*. The conservative *Bibliotheca Sacra*, too, began to see the writing on the wall. In 1897, it ran a long piece by Henry Morton, President of the Stevens Institute of Technology in Hoboken, New Jersey. Morton declared that, when the Bible was understood as inspired on moral but not factual matters, 'the problem worked on by the reconcilers simply vanished'.[155]

Although the controversy retreated from high culture, the logistical and literary problems of the Genesis visions were still being addressed. The author of one ambitious twentieth-century contribution to the field, David L. Holbrook, was a Congregational minister. The title of his book *The Panorama of Creation* (1908), invoking the technologies of visual spectacle, suggested that Holbrook would be working along similar lines to Miller's idea of the Genesis visions as a kind of miraculous diorama. In fact, Holbrook's thesis, combining half a century of figurative day-age devices, was multifaceted to the point of bewilderment. Firstly, the six days of creation, or rather 'panels in the biblical panorama', were to be 'regarded as so many paintings of geological landscapes, such as may be seen on the walls of a college museum'.[156] Holbrook may even have had in mind Hawkins' paintings at Princeton, already associated with day-age schemes in Guyot's *Creation*. We have

---

[152] Gladstone, 'The Creation Story', 305–6.
[153] John Stuart Buchan to Dawson, 15 February 1887, Dawson-Harrington Families Fonds, 1022-2-1-228-0022.
[154] John William Dawson, *The Meeting-Place of Geology and History* (Chicago: Fleming H. Revell, 1894), 18, 20.
[155] Henry Morton, 'The Cosmogony of Genesis and Its Reconcilers', *Bibliotheca Sacra*, 54 (1897), 264–92, 436–68 (270).
[156] David L. Holbrook, *The Panorama of Creation as Presented in Genesis Considered in Its Relation with the Autographic Record as Deciphered by Scientists* (Philadelphia, PA: Sunday School Times, [1908]), v. Subsequent references included in text.

seen various instances of scenes from deep time being used to illustrate passages of Genesis, but here, strikingly, Genesis *became* scenes from deep time, the conventions and display conditions of the latter shedding light on the interpretation of former.

Holbrook took the painterly metaphors of day-age theory more seriously than his predecessors. If Genesis verses appeared inaccurate, Holbrook argued, it was because they were 'not diagrammatic, but pictorial', unmeasurable 'by any historical or scientific scale' because '[o]bjects in the foreground' were 'proportionately larger than those in the background' (9). The Bible's apparent anachronism of modern plants emerging on the third day-age, during a geological age containing only their archaic ancestors, resulted from the text's figuratively 'foreshortened view' (36). Establishing the perspectival conventions of the Hexaemeral visions was necessary, as Holbrook's book was yet another virtual experiment in 'holding the reader long enough in the true position to enable him to see it for himself' (vi). Given that Holbrook believed that Genesis described geohistory phenomenologically, from the perspective of '*an ordinary human observer at the surface of the earth*' (12), this might initially appear a relatively straightforward task. However, not only were Holbrook's observers to take into metaphorical account the laws of perspective, but they also were advised not to imagine a literal viewpoint. Rather than a 'narrowly visual' approach, that employed by 'THE STANDPOINT OF THE BIBLICAL NARRATIVE OF CREATION IS THAT OF A SPECTATOR IDEALLY PRESENT AT THE SURFACE OF THE EARTH DURING GOD'S CREATIVE WORK' (19). This was probably a way of explaining that, like Riou's and Hawkins' images, the Genesis 'panels' or 'paintings' clustered together relevant ecological details into an ideal scene.

With readers now prepared, Holbrook tested the theory. 'Let one now', he declared, 'by an act of scientific imagination, place himself on some lone igneous rock in the primeval sea' (23). The book takes readers from one semi-ideal spot to the next, comparing the details described in the Bible with the facts of palaeoscience. At one point the reader's mental eye even becomes a camera lens, when Holbrook's information on Carboniferous flora 'sharpens the picture' (33). After this imaginative voyage is complete, the reader is presented with the information to make an informed decision about the concordance described. Holbrook was scrupulous in ensuring that the jury had the facts. Noting that a two-column 'tabular view' of 'selected words and phrases' from the Bible and geology was the 'usual method' of illustrating reconciliation, Holbrook wished to avoid the 'suspicion of manipulation' accompanying acts of 'selection' (55). He thus printed a triune table containing not just the entirety of the Genesis creation text but also two alternative scientific theories of creation (Figure 1.5). With this balanced presentation, Holbrook declared, 'the reader may have practically all the facts under his eye, and judge for himself as to their relation and significance' (55).

58   *Panorama of Creation*

Parallel Records of Creation in Its Visible Aspects

Autographic record as interpreted on [1]the nebular theory.	Biblical record as rendered in the American Standard Revision. Copyright, 1901, By Thomas Nelson and Sons.	Autographic record as interpreted on the planetesimal theory.

[1] For a fuller statement of the relation of the nebular theory to this view see Appendix II.

ORIGIN OF THE WORLD AND INITIAL CONDITION OF THE EARTH.

| As to the origin of the universe the nebular hypothesis is silent. It implies that the earth, in its earliest, non-luminous stage, was in darkness at the surface, drenched under the weltering deep of a vast and vapor-laden atmosphere. | 1 In the beginning God created the heavens and the earth. 2 And the earth was waste and void; and darkness was upon the face of the deep: and the spirit of God [1]moved upon the face of the waters.
[1] Or, *was brooding upon* | As to the origin of the universe the planetesimal hypothesis is silent. At the earliest stage at which it takes cognizance of the [1]materials which now constitute this earth, it postulates molecular activity |

[1] For other points in this connection see Appendix III.

LIGHT, AND THE PHENOMENA OF DAY AND NIGHT.

| To an observer in such conditions, at the surface of the earth, the first notable phenomenon would be simply light. Hav- | 3 And God said, Let there be light: and there was light. 4 And God saw the light, that it was good: and God divided the | manifesting itself in light. The earth's separate existence was begun by matter shot forth from an ancestral sun, under |

Figure 1.5 David L. Holbrook, *The Panorama of Creation as Presented in Genesis Considered in Its Relation with the Autographic Record as Deciphered by Scientists* (Philadelphia, PA: Sunday School Times, [1908]), 58. Holbrook presents a tabular summary of two scientific theories of Earth's origins, both running alongside the Genesis creation story's text.

The reader's opinion was consequential: after all, Genesis was emotive 'literature' rather than cold, hard 'science' (v), having been written not for 'the professional scientist' but rather, as Gladstone had observed, for 'the unscientific reader' (69). Undecided members of the 'unscientific' jury approaching Holbrook's monograph, however, were likely to be few. Most reviewers gathered in one of two camps: those who, like the critic in the Boston Methodist *Zion's Herald*, agreed that the Mosaist 'must have been supernaturally, divinely guided' and those who, with the non-denominational *Hartfield Seminary Record*, insisted that 'discussions of this sort have lost their significance and their interest for our generation'.[157] It appeared that, no matter how ingenious the analogy, the question was unlikely to be settled by a jury after all.

Conclusion

Back in the 1850s, Hugh Miller had triumphantly tested the principles of day-age theory on working-class visitors to the British Museum. Several decades later, Kinns had preached the theory to children in the Museum's halls, airing the content that later became *Moses and Geology*. In the twentieth century, the theory would again perform paternalistic service in a branch of the British Museum, although this time it was wielded by a very different figure: the birth control campaigner Marie Stopes. In her novel *Love's Creation* (1928), Stopes's protagonists visit the Natural History Museum, where idealistic heroine Rose Amber learns that a security guard's literalistic Christian faith has been shaken by the labels of palaeontology exhibits. Rose reassures the man, in Dawsonian rhetoric, that, in Hebrew, the word day

> might just as well have been translated as periods, and then the first chapter of Genesis would read in your Bible 'And God made the world in seven periods,' and that is just what these labels are saying! These geological records really do correspond very well to short poetical descriptions of the seven periods in the beginning of the Bible.[158]

She confides with her friends the utility of this white lie: 'Why should some cruel bits of truth be rammed down their throats and none of the beautiful, comforting bits that would help them to keep their God and learn something as well?'[159] If

---

[157] Untitled review of *The Panorama of Creation*, by David L. Holbrook, *Zion's Herald* [Boston, MA], 26 May 1909, 660; L. B. P., untitled review of *The Panorama of Creation*, by David L. Holbrook, *Hartford Seminary Record*, 19 (1909), 311–12 (311).
[158] Marie Stopes, *Love's Creation: A Novel* (London: John Bale, Sons & Danielsson, 1928), 140.
[159] Stopes, 141.

Miller's successors had sought to bring intellectual respectability to day-age theory, to Stopes it was a mere sop for passive, unquestioning members of society. Unlike Stopes, however, these day-agers would have known that explaining the double meaning of the Hebrew word *yom* was just the start. More than philological technicalities, persuading believers required literary technologies that could simulate and attest to biblical events no modern humans would ever see.

The day-agers I have examined were, for the most part, conservative Nonconformists and idiosyncratic Anglican laymen. Although many held privileged connections in the worlds of literature and science, these were strained by their increasingly unfashionable views on biblical hermeneutics. To communicate these day-age views, men like Collingwood drew on old, tried-and-tested genres, including familiar conversations, Miltonic epics, and didactic miscellanies, although others, like Dawson, employed the strategic opportunities available in more timely formats like popular science articles. Eschewing the encroachment of technical specialization, literary utilitarianism, and scientific naturalism, they invoked time-honoured forms of authority like vatic poetic inspiration, esoteric intuition, patronage, testimonials, and the freedoms of the public sphere. In so doing they brought the maximum scientific authority to the text of Genesis, along with the maximum independence from scientific evaluation. This tactic extended an appropriately Protestant stress on individual interpretation, rather than acquiescence to authority, into a palaeoscientific context. Naturally, individual interpretation could be given a nudge in the right direction through provision of the right literary technology with which to parse the Hexaemeron's sparing text. After all, when deciding whether the creation story represented a vision, readers' imaginative faculties were being exercised just as much as their ability to evaluate disputable geological and theological evidence.

The result of day-agers' efforts was that the theory was promoted at North American universities and debated in cutting-edge periodicals decades after progressives had declared it moribund. It was by no means without its converts: as we shall see in chapter 5, the theory remained a difficult obstacle for young-earth creationists to dislodge within Christian fundamentalist circles deep into the twentieth century. Nonetheless, in most spheres it remained not just borderline palaeoscience but borderline theology. Across both centuries, there was also competition in the visionary department from an unexpected direction. As day-agers attempted to show exactly how the Mosaist had gazed upon the Mesozoic seas, their contemporaries in the occult community were devolving the task of godlike clairvoyance upon themselves. Unlike in the masculine world of evangelical geology, moreover, women were to play central roles in this new time-transcending project.

2
Deep Clairvoyance

The Tĕx'rŏnz, the Nŏl-ō-kā-thē'rŭm, and the Skär'dĕnt: these extinct genera native to the lost continent of Atlantis were, respectively, ancient descendants of the camel, crocodile, and crow. We know this, and the correct pronunciation of these names (indicated diacritically), thanks to *Submerged Atlantis Restored* (1911), an encyclopaedic tome by American Spiritualists J. Ben Leslie and Carrie C. Van Duzee. With Van Duzee acting as medium, the two investigators detailed the natural history of Atlantis through consultation with its ghostly denizens.[1] Their use of psychic communion to learn about the prehistoric past was no novelty. The ability purportedly to obtain secret geohistorical knowledge through occult means emerged back in the mid-nineteenth-century United States, originating among the varied Spiritualist phenomena that subsequently also took Europe by storm.[2] At febrile, utopian intersections of research on the frontiers of science, the idea had begun to circulate not just that the living might communicate with the dead, but also that little-understood powers of perception held the potential to transcend time, bypassing the untidy speculation so necessary to palaeontology, geology, and other historical sciences. This superhuman capability would further reveal the limitations of a purely materialist worldview. For the cautiously optimistic, this was merely a tantalizing possibility. Among radical clairvoyants, it was already being demonstrated.

Previously, we heard that, after he embraced esoteric Christianity, Cuthbert Collingwood lost interest in how the author of Genesis had actually seen into the prehistoric past. In contrast, the occult subjects of this chapter, visionary 'psychometers' and members of the arcane Theosophical Society, became even more obsessed with this question of seeing into the past. Training their inner eyes on ethereal remnants of prehistoric happenings, they employed what I call 'deep clairvoyance' to actually *see* mastodons, dinosaurs, and ancient catastrophes. Their practice was inspired by what Gowan Dawson calls 'the occult self-fashioning' of elite palaeontologists—the clairvoyants 'making literal what was initially only used figuratively'.[3] Here, the vivid language used by scientific practitioners to

---

[1] J. Ben Leslie, *Submerged Atlantis Restored; or, Rĭn-Gä'-Sĕ Nud Sī-ī-Kĕl'Zē (Links and Cycles)* (Rochester, NY: Austin Publishing, 1911), 131, 134, 169.

[2] Ann Taves, *Fits, Trances, and Visions: Experiencing Religion and Explaining Experience from Wesley to James* (Princeton, NJ: Princeton University Press, 1999), esp. part two.

[3] Gowan Dawson, *Show Me the Bone: Reconstructing Prehistoric Monsters in Nineteenth-Century Britain and America* (Chicago: University of Chicago Press, 2016), 356.

market palaeontology to general readerships was repacked as the register for occult literary technologies designed to engender virtual witnessing. The earliest clairvoyants had reason to think their project not doomed to the fringes, given the enduring currency of the similarly visionary day-age theory and the endorsement granted to Spiritualist and psychic phenomena by many eminent savants.[4] Nonetheless, accounts of clairvoyant time travel faced extreme scepticism from the establishment. By the century's final decades, clairvoyant palaeoscience was chiefly addressed to what Mark S. Morrison describes as the 'counter-public spheres' of occult publishing.[5] From this counter-establishment vantage point, however, eyewitness reports of the primeval world proliferated, engrossing believers and non-believers on both sides of the Atlantic.

By surveying the literary stylings of deep clairvoyance, this chapter explores a suggestive overlap between literature and science studies, Earth science history, and esotericism studies. Scholars in the latter field are already exploring the role of esoteric texts in the emergence of science fiction.[6] It should be no surprise, then, that the techniques and conventions of the former could also overlap with those of palaeoscientific non-fiction, given their shared dedication to visualizing unseen but scientifically explicable worlds. The occult register sometimes employed by geologists and palaeontologists has not passed unnoticed, but it has usually been discussed only with reference to the culture of the early nineteenth century.[7] Moreover, perhaps because connections between occultism and palaeoscience are most often found only in dispersed passages in texts focused on other matters, these connections have simply not been the primary and sustained focus of researchers, excepting those working specifically on enchanted lost continents like Atlantis.[8] Scholars of occult science looking at this period have usually focused on the human and physical sciences instead.[9] As I demonstrate, Earth's deep history played a far stronger role in nineteenth- and early twentieth-century occultism than has

---

[4] For instance, see Efram Sera-Shriar, *Psychic Investigators: Anthropology, Modern Spiritualism, and Credible Witnessing in the Late Victorian Age* (Pittsburgh, PA: Pittsburgh University Press, 2022).

[5] Mark S. Morrison, 'The Periodical Culture of the Occult Revival: Esoteric Wisdom, Modernity and Counter-Public Spheres', *Journal of Modern Literature*, 31 (2008), 1–22. See also Christine Ferguson, 'The Luciferian Public Sphere: Theosophy and Editorial Seekership in the 1880s', *Victorian Periodicals Review*, 53 (2020), 76–101.

[6] Aren Roukema, 'The Esoteric Roots of Science Fiction: Edward Bulwer-Lytton, H. G. Wells, and the Occlusion of Magic', *Science Fiction Studies*, 48 (2021), 218–42; Christopher Keep, 'Life on Mars?: Hélène Smith, Clairvoyance, and Occult Media', *Journal of Victorian Culture*, 25 (2020), 537–52.

[7] For examples of geologists' and palaeontologists' occult and necromantic metaphors in the eighteenth and early nineteenth centuries, see Martin J. S. Rudwick, *Bursting the Limits of Time: The Reconstruction of Geohistory in the Age of Revolution* (Chicago: University of Chicago Press, 2005), 301, 413; and Ralph O'Connor, *The Earth on Show: Fossils and the Poetics of Popular Science, 1802–1856* (Chicago: University of Chicago Press, 2007), 56–58, 82–87, 90–92, 95–98, 100–15, 128–29. The aforementioned discussion in Dawson, *Show Me the Bone*, 352–57, refers to the mid-century.

[8] Sumathi Ramaswamy, *The Lost Land of Lemuria: Fabulous Geographies, Catastrophic Histories* (Berkeley: University of California Press, 2004), esp. chapter 3; Joscelyn Godwin, *Atlantis and the Cycles of Time: Prophecies, Traditions, and Occult Revelations* (Rochester, VT: Inner Traditions, 2011).

[9] For example, see Sumangala Bhattacharya, 'The Victorian Occult Atom: Annie Besant and Clairvoyant Atomic Research', in *Strange Science: Investigating the Limits of Knowledge in the Victorian Age*,

hitherto been recognized. Even Wouter J. Hanegraaff's authoritative examination of psychometric clairvoyance does not consider how the geological career of William Denton, a significant figure in his article and one of the main subjects of my chapter, may have shaped the paranormal practices of his family circle.[10]

The fascinating palaeoscientific works of the Denton circle, and of other occult authors like H. P. Blavatsky and William Scott-Elliot, reward more sustained attention. I argue, firstly, that literalized or augmented versions of geological and palaeontological metaphors and generic conventions infused their clairvoyant narratives, shaping the ways they psychically experienced, and subsequently represented, the prehistoric past. Their maverick activities were, therefore, products of the success of palaeoscience's popularizers in helping wide audiences to visualize extinct animals and ancient landscapes. Secondly, I show how these clairvoyants used their powers to undertake otherwise impossible research on beings and events upon which the fossil record was utterly silent. By so doing, they could reconfigure the story of life's evolution without the need for specialist training, expensive facilities, or painstaking fieldwork—sometimes even from the comfort of their drawing rooms. These authors demonstrated not a wholesale rejection of mainstream interpretations of Earth's history, as did some of the figures I discuss later in this book. Rather, espousing an ultra-heightened faith in the imaginatively perceptive faculties by which savants had reconstructed prehistoric worlds, they proposed to show in more detail than ever before what had happened in the darkest depths of time's abyss.

The Vision of the Mystery

French savant Georges Cuvier, one of the founders of palaeontology, memorably declared that the new sciences of the Earth would 'burst' the limits of time. As Martin J. S. Rudwick reminds us, Cuvier's declaration referred less to expanding the conventionally brief biblical timescale of 6,000 or so years than it did to expanding the *knowability* of the deep past.[11] The proper role of imagination and speculation in this project was controversial, especially in the fiercely empirical culture of early nineteenth-century geology.[12] Even in the 1830s, when the project of reconstructing Earth's deep history was well underway, Charles Lyell questioned the advisability of pinning too much faith in geohistorical grand narratives: humans, who inhabit only the Earth's surface, Lyell argued, have just as incomplete a notion of geological processes as would be available to a 'dusky' subterranean gnome

---

ed. by Lara Karpenko and Shalyn Claggett (Ann Arbor: University of Michigan Press, 2017), 197–214; and Sera-Shriar, *Psychic Investigators*.

[10] Wouter J. Hanegraaff, 'The Theosophical Imagination', *Correspondences*, 5 (2017), 3–39 (22, 24).
[11] Rudwick, *Bursting the Limits of Time*, 506.
[12] O'Connor, *Earth on Show*, 18–19, 207.

(to be specific, Lyell referred to the Rosicrucian spirit Umbriel, from Alexander Pope's mock-epic poem *The Rape of the Lock*).[13] He nonetheless exalted the mental muscle used by cosmopolitan savants, in the face of all these difficulties, to conceptualize the deep timescales over which geological processes took place. As Gillian Beer points out, Lyell, despite his decentring of humanity as the measure of things still insisted on the time-transcending 'power of man's imagination' at the 'humanistic core' of his work.[14] Nor was the idea that we might only possess fragmentary knowledge of the events of the deep past necessarily a deflating one. For some, the nebulous nature of geologists' access to the obscure deep past generated a thrilling sense of the sublime.[15]

Creative literature and art provided a way to see into the past more freely, enabling the construction, in Virginia Zimmerman's words, of 'a path for the individual through time's roar'.[16] Imaginative depictions of prehistoric worlds were at their most engaging in material addressed to general audiences, appearing not just in books but also in lantern slides, models, posters, and panoramas. Authors generated a legion of genres, modes, and analogies for conceptualizing literary forms of time travel, often promiscuously adopted from other discourses. In addition to becoming linked with the languages of epic, painting, and theatrical spectacle we saw on show in chapter 1, seeing through time was framed as reading nature's book, translating hieroglyphs, shining light into darkness; resurrecting the dead, travelling to the underworld, experiencing reincarnation; rambling, voyaging, exploring; and experiencing phantasmagorical dreams, nightmares, and drug-induced hallucinations.[17]

Nonetheless, in certain quarters, a merely deductive or mediated ability to explore prehistory seemed insufficiently direct. Savants toyed with the notion that their powers went further: Cuvier and successors like Richard Owen, Oxford geologist William Buckland, and Harvard palaeontologist-polygenist Louis Agassiz relished their reputation as pseudo-necromancers and prophets resembling the biblical Ezekiel, especially regarding their purported ability to identify an animal from a single bone (the subject to which Dawson's discussion of 'occult self-fashioning', referred to above, relates).[18] Some members of the occult community took the bait, contending that this line was not mere rhetoric. For George

---

[13] Adelene Buckland, 'The World beneath Our Feet', in *Time Travelers: Victorian Encounters with Time & History*, ed. by Adelene Buckland and Sadiah Qureshi (Chicago: University of Chicago Press, 2020), 42–64 (45–46).

[14] Gillian Beer, *Darwin's Plots: Evolutionary Narrative in Darwin, George Eliot and Nineteenth-Century Fiction*, 3rd edn (Cambridge: Cambridge University Press, 2009), 17, 39.

[15] O'Connor, *Earth on Show*, 439–43.

[16] Virginia Zimmerman, *Excavating Victorians* (Albany: State University of New York Press, 2008), 23.

[17] O'Connor, *Earth on Show*, passim.

[18] Dawson, *Show Me the Bone*, 351–52. For a sketch of Owen with a magic wand, see Irina Podgorny, 'Fossil Dealers, The Practices of Comparative Anatomy, and British Diplomacy in Latin America, 1820–1840', *British Journal for the History of Science* 46 (2013), 647–74 (664).

Winslow Plummer, Imperator and Supreme Magus of the Societas Rosicruciana in America, such uncanny abilities could only be an 'inherited cosmic memory' from the 'Atlantean Epoch' when humans lived alongside 'gigantic reptilia'.[19]

Owen and Agassiz also had more tangible connections to esotericism, both holding sympathies with the Romantic, anti-materialistic school of German science known as *Naturphilosophie*.[20] Strikingly, Agassiz claimed that dreams guided his ability to reconstruct fossil fish.[21] Owen's traffic with transcendentalism was more guarded, but his German-inspired morphological theories, mostly published in the 1840s, appealed to later occultists long after they fell from favour among anatomists. Recalling his theory about the divinely designed archetypal form underlying all vertebrate skeletons, the Theosophist Annie Besant observed in *The Pedigree of Man* (1904) that Owen 'builded truer than he knew'. His ideas, she explained, were foreshadowed by discussions of 'archetypal forms' of life in the ancient Hindu Puranas, texts that, while superficially 'dim', were actually 'the best description that human language is able to give' of these primordial archetypes.[22]

Besant's claims, as we shall see in more detail in a later section, were based on the esoteric tenet that an immensely ancient scientific wisdom was kept alive in present-day Asia, especially in India. Again, these ideas were fuelled by the Orientalist language used by accredited savants. Palaeoscientific imagination had long exhibited a similar vein of Orientalist Romanticism, drawing freely upon imagery from Asian religion and myth to express the mysterious strangeness of prehistory. Comparisons between the wonders of palaeontology and those described in the *Arabian Nights* were commonplace.[23] Perhaps most strikingly, Austrian geologist Eduard Suess compared the strenuous global project of bridging deep time to the bridging of the ocean by Rama, an avatar of the god Vishnu, in the Sanskrit Hindu epic *Ramayana*.[24] By verbally donning Orientalist attire, geologists could become like unto gods, performing seemingly supernatural reconstructive feats of the mind, with the stated or implicit caveat that rational Western science underlay all.

More frequently, transcendent insight into the deep past was framed within the Christian framework of day-age theory, but the Mosaist's God-given ability to

---

[19] 'Khei X' [George Winslow Plummer], *Rosicrucian Fundamentals: An Exposition of the Rosicrucian Synthesis of Religion, Science and Philosophy in Fourteen Complete Instructions* (New York: Flame Press, 1920), 226.

[20] Olaf Breidbach and Michael Ghiselin, 'Lorenz Oken and *Naturphilosophie* in Jena, Paris and London', *History and Philosophy of the Life Sciences*, 24 (2002), 219–47.

[21] Dawson, *Show Me the Bone*, 352–53.

[22] Annie Besant, *The Pedigree of Man* (Benares [Varanasi]: Theosophical Publishing Society, 1904), 55–56.

[23] O'Connor, *Earth on Show*, 97, 197, 341, 362–63, 366.

[24] James A. Secord, 'Global Geology and the Tectonics of Empire', in *Worlds of Natural History*, ed. by H. A. Curry, N. Jardine, J. A. Secord, and E. C. Spary (Cambridge: Cambridge University Press, 2018), 401–17 (405). See also Pratik Chakrabarti, *Inscriptions of Nature: Geology and the Naturalization of Antiquity* (Baltimore, MD: Johns Hopkins University Press, 2020).

see behind the veil was different from esoteric experiences in degree rather than kind. That degree could become remarkably minute. *Panthea* (1849), a philosophical novel by the polymathic savant Robert Hunt, tells the story of a young Lord, Julian Altamont, who befriends a Rosicrucian mystic. Empowered by the mystic to see through geohistory, Altamont experiences pseudo-day-age visions: he gapes at 'frog-like monsters' and 'rapacious Saurians', and, when each phase ends, '[a]ll things passed into night, and another day at length broke upon Julian's astonished vision'.[25] '[T]he vision of the mystery' is compared to the Eleusinian Mysteries, a secretive ancient Greek ritual beloved of occultists.[26] Given that the biblical creation story was sometimes explained as deriving from Moses's esoteric Egyptian knowledge, a surprising number of different religious traditions could be syncretized in the same occult geohistorical package. Some, including Plummer, even argued that day-age theory was compatible with the Theosophical cosmogony, an elaborate narrative structured around the cyclical rise and fall of human cultures on fabled continents like Atlantis (Figure 2.1). This surprising compatibility, it was argued, reflected the shared ancient origins of these notions.

This section has briefly established some of the overlaps between occult thought and visionary or theistic currents across nineteenth-century palaeoscience, but now it is necessary to walk a narrower path. The remainder of the chapter will follow a chronological narrative of the previously overlooked development of deep clairvoyance and its concomitant literary technologies, beginning with the mid-nineteenth-century context in which visions of the past moved beyond the language of metaphor and into the realm of truth claims.

Mental Fossils

Edward Hitchcock, Congregationalist pastor of Amherst College, Massachusetts, was famous for research on the mysterious fossil footprints of the Connecticut valley, all that remained of what had apparently been giant Triassic birds. Drawing on palaeontology's necromantic connotations, Hitchcock had even tried his hand, in 1836, at a mock-Romantic poem in which a sorceress not unlike the biblical Witch of Endor revives one of these titanic avians.[27] However fanciful, the poem's conceit was little more than a supernaturalised precursor to its author's most ambitious hypothesis. This hypothesis, inspired by combining his interest in recovering geology's faintest traces with his enthusiasm for the new technological

---

[25] Robert Hunt, *Panthea, the Spirit of Nature* (London: Reeve, Benham, and Reeve, 1849), 81–82. For *Panthea*'s origins, see James A. Secord, *Victorian Sensation: The Extraordinary Publication, Reception, and Secret Authorship of* Vestiges of the Natural History of Creation (Chicago: University of Chicago Press, 2000), 467, 469.
[26] Hunt, 188.
[27] 'Poetaster' [Edward Hitchcock], 'The Sandstone Bird', *Knickerbocker*, 8 (1836), 750–52.

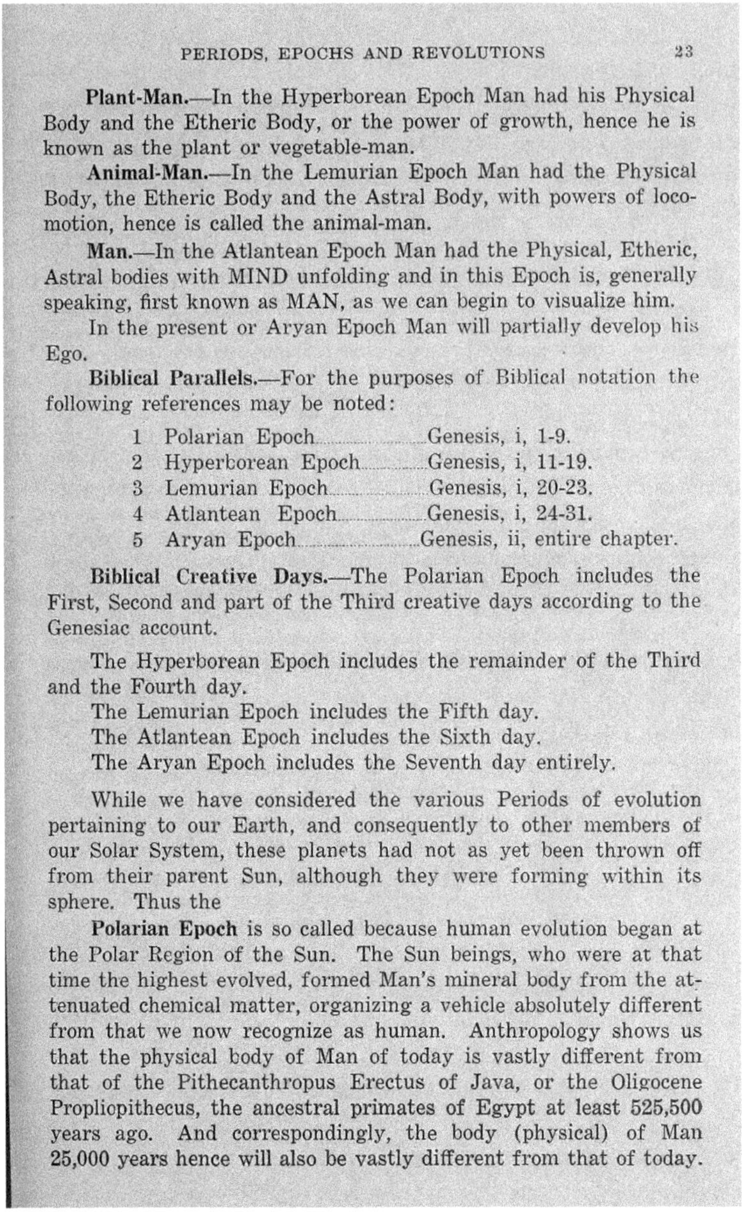

> PERIODS, EPOCHS AND REVOLUTIONS 23
>
> **Plant-Man.**—In the Hyperborean Epoch Man had his Physical Body and the Etheric Body, or the power of growth, hence he is known as the plant or vegetable-man.
> **Animal-Man.**—In the Lemurian Epoch Man had the Physical Body, the Etheric Body and the Astral Body, with powers of locomotion, hence is called the animal-man.
> **Man.**—In the Atlantean Epoch Man had the Physical, Etheric, Astral bodies with MIND unfolding and in this Epoch is, generally speaking, first known as MAN, as we can begin to visualize him.
> In the present or Aryan Epoch Man will partially develop his Ego.
> **Biblical Parallels.**—For the purposes of Biblical notation the following references may be noted:
>
> 1 Polarian Epoch............Genesis, i, 1-9.
> 2 Hyperborean Epoch............Genesis, i, 11-19.
> 3 Lemurian Epoch............Genesis, i, 20-23.
> 4 Atlantean Epoch............Genesis, i, 24-31.
> 5 Aryan Epoch............Genesis, ii, entire chapter.
>
> **Biblical Creative Days.**—The Polarian Epoch includes the First, Second and part of the Third creative days according to the Genesiac account.
> The Hyperborean Epoch includes the remainder of the Third and the Fourth day.
> The Lemurian Epoch includes the Fifth day.
> The Atlantean Epoch includes the Sixth day.
> The Aryan Epoch includes the Seventh day entirely.
> While we have considered the various Periods of evolution pertaining to our Earth, and consequently to other members of our Solar System, these planets had not as yet been thrown off from their parent Sun, although they were forming within its sphere. Thus the
> **Polarian Epoch** is so called because human evolution began at the Polar Region of the Sun. The Sun beings, who were at that time the highest evolved, formed Man's mineral body from the attenuated chemical matter, organizing a vehicle absolutely different from that we now recognize as human. Anthropology shows us that the physical body of Man of today is vastly different from that of the Pithecanthropus Erectus of Java, or the Oligocene Propliopithecus, the ancestral primates of Egypt at least 525,500 years ago. And correspondingly, the body (physical) of Man 25,000 years hence will also be vastly different from that of today.

Figure 2.1 [George Winslow Plummer], 'Khei X', *Rosicrucian Fundamentals: An Exposition of the Rosicrucian Synthesis of Religion, Science and Philosophy in Fourteen Complete Instructions* (New York: Flame Press, 1920), 23. Plummer correlates the epochs of the occult cosmos, mostly named for sunken continents like Atlantis, with the day-age interpretation of Genesis.

developments of photography and telegraphy, was his 'Telegraphic System of the Universe', described in *The Religion of Geology and Its Connected Sciences* (1851).[28] Since rocks could preserve footprints, photographs capture moments, and telegraphs link distant individuals, Hitchcock mused, might invisible forces pervading the world be recording every moment in a 'vast panorama'?[29] And might these superfine recordings be accessible? In support of the idea, Hitchcock cited a recent thought experiment by German astronomer Felix Eberty. Given the limitations of the speed of light, Eberty observed, individuals on very distant stars must be seeing our Earth in its past state. Using this '*microscope for time*', an alien could, in theory, answer all our questions about 'geology and the creation'.[30] The provisional aspect of geohistorical reconstruction could, theoretically, be bypassed through entirely scientific means.

Although Hitchcock's substantial scientific clout would prove useful to clairvoyants citing his ideas in subsequent decades, paranormal interest in prehistory was already in the air. It had begun to germinate in 1840s New York state, when a heady atmosphere of Jacksonian democracy was fostering a climate of scientific and religious individualism.[31] These individualists are best understood with reference to what Egil Asprem calls 'open-ended naturalism', a term he uses to characterize the methodology of practitioners of theistic and heterodox science who insist that their dramatic claims are empirically founded.[32] At the centre of the first wave was young mesmeric subject (and subsequently leading Spiritualist) Andrew Jackson Davis, the Poughkeepsie Seer. In 1845, even before the now-infamous 'Rochester Rappings' of 1848 (the ghostly occurrences that provoked the emergence of modern Spiritualism), Davis began delivering a detailed divine revelation while in clairvoyant trance.

Davis's insights were published as *The Principles of Nature* (1847), a magisterial synthesis of scientific knowledge. As Dana Luciano has pointed out, geology held thematic pride of place in the *Principles of Nature* as the exemplary modern science, epitomizing science's ability to shed light on the darkness of time itself.[33] Endowed with heightened lucidity, Davis revealed much more

---

[28] Edward Hitchcock, *The Religion of Geology and Its Connected Sciences* (Boston: Phillips, Sampson, 1851), 409

[29] Hitchcock, 417. For panorama as occult metaphor, see Erkki Huhtamo, *Illusions in Motion: Media Archaeology of the Moving Panorama and Related Spectacles* (Cambridge, MA: MIT Press, 2013), 348–49.

[30] [Felix Eberty], *The Stars and the Earth; or, Thoughts Upon Space, Time, and Eternity, Part II* (London: H. Bailliere, 1847), 12, 14. See also O'Connor, *Earth on Show*, 429–30.

[31] Daniel Patrick Thurs, *Science Talk: Changing Notions of Science in American Popular Culture* (New Brunswick, NJ: Rutgers University Press, 2007), 30–31, 33–41.

[32] Egil Asprem, *The Problem of Disenchantment: Scientific Naturalism and Esoteric Discourse, 1900–1939* (Leiden: Brill, 2014), 79.

[33] Dana Luciano, 'Sacred Theories of Earth: Matters of Spirit in *The Soul of Things*', *American Literature*, 86 (2014), 713–36 (719).

about prehistory than what was found in merely material fossils: the English dinosaur *Megalosaurus*, for instance, unbeknownst to palaeontologists, was actually equipped with 'wings', 'fins', and 'a tortoise shell-like coating', while the wildlife of the planet Saturn included an animal resembling another dinosaur, *Iguanodon*.[34] Although the seer claimed to be poorly read, and the winged *Megalosaurus* and Saturnian *Iguanodon* numbered among his original contributions, the New York *Christian Examiner* suspected that 'the bulk of the work' was a 'generalization' of Scottish journalist Robert Chambers' anonymously published evolutionary epic, *Vestiges of the Natural History of Creation*.[35] Both the *Vestiges* and the *Principles* presented revolutionary new syntheses of the sciences, and the occult origins of the latter potentially intrigued, rather than repulsed, audiences receptive to philosophical radicalism. In 1848, one reader, the liberal British intellectual John Chapman, unsuccessfully attempted to convince Owen of the value of Davis's weighty psychic volume.[36]

Indeed, although Bernard Lightman places the birth of the Spiritualist evolutionary epic genre later in the century, it seems to have been an element of Spiritualist science from the beginning.[37] Davis's tome was followed by another vast scientific synthesis, *The Arcana of Nature* (1859), this time produced using the mortal frame of the Ohio medium Hudson Tuttle. Even more striking than the claims made in Tuttle's prose was an accompanying panorama hundreds of feet in length, painted some time before 1855, in which he depicted the geohistory of the planet. As Tuttle later recalled, he had never even 'seen a panorama' before, but a spiritual 'guide' exhibited a succession of moving images that he mechanically reproduced:

> The first scene was a glowing fire mist, the next a condensing nebula. Then, the molten, heaving, lava surface was represented, and the painting was thenceforth continuous to the end. The surface grew dark, black clouds appeared, watery vapors condenst [sic], falling into boiling seas; vegetation came on the coast lines, wonderful forests of the Coal Age covered land and water; the atmosphere cleared and reptiles came, gigantic saurians of hideous shape basked on the shores and sported in the waves; these again giving place to equally huge animals of the Tertiary Age.[38]

---

[34] Andrew Jackson Davis, *The Principles of Nature, Her Divine Revelations, and A Voice to Mankind* (New York: S. S. Lyon and William Fishbough, 1847), 179, 261, 263–64.

[35] W. S., untitled review of *The Principles of Nature*, by Andrew Jackson Davis, *Christian Examiner*, 43 (1847), 452–55 (452).

[36] Secord, *Victorian Sensation*, 486.

[37] Bernard Lightman, *Victorian Popularizers of Science: Designing Nature for New Audiences* (Chicago: University of Chicago Press, 2007), 239.

[38] Hudson Tuttle, *Arcana of Nature*, 2nd rev. edn (New York: Stillman Publishing, 1909), 60. See also Hudson Tuttle, *Scenes in the Spirit World; or, Life in the Spheres* (New York: Partridge and Brittan, 1855), 8.

The automatically produced (and now lost) painting was so convincing that it was loaned from Tuttle for use in scientific lectures, although its paranormal origin was kept quiet.[39]

Unlike the avowedly unlettered Davis, however, Tuttle hinted at his textual sources. The original edition of the *Arcana of Nature* cited the *Vestiges* among the books 'consulted' by the author, alongside references to other works, such as those of Hugh Miller, in order to 'trace [clairvoyantly] *received* facts to their legitimate sources'.[40] Although his retrospective phrasing implied the happy coincidence of facts rather than intellectual debts, palaeo-artistic conventions shaped the form of Tuttle's painting. Despite Tuttle's claim never to have seen a panorama of any kind, geohistorical panoramas of a distinctly similar format were coming into use in the years before his clairvoyant experience.[41] The illustrated ideal scenes of prehistory adorning the frontispieces of many geological tomes would also have provided blueprints for the art of fitting representative sketches of different prehistoric ecosystems into a continuous image. As Samuel Phelps Leland, one of the savants who lectured using the panorama, explained, Tuttle's painting presented not naturalistic glimpses into the past, but rather a more artfully 'connected story of creation, each great period blending into another without a break', geological periods being 'shown by the characteristic animal and vegetable forms' of the time (for a comparable image from a later decade, see Figure 1.1).[42]

Another American maverick developed a theory, or method, that allowed one to transcend time more proactively than the passive mediumship of Spiritualism could enable. It would cast a long shadow. In 1849, the physician Joseph Rodes Buchanan announced that, seven years prior, he had invented a new science called 'psychometry' or 'mind-measuring'.[43] Working at the anti-establishment Eclectic Medical Institute in Cincinnati since 1846, Buchanan's wide-ranging interests included mesmerism and phrenology.[44] Thinking along the same lines as Hitchcock, Buchanan's psychometry was based on the idea that we unknowingly impress traces of our minds and actions upon physical objects, like a '*mental daguerreotype*', and that a person of psychic sensitivity can recover these invisible impressions.[45]

---

[39] Tuttle, *Arcana of Nature* (1909), 466–67.

[40] Hudson Tuttle, *Arcana of Nature; or, the History and Laws of* Creation, 2 vols (Boston: Colby & Rich, 1859), I, 5, 15, 73–74.

[41] Martin J. S. Rudwick, *Scenes from Deep Time: Early Pictorial Representations of the Prehistoric World* (Chicago: University of Chicago Press, 1992), 92–95.

[42] Tuttle, *Arcana of Nature* (1909), 467.

[43] Joseph Rodes Buchanan, 'Psychometry', *Buchanan's Journal of Man*, 1 (1850 [1849]), 49–62, 97–113, 145–56, 208–27 (62).

[44] Greg L. Hester, 'Into the Celestial Spheres of Divine Wisdom: Joseph Rodes Buchanan and Nineteenth-Century Esotericism', unpublished MA thesis, University of Amsterdam (2015), 9–10, 33, 37.

[45] Buchanan, 58. See also Cameron B. Strang, 'Measuring Souls: Psychometry, Female Instruments, and Subjective Science, 1840–1910', *History of Science*, 58 (2020), 76–100.

Buchanan's psychometers chiefly investigated the personalities of the senders of unopened letters, but his most glamorous rhetoric drew upon the register of geological necromancy. The geologist's 'magic power' had allowed 'huge Saurian monsters' to 'rise before the eye', and now psychometry's 'mental telescope' would similarly 'pierce the depths of the past' to extract 'mental fossils'.[46] Buchanan thus implied that psychometric discoveries were no harder to believe in than the astonishing narratives geologists had already extracted from the rocks trodden underfoot. Science, it seemed, was undergoing a revolution, in which geology's almost magical insights heralded the arrival of ever more powerful tools for exposing nature's secrets. Even half a century later, when chastising conservative 'hostility to Homeopathy, Geology, Phrenology, American Spiritualism', the elderly Buchanan evidently still counted palaeoscience among those heterodox fields used to investigate hidden worlds.[47]

Enter into the Soul of Things

The nineteenth century's most prolific psychometers, the family circle surrounding geologist William Denton, took Buchanan's geological rhetoric seriously. Denton was born, in 1823, to an English Methodist family of what a sympathetic biographer called 'comparatively humble circumstances' in Darlington.[48] The young Denton was a scientific autodidact, devouring cheap educational periodicals like the *Penny Magazine*, attending the local Mechanics' Institute, and finding inspiration, as his American Spiritualist contemporaries did, in Chambers' *Vestiges*.[49] His particular enthusiasm for geology would have been encouraged by that science's reputation as one in which proletarian practitioners like himself could make meaningful contributions to knowledge, thanks to the ubiquity of geological specimens.[50] Hugh Miller's popular book, *The Old Red Sandstone* (1841), had called for working men to employ their time, as he did, in geological pursuits, rather than anti-establishment politics.[51] Denton was evidently convinced that the two activities could be combined. Frustrated by British intolerance of his social radicalism and freethinking religious heterodoxy, he emigrated

---

[46] Buchanan, 147–48.
[47] Joseph Rhodes Buchanan, 'The New World of Science—for 1898', John Uri Lloyd Papers, 1849–1936, unpaginated booklet, series VII.6, folder 80, Lloyd Library and Museum.
[48] J. H. Powell, *William Denton, the Geologist and Radical: A Biographical Sketch* (Boston, MA: J. H. Powell, 1870), 5. For the Dentons, see Luciano; Dawson, 354–57; and Hanegraaff, 22–26.
[49] Powell, 8, 12, 15–19.
[50] Simon J. Knell, *The Culture of English Geology, 1815–1851: A Science Revealed through Its Collecting* (Aldershot: Ashgate, 2000), 36–37. For labourers' access to geology, see William Denton, *Our Planet, Its Past and Future; or, Lectures on Geology* (Boston: William Denton, 1868), 14.
[51] Hugh Miller, *The Old Red Sandstone; or, New Walks in an Old Field, Edited with a Critical Study and Notes*, 2 vols, ed. by Ralph O'Connor and Michael A. Taylor (Edinburgh: National Museums Scotland Publishing, 2023), II, 1–3.

to the United States in 1848, where he was subsequently joined by his family.[52] There, Denton maintained a precarious living by lecturing on geology and promoting controversial causes like Spiritualism, abolitionism, evolution, and mesmerism.

Buchanan's psychometry was soon added to his repertoire and, although the expatriate Denton did not himself possess psychometric powers, such powers, it emerged, were widespread. In Dentonian practice, their reach became far more ambitious than merely learning about the personalities of the senders of letters. Just by holding an object, those with psychic sensitivity might understand and even see its past, including events that had taken place aeons ago. Many members of their family and local circle participated, especially women and children, but William's most receptive psychometers were his American second wife Elizabeth Melissa Foote Denton, a typesetter and proto-feminist campaigner, and, later, their artistic teenage son Sherman.

After years of psychometric research, William and Elizabeth compiled their findings into *The Soul of Things* (1863), a monograph published by the liberal Unitarian firm Walker and Wise of Boston. The epigraph, 'Enter into the soul of things', attributed to the poet William Wordsworth, adapts a line from his poem *The Excursion* (1814), in which the speaker predicts that the formerly 'dull Eye' of 'Science' will one day enhance 'the Mind's *excursive* Power', thereby 'deeply drinking in the Soul of Things'.[53] These notions of a Romantic, intuitive connection with nature would have found sympathetic auditors around Wellesley, Massachusetts, the town near Boston where the Dentons eventually settled. After all, Massachusetts was the home territory of the transcendentalist Ralph Waldo Emerson, beside whom William would lecture in 1869.[54] The Dentons thus bridged American and British Romanticisms. Wordsworthian-Emersonian pantheism had even been united with a yet more controversial Romantic ingredient, levelling Shelleyan politics, in William's *Poems for Reformers* (1856), reprinted in 1871 as *Radical Rhymes*.[55] While the Dentons made no attempt to play down their manifold radicalism in America, attempted concessions to respectable readers were made in the British edition of *The Soul of Things* (retitled *Nature's Secrets*). This edition was edited by 'A CLERGYMAN OF THE CHURCH OF ENGLAND' and

---

[52] Powell, 21–22, 25.
[53] William and Elizabeth M. Foote Denton, *The Soul of Things; or, Psychometric Researches and Discoveries* (Boston: Walker, Wise and Company, 1863), title page. Subsequent references included in text. William Wordsworth, *The Excursion, being a portion of The Recluse, a Poem* (London: Longman, Hurst, Rees, Orme, and Brown, 1814), 197. For Wordsworthian pantheism, see Robert M. Ryan, *Charles Darwin and the Church of Wordsworth* (Oxford: Oxford University Press, 2016), 57–59, 65–67, 71.
[54] 'Address of Professor William Denton', *Proceedings at the Second Annual Meeting of the Free Religious Association* (Boston: Roberts Brothers, 1869), 37–42. See Patrick J. Keane, *Emerson, Romanticism, and Intuitive Reason: The Transatlantic 'Light of All Our Day'* (Columbia: University of Missouri Press, 2005), esp. chapters 1 and 2.
[55] William Denton, *Poems for Reformers* (Dayton, OH: William and Elizabeth Denton, 1856).

William Denton was reassuringly but inaccurately described as having attended an English university.[56]

William's theorization of the workings of psychometric clairvoyance in *The Soul of Things* drew on typical sources used by Spiritualists to evidence the survival of beings, in some form, after death. It also cited the work of geologists. These citations included Miller's ruminations on the mind's ability to store photographic records of experience that re-emerged in states of disorientation (17–20) and Hitchcock's accounts of vivid phantasmagorical hallucinations (263). The latter man, before his death in 1864, was the Dentons' most promising link to the international geological community: William pointed to a personal communication from Hitchcock's son, the geologist Charles Henry Hitchcock, claiming that, while delirious, his father had accurately visualized the Connecticut valley sandstone, enabling him 'to clear up some doubtful points' about its fossil bird footprints (264–65). Tapping into literary associations between the marvellous subjects of palaeoscience and those of dreams or hallucinations, William implied that geologists' vaunted reconstructive power was a form of unconscious psychometry.[57] Elizabeth, in her section of the book, supported the notion with reference to Agassiz's aforementioned intuitions about fossil fish (329–30).

Beyond these methodological sections, *The Soul of Things* consisted of lightly edited transcriptions of numerous psychometric experiments, followed by William's commentaries. Typically, psychometers held an object, preferably one unknown to them, to their head, and described the visual memories and sensations they found implanted upon it. Although all manner of objects were tested, William, whose presiding passion remained geology, preferred to give psychometers fossils and rocks from his own humble collection. Object in hand, 'the history of its time passed before the gaze of the seer like a grand panoramic view; sometimes almost with the rapidity of lightning' (36). With mental training, Elizabeth, the primary psychometer, became able to 'pause' the 'flying scenes', transforming their 'fragmentary' nature into continuity (313). Whereas the evangelical day-age vision theorists I discussed in chapter 1 were uniformly men, Elizabeth's powers reflected the pivotal role women played in the occult community, usually acting as medium or seer, even if her scientifically well-read husband remained the authoritative focal point of their psychometric circle.[58]

The earliest psychometric visions resembled pictures, but these scenes soon became three-dimensional and even explorable, not unlike the island of prehistoric animal models then-recently erected outside London's Crystal Palace by

---

[56] William and Elizabeth M. Foote Denton, *Nature's Secrets or Psychometric Researches*, ed. by [W. L. Thompson] (London: Houlston and Wright, 1863), viii. The editor's identity is provided in Powell, 33.

[57] For geological dreams, see O'Connor, *Earth on Show*, esp. 180–81, 72–73, 440.

[58] For women and Spiritualism, see Alex Owen, *The Darkened Room: Women, Power and Spiritualism in Late Victorian England* (London: Virago, 1989).

Benjamin Waterhouse Hawkins.[59] Elizabeth guided listeners through her experiences in these scenes in real time, sometimes losing herself in the object of examination. In the following example, she reads the impressions of a glacially striated pebble, its nature purportedly unknown to her:

> I feel as if I were below an immense body of water, – so deep that I cannot see down through it, and yet it seems as if I could see upward through it for miles. Now I am going, going, and there is something above me, I cannot tell what. It is pushing me on. It is above and around me. It must be ice; I am frozen in. (51)

The content of Elizabeth's adventures often triumphantly tallied, as in this case, with the nature of the object provided by William. When even William did not know the examined object's identity, he verified the psychometers' descriptions, in true autodidact style, with the *Encyclopædia Britannica* (170–71). As his family's visions became increasingly impressive, William's adopted home in the New World seemed to promise not only more religious freedom than had the Old, but also even more spectacular scientific insights.

The financially unstable Dentons were eager to extend the reach of *The Soul of Things*, its two self-published sequels, and their numerous pamphlets on Spiritualism and freethought. They were, in Ann B. Shteir's terms, a scientific 'family firm', and thus bookselling was one of the jobs of their son Sherman, who hawked *The Soul of Things* to uncooperative butchers in the winter cold.[60] To lubricate the process, William circulated a form for agents to sell the book at a substantial profit to themselves.[61] One respondent from Waukegan, Illinois, replied that '[i]f the sale is fair I shall be induced to do all I can to spread such useful knowledge before the people'.[62]

Zeal for the Dentons' self-empowering psychometric message manifested in a crowd of loyal supporters, many of whom did not, like the prospective Waukegan agent, require financial incentives. Fans penned laudatory poems to William: after his death, Emma Train roused the Spiritualist community with an ode to one '[w]ho found a sermon all unknown/Within each pebble, rock, and stone'.[63] William's allies included eminent figures within American freethought. A representative list of names attested to his selfless 'services to mankind' following

---

[59] James A. Secord, 'Monsters at the Crystal Palace', in *Models: The Third Dimension of Science*, ed. by Soraya de Chadarevian and Nick Hopwood (Stanford, CA: Stanford University Press, 2004), 138–69.

[60] Sherman F. Denton to William Denton, 31 December 1871, William Denton Papers, Box 7, Folder 3, Denton Family Papers, Wellesley Historical Society, Wellesley, MA; Ann B. Shteir, 'Botany in the Breakfast Room: Women and Early Nineteenth-Century British Plant Study', in *Uneasy Careers and Intimate Lives: Women in Science 1789-1979*, ed. by Pnina G. Abir-Am and Dorinda Outram (New Brunswick, NJ: Rutgers University Press, 1987), 31–43 (34).

[61] 'Confidential and Special Terms to Agents', William Denton Papers, Box 18, Folder 4.

[62] Completed agent form, n.d., William Denton Papers, Box 18, Folder 4.

[63] Emma Train cited in Warren Chase, *Forty Years on the Spiritual Rostrum* (Boston: Colby & Rich, 1888), 268–69 (268).

an 1880 series of geological lectures in New York, including those of Buchanan, statesman John L. O'Sullivan (the Spiritualism sympathizer who coined the term 'manifest destiny'), sexologist Edward Bliss Foote, and Theosophist Henry J. Newton.[64] Although scorned in London, *Nature's Secrets* received positive press in industrial Newcastle, not far from William's birthplace, while, in the United States, Charles Henry Hitchcock implied that his famous geologist father would share his own enthusiasm for the book.[65]

The Strangest-Looking Being I Ever Saw

The psychometric narratives related in the three volumes of *The Soul of Things* present fascinating case studies for the relationship between occult experience and palaeoscientific writing. Given that Elizabeth and her collaborators psychically visited all the conventional showpieces of the geological picturesque, such as Mount Vesuvius (187), the British *Athenæum* scathingly suggested that the psychometers' descriptions were based not on experience but on engravings and novels like Edward Bulwer-Lytton's *The Last Days of Pompeii* (1834).[66] The *Athenæum* critic, oddly, ignored the more complicated priming required for describing prehistoric scenes even less accessible than the Bay of Naples was to an American. It is significant that the psychometric visionary experience was routinely characterized by disjointed perspectival and temporal transitions, recalling the 'shifting scene' passages penned by geological writers and lecturers. This technique, usually delivered in the present tense and inspired by theatrical scene changes, diorama effects, and magic lantern dissolving views, was used by palaeoscientific authors to dazzle, to disguise gaps in geohistorical knowledge, and to replicate the transitions between day-ages experienced by the Mosaist seer.[67] The psychometer's experience could be similarly bewildering: exploring Kentucky's Mammoth Cave, Elizabeth found herself '[a]ll at once' teleported to 'the surface' (84); exposed to conflicting psychometric influences, she seemed 'to oscillate between the far past and a more recent period' (203). If disorientation was potentially convenient when performing cold readings on difficult geological objects, it also notably reflected the disjointed manner in which mid-century audiences of popular geology had typically glimpsed into prehistory.

To investigate the influence of palaeoscientific prose on Dentonian visions, we can compare *The Soul of Things* to the works of Miller, probably the most popular geological author in Britain and the United States during the 1850s, when the Dentons began their research. At the end of an Edinburgh lecture posthumously

---

[64] Excised testimonial, c. November 1880, Willian Denton Papers, Box 18, Folder 4.
[65] Dawson, *Show Me the Bone*, 356–57.
[66] 'The Soul of Things', *Athenæum*, 1871 (1863), 295–97.
[67] O'Connor, *Earth on Show*, e.g. 281, 397, 403, 406, 414; Adelene Buckland, *Novel Science: Fiction and the Invention of Nineteenth-Century Geology* (Chicago: University of Chicago Press, 2013), 203–5.

published in 1859, Miller took his listeners on 'a short walk' in the Jurassic.[68] In what O'Connor calls a passage of unusual 'immediacy', even for Miller, the Scottish geologist situates the audience spatially ('We stand on an elevated wood-covered ridge'), cautiously deciphers the sights ('cycadaceæ, whose leaves seem fronds of the bracken'), gestures to points of interest ('there is a noble Araucarian'), simulates soundscapes ('Tramp, tramp, tramp,—crash, crash'), details anatomy (*Iguanodon* 'has his jaws thickly implanted with saw-like teeth'), or finds it obscured ('The body is but dimly seen'), surrounds the audience ('Reptiles, reptiles, reptiles,—flying, swimming, waddling, walking'), and even seems to threaten its safety ('the night grows dangerous').[69] Elizabeth's accounts are similarly immediate, phenomenological, deictic, and descriptive:

> I begin to get the outline of objects moving, some on this flat and some among bushes that grow near there. One that I see attracts my attention much by its great singularly; it is without exception the strangest-looking being I ever saw. (When I go back so far, there is a difficulty in seeing objects at a distance, which I think is owing to the thick, heavy atmosphere of those early times.) (57–58)

> It is a reptile, with a head like a crocodile, but larger. It has enormous jaws, large eyes, small neck, and broad shoulders. It is looking at the other animals, and crawling softly toward them. It has a sly look. It is crested with an edge of thick points all along the back. (I feel as if I should be swallowed alive, with so many rapacious monsters around me.) (61)

Here, Elizabeth's vision almost encroached upon her just as Miller's word-painted reptiles had virtually menaced his Edinburgh auditors.

Immersive techniques did not, of course, belong exclusively to popular geology, even if this was a likely place for the Dentons to encounter them. On occasion, however, the psychometer's appropriations were explicit. Miller had opined about the literary possibilities of an 'autobiography of a single boulder' and Elizabeth obligingly provided '*The Autobiography of a Boulder*', a sensory narrative of her time spent as a volcanic boulder absorbed in glacial ice (114–21).[70] The inanimate object biography was becoming a classic genre of popular science, as seen also in geologist Archibald Geikie's *The Story of a Boulder* (1858).[71] The psychometer's occasional and unexplained ability to feel 'all that was felt' by a specimen

---

[68] Hugh Miller, *Sketch Book of Popular Geology: A Series of Lectures Read before the Philosophical Institution of Edinburgh*, ed. by Lydia Miller (Boston: Gould and Lincoln, 1859), 198.

[69] Miller, *Sketch Book of Popular Geology*, 198–202; O'Connor, *Earth on Show*, 404. Also compare with William Denton, *Our Planet*, 136–38, 162–63.

[70] Hugh Miller, *The Cruise of the Betsey; or, A Summer Ramble among the Fossiliferous Deposits of the Hebrides* (Edinburgh: Thomas Constable, 1858), 322.

[71] For the major study of this subject, see Melanie Keene, 'Object Lessons: Sensory Science Education, 1830–1870', unpublished PhD thesis, University of Cambridge (2008), especially chapter 1.

(50) was likely indebted to the innovations of geological authors too. Elizabeth's experience of life as a mastodon had precedent, for instance, in Charles Kingsley's novel *Alton Locke* (1850), the protagonist of which hallucinates that he is a giant ground sloth, and in John Mill's *The Fossil Spirit* (1854), a children's book narrated by a Hindu holy man who recalls being reincarnated as various extinct animals and relates what occurred to him.[72] Both fiction and non-fiction dealing with geological subject matter presented useful literary exemplars for how one might experience transportation into the objects and animals of the prehistoric past.

If Elizabeth Denton's visionary narratives hint at frameworks fashioned by prior reading, this reading was obligatory for her geologist husband. One sceptic in *Scientific American* even complained that the self-taught William's knowledge was 'gained mainly by reading'.[73] His lectures, published in the ostensibly non-psychometric book *Our Planet, Its Past and Future* (1868), cited the likes of 'Lyell', 'Owen', and '[Gideon] Mantell', and was punctuated with virtual tours through prehistory, including an invitation to take 'the wing' of a distinctly Lyellian 'dusky demon' and 'descend with me into the nether regions' to 'let us' uncover the origin of earthquakes.[74] The last-quoted phrase suggests a more subtle debt to Lyell, who, as J. M. I. Klaver has observed, habitually adopted the grandiose biblical 'let' of Genesis I in his own thought experiments.[75] William's lectures were similarly demiurgic: 'Let water be poured on it'; 'Let the earth be gradually heated'; 'Let the rocks be represented'; 'Let us transport ourselves to the carboniferous times'.[76] This phraseology may, of course, simply represent the impious William's mischievous appropriation of the language of Genesis itself. This would be unsurprising, as, in 1872, he ruthlessly satirized Miller's notion of God as prehistoric diorama showman.[77] By this point an expert in the mechanics of deep clairvoyance, he was perhaps eager to discard its connection to metaphorical scaffolding. He had also become skilled in describing prehistoric scenes, even if he could not, like his wife, actually observe them. As the *Boston Traveler* attested, William 'takes his audience with him in all his descriptions, and they seem to see everything with their own eyes'.[78]

The Dentons' experiments were intended to advance scientific knowledge. When Elizabeth spotted a large Tertiary mammal capable of 'lengthening its neck at will' (105), her husband examined William Buckland's 'engraving of the skeleton of the megatherium' and concluded that this extinct ground sloth, unbeknownst to savants, 'had the power of protruding the head' (107). Psychometry was, moreover,

---

[72] Buckland, *Novel Science*, 204–7; O'Connor, *Earth on Show*, 253–54.
[73] Untitled review of *Is Darwin Right? Or, the Origin of Man*, by William Denton, *Scientific American*, 44 (1881), 250.
[74] William Denton, *Our Planet*, 3, 41, 51.
[75] J. M. I. Klaver, *Geology and Religious Sentiment: The Effect of Geological Discoveries on English Society and Literature between 1829 and 1859* (Leiden: Brill, 1997), 45.
[76] William Denton, *Our Planet*, 22, 52, 41, 136, 170.
[77] Huhtamo, 347.
[78] Advertisement for 'Six Lectures on Geology' (1871), William Denton Papers, Box 18, Folder 2.

happily unrestrained by the surface bias that Lyell had diagnosed in geologists, and Elizabeth actually found it easier to travel 'under the surface through the rock' than above it (141). These voyaging powers helped her to develop a theory on the origin of petroleum, which she explained to conchologist Isaac Lea, President of the American Association for the Advancement of Science, in 1860. Rejecting 'the generally received idea' that petroleum came from 'the vegetation of which coal was formed', she discovered that the source was, rather 'ancient coral'.[79] She concluded the letter with the hope that her contribution to 'the accumulating treasure of scientific information' would justify the 'liberty' she had taken in reaching out to such an eminent man of science.[80]

Elizabeth was used to writing unsolicited letters to famous savants, including John William Dawson in Canada, to promote her work and her husband's career.[81] The true origin of petroleum made for a bravura scene in *The Soul of Things*, as Elizabeth was transported underground into 'a 'coralline forest' shining with 'all the colors of the rainbow' and oozing with oily 'liquid' (226). William was pleased to remark in the 1866 edition of *The Soul of Things* that his wife's claims about 'the animal origin of petroleum' had since been 'strengthened' by 'discoveries of fossils in the Laurentian rocks'.[82] He reinforced this idiosyncratic theory in *Our Planet*, albeit alluding only to non-psychometric evidence.[83]

Similarly empowering was the psychometer's ability to fill in the fossil record's imperfections, famously described by Charles Darwin using the metaphor of a frustratingly shredded book.[84] This was chiefly a project of father and son: it provided an ideal opportunity for the young clairvoyant Sherman, whose artistic skills allowed him to sketch what he saw psychometrically, often in real time. (Figure 2.2). Numerous engravings described as '*facsimiles* from his drawings' were printed in the third volume of *The Soul of Things* (1874).[85] The Dentons having by then taken publication of their works upon themselves, this latest instalment, following on from the second in 1873, was more ambitious than ever. Throughout the book, Sherman drew evolutionary missing links back into sparse lineages like those of birds, including not just the notoriously elusive species that left disembodied footprints in Hitchcock's Connecticut valley sandstone, but also unexpected hybrids like the 'Bird-reptiles of Lebanon' (Figure 2.3). As Sherman would have known, links between birds and reptiles, rarely preserved in fossil form, were among the most sought-after (and financially valuable) specimens in palaeontology.

[79] Elizabeth M. F. Denton to Isaac Lea (copy), 22 May 1860, Elizabeth M. F. Denton Papers, Box 11, Folder 3, Denton Family Papers, Wellesley Historical Society, Wellesley, MA.
[80] Elizabeth Denton to Lea, 22 May 1860.
[81] Elizabeth M. F. Denton to John William Dawson (copy), 6 December 1862, Elizabeth M. F. Denton Papers, Box 11, Folder 3.
[82] William Denton, 'Preface', in William and Elizabeth M. F. Denton, *The Soul of Things; or, Psychometric Researches and Discoveries*, 3rd edn (Boston: Walker, Wise, 1866), iii–iv (iv).
[83] William Denton, *Our Planet*, 119.
[84] Zimmerman, 31.
[85] William Denton, *The Soul of Things; or, Psychometric Researches and Discoveries*, vol. III (Boston: William Denton, 1874), 36.

Figure 2.2 Photograph of Sherman Foote Denton, n.d., Photograph Collection, Denton Family Papers, Wellesley Historical Society, Wellesley, MA. Courtesy of Wellesley Historical Society, MA. Sherman was a psychometric illustrator during his early-to-mid teens.

DEEP CLAIRVOYANCE 83

(A)

BIRDS OF THE CONNECTICUT VALLEY.   79

"These birds go strutting along, not hopping like the kangaroo-reptiles."

FIG. 66.—Larger web-footed bird of the Connecticut.

("See where the water from those ponds flows, and examine the country.") "It is very flat. On each pond is a kind of beach; and between them prickly bushes grow very thick. It would be impossible for a man to make his way through them.

Figure 2.3 (A and B) William Denton, *The Soul of Things; or, Psychometric Researches and Discoveries*, vol. III (Boston: William Denton, 1874), 79, 120. Reproduction of Sherman's sketches, based on his psychometric readings of the fossil footprints of the Connecticut valley and limestone from Mt. Lebanon. (C) Untitled sketchbook, Sherman F. Denton Papers & Artwork, Box 1, Folder 9, Denton Family Papers. Sherman's sketches in the process of refinement for publication.

(B)

120   THE SOUL OF THINGS.

tips; and there are claws on both feet. Their bodies are thickly covered with stout bristles. The skin is black, and the bristles not so dark. They have no teeth, but a bill like a bird (1). (See Fig. 101.)

FIG. 101.— Bird-reptiles of Lebanon.

"Now I see some curious reptiles with small bodies and enormous heads. They are about seven feet long, and of a dirty color; and, like the frog,

> 'For convenience' sake they wear
> Their eyes on the top of their heads.'

They have a bunch of stout bristles under the throat, — a kind of beard. (Draws Fig. 102.)

"They are good swimmers. There is no web uniting the toes; but each toe has a web like a dabchick's.

"They seem to be friendly with the bird-reptiles.

"The young bird-reptiles look very odd, quite white, and like a chicken with its feathers picked off. The fore-feet look like a wing when it is picked, or a turtle's

Figure 2.3 Continued

Figure 2.3 Continued

A psychometrically equipped individual could solve palaeontological problems where Darwin feared to tread, but this boldness was imaginatively scaffolded by familiarity with prior scenes from deep time. Regarding his son's cutting-edge depiction of dinosaurs as bipedal, William admitted the precedent of Edward Drinker Cope's recent restoration of the kangaroo-like *Lælaps aquilunguis* but thought it unlikely Sherman had seen this image in the *American Naturalist*.[86] The similarity between Sherman's depiction of a pterodactyl and that drawn by Édouard Riou in Louis Figuier's *The World before the Deluge*—a book we have already seen inspiring day-age theorists—is even closer.[87] Sherman's sketchbook, from which the images reproduced in *The Soul of Things* originate, shows the teenager polishing his initial, more rudimentary, sketches of the objects of his visions. Refining his plesiosaur, Sherman made the aquatic reptile look significantly more like the model erected by Hawkins at the Crystal Palace—a model familiar to Americans through engravings and appropriations (Figure 2.4). William was even able to exhibit his son's findings to Hawkins himself, who was then working in the United States. Visiting Boston in June 1873, William was introduced to 'the man who made the restorations of the iguanodon & other animals

---

[86] *Soul of Things*, III, 85.
[87] Compare Louis Figuier, *The World before the Deluge*, trans. by W. S. O. (London: Chapman and Hall, 1865), facing 210, with *Soul of Things*, III, 88.

for the crystal palace [sic]', upon which he whipped out Sherman's sketches and 'surprised him very much'.[88]

Despite these apparent triumphs, a frustrated teenage Sherman soon renounced his irregular powers. In November 1874, he told his mother that 'I will never try another specimen for any soul on this planet'.[89] Tensions were already visible in the later volumes of The Soul of Things, although the most flagrant of these were diplomatically omitted from publication. In an excised paragraph in the manuscript draft of the third volume, William attempted to direct Sherman's examination of the Tertiary period, asking his son to '[g]o up in the air and see the country', to which Sherman replied, 'I do not wish to'.[90] Although Sherman

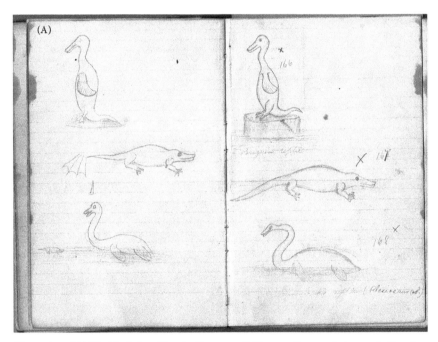

Figure 2.4 (A) Untitled sketchbook, Sherman F. Denton Papers & Artwork, Box 1, Folder 9. Sherman's revisions of his psychometric sketch of a long-necked reptile suggest influences from Benjamin Waterhouse Hawkins's model of the Plesiosaurus at the Crystal Palace Park in London. Courtesy of Wellesley Historical Society, MA. (B) Hawkins's widely imitated design. Photograph by Chris Sampson, CC BY 2.0.

---

[88] William Denton to Elizabeth M. Foote Denton, 2 July 1873, William Denton Papers, Box 9, Folder 3.

[89] Sherman Foote Denton to Elizabeth M. Foote Denton, 5 November 1874, Elizabeth M. F. Denton Papers, Box 9, Folder 2.

[90] Manuscript for The Soul of Things, vol. 3, numbered 441/7, William Denton Papers, Box 14, Folder 5.

Figure 2.4 Continued

(and, after William's death in 1883, seemingly the entire family) abandoned psychometry, he did not lose his passion for capturing transient glimpses of the natural world. He and a brother established a successful natural history business that became renowned for preserving specimens as if they were alive. His entomological photographic reproduction technique, 'Nature Prints', took 'direct transfers from the insects themselves' in order to present lepidoptera 'not as dissected fragments for scientific classification, but as one sees them in our woods, fresh and lovely'.[91]

The Past Is the Present

In the final quarter of the century, the psychometric framework pioneered by Buchanan and the Dentons was metamorphosed into a tool for writing the deep history of humanity itself. While palaeoscience remained an important ingredient, its role and literary expression decisively changed. The Theosophical Society, an organization dedicated to the syncretism of science, esoteric lore, and world religions, was behind this new clairvoyant agenda. Established in New York in 1875 by the Russian occultist Helena Petrovna Blavatsky and her American companion Henry Steel Olcott, the Society soon formed bases in Adyar and London.[92] Theosophists declared that the diverse public (exoteric) content of all religions disguised one unified but secret (esoteric) truth, although the Society came increasingly to focus on the insights of Buddhism and Hinduism. The leadership of charismatic Theosophists like Blavatsky was predicated on their extraordinary ancient knowledge, much of it absorbed clairvoyantly, both through access to immaterial tomes and through time-transcending visions.

Much of the Society's information about geohistory first emerged in the 'Mahatma Letters'. This series of mysterious communications about the secrets of science and religion, purportedly authored by reclusive Himalayan adepts, was received by Anglo-Indian journalist Alfred Percy Sinnett in the early 1880s and publicized in his book *Esoteric Buddhism* (1883). The visionary method of expanding Theosophical knowledge of the cosmos, however, took off after Sinnett sought those who might continue the stream of revelations that had ceased around 1884, when the Mahatma communications dried up. The most successful candidate was former Anglican priest Charles Webster Leadbeater, who proved capable of astonishing clairvoyant exertions. Upon Blavatsky's death in 1891, the already multifarious movement split, the leadership of its Adyar wing being domineered by Leadbeater and his companion, the radical campaigner Annie Besant, until the end of the 1920s. A third figure also became integral to this wing of Theosophy: born in Ceylon (Sri Lanka), the young Curuppumullage Jinarajadasa was brought to Europe by Leadbeater in 1889 under hazy and suspicious circumstances. Educated at the University of Cambridge, by 1921 Jinarajadasa was Vice President of the Theosophical Society. Although diverse branches of Theosophy spread across

---

[91] Sherman F. Denton, *As Nature Shows Them: Moths and Butterflies of the United States East of the Rocky Mountains*, 2 vols (Boston, MA: Bradlee Whidden, 1900), I, v.
[92] An overview of Theosophy is provided in Joscelyn Godwin, *The Theosophical Enlightenment* (Albany: State University of New York Press, 1994). A summary of the Theosophical cosmogony is found in Godwin, *Atlantis*, 77–87.

the world, I will focus on the milieu surrounding Blavatsky and her London-Adyar successors.

In Wouter Hanegraaff's words, the 'Theosophical clairvoyance' practised by Blavatsky, Leadbeater, and others was 'just another word for psychometry'.[93] These origins were somewhat occluded: in line with the Society's increasingly Orientalist vocabulary, what began as Dentonian psychometric readings had, by the 1890s, become communion with the *akashic* records, an ethereal archive of everything that has ever happened. A suitably powerful Theosophist could consult these records and thereby gaze upon the past without even needing to examine a relevant object (a development foreshadowed by Sherman Denton's precocious ability to visit other planets merely by sight).[94] This was the ultimate freedom from the material and logistical constraints of nineteenth-century science. If an advanced clairvoyant 'desires to see the earth before its crust has solidified', explained Jinarajadasa, '[t]he Book of Time is spread out before him' and 'the past is the present'.[95] Olcott, at least, was honest about the *akashic* records' conceptual origins. He advised anyone who desired 'a complete understanding' of Theosophy's 'revelations' to 'familiarise himself with the principles and history of psychometry', not least because Buchanan, an early if uneasy convert to Theosophy, had been Olcott's friend since the 1850s.[96] The Dentons' work was at first openly esteemed by the Society: in 1883, the official journal *Theosophist* boasted of encouraging psychometric research by 'putting more than seventy copies of the *Soul of Things* [sic] into circulation in India' and it hoped to circulate 'seven hundred more' as well as to invite William Denton to lecture.[97] This latter arrangement was prevented by his premature death the same year.

While it emerged in large part from the Dentons' palaeoscientific psychometry, Theosophical clairvoyance was used mainly for a different end: placing humans and ethnology at the centre of cosmic history. Consulting *akashic* recordings that preserved every moment since the dawn of time, Theosophical clairvoyants, with the help of the Mahatma Letters, revealed that the geohistorical narrative known to naturalists was just a fraction of the truth. Not only were humans capable of mentally travelling deep into the past, but humans had *existed* in these remote periods too. In the elaborate Theosophical cosmogony, human beings—divided into consecutive series called 'Root Races'—were believed to have taken many forms. We were ethereal invertebrates in the Earth's earliest epochs, first materializing as three-eyed giants during the Jurassic period on the Indo-Pacific continent

---

[93] Hanegraaff, 35. See also Hester, 41.
[94] *Soul of Things*, III, 147.
[95] Curuppumullage Jinarajadasa, *First Principles of Theosophy*, 2nd edn (Adyar: Theosophical Publishing House, 1922), 30.
[96] Henry Steel Olcott, *Old Diary Leaves: The Only Authentic History of the Theosophical Society, Fifth Series: January, 1893–April, 1896* (Adyar: Theosophical Publishing House, 1932), 398–99; Hester, 37.
[97] '*The Soul of Things*', *Theosophist*, 4 (1883), 239–40 (240).

of Lemuria. As I mentioned in the introduction, this was a theorized location described by Ernst Haeckel as the birthplace of anthropoid apes.[98] Humanity became anatomically recognizable around the time of the Cenozoic Era, far earlier than any mainstream palaeontologists claimed, when the continent of Atlantis was the centre of a wizardly super-civilization. This cosmogony involved a significant revision of evolutionary theory: rather than humans evolving from animals, animals were the 'cast-off clothes' of earlier embodiments of humanity.[99]

Even this expanded, anthropocentric history of Earth was revealed to be just the latest in an infinite series of septenary cycles throughout which humans have always existed. Such sublime depths of time, while repugnant to the young-earth creationists I will discuss in chapter 5, were commonplace to Theosophists, who took their chronological frameworks from the immensely long units of inhabited time laid out in Hindu cosmology. The apparent geological insights of Hindu conceptions of deep time had already been dutifully qualified and appropriated by evangelical day-agers: John William Dawson, for one, noted that some supposed the Hindu cyclical period, the *manvantara*, 'to represent the Mosaic day', with a duration of '308,571 years', and, although these numbers were 'probably conjectural', they demonstrated 'that the idea of long creative periods' had 'in the infancy of the postdiluvian world ... been very widely diffused'.[100] Unlike Theosophists, who found ancient wisdom surviving in the modern Hindu cosmogony, Dawson viewed this cosmogony as a corruption of knowledge revealed by God in humanity's earliest ages.

The Dentons had, indeed, used psychometry to visit ancient human cultures in some of their visions, but this interest in human prehistory did not overshadow their similarly careful investigations into matters like the evolution of birds.[101] In contrast, throughout Blavatsky's monumental work, *The Secret Doctrine* (1888), profuse palaeontological data took on a more subservient role as collateral evidence for the esoteric truths about human history hidden in exoteric religion. Palaeontology was at its most useful when propping up the Theosophical cosmogony by providing support for the existence of human civilizations far antedating recognized history. For instance, when the Kabbalistic Jewish *Zohar* obscurely stated that 'a kind of *flying camel*' had tempted Eve in Eden, the fact that Cuvier himself had described 'a *flying* saurian, "the Pterodactyl"', was shown to have 'vindicated' the *Zohar*'s pre-scientific intent, suggesting that this extinct reptile was known to ancient humans.[102] This logic may appear tenuous, but the apparently non-sequitous comparison between pterodactyl and Satan was a

---

[98] Ramaswamy, chapter 3.
[99] Helena Petrovna Blavatsky, *The Secret Doctrine: The Synthesis of Science, Religion, and Philosophy*, 2 vols (London: Theosophical Publishing Company, 1888), II, 290.
[100] John William Dawson, *The Origin of the World, According to Revelation and Science* (New York: Harper & Brothers, 1877), 151–52.
[101] See *Soul of Things*, III, 305–8.
[102] Blavatsky, *Secret Doctrine*, II, 205.

long-standing literary convention. It had originated in William Buckland's oft-quoted comparison between a fossil unearthed by Mary Anning in 1828 and John Milton's description of the airborne Lucifer in *Paradise Lost*.[103] Here, as elsewhere, the figurative was literalized, an analogy intended to glamorize fossil discoveries transforming into a startling truth claim.

The circumstantial nature of this kind of evidence was freely confessed by Theosophists. This was apparent in an exchange about the third eye, a traditional Asian symbol of clairvoyance supposedly possessed by the giant Lemurian humans of the occult Jurassic. Various mainstream naturalists, including Owen, had detected a third eye in certain reptiles and amphibians. Blavatsky thus triumphantly cited the case of 'the antediluvian *Labyrinthodon*', an extinct amphibian in which 'naturalists' suspected that the vestigial third eye 'was a real organ of vision'.[104] More tangible evidence of the Lemurian's third eye, however, was not forthcoming. In 1889, anthropologist Charles Carter Blake wrote to Blavatsky's periodical *Lucifer* to celebrate 'the light that comparative anatomy and palæontology are beginning to throw on the formation of the Third Eye', while appealing to the editor for further evidence.[105] Blavatsky's reply was irritable: 'As three-eyed men are no longer extant, what evidence can be expected other than of a circumstantial character?'[106]

Regular citations of Owen's work demonstrate that early Theosophical interest in palaeontology, while subservient to the study of humans, was wide-ranging. In her earlier work, *Isis Unveiled* (1877), Blavatsky pointed triumphantly to Owen's research into the gigantic New Zealand moa as validating the existence of 'the *Ruc* (or Roc)', that 'monstrous bird of the *Arabian Nights*'.[107] This particular observation was less a literalization of figurative language than it was a confident repetition of a claim more tentatively voiced by respected savants.[108] The use of fossil evidence to scientize Asian lore had ample precedent in the Indian research of British geologists like Hugh Falconer, who had sought physical origins for mythic animals and locales referred to in Hindu texts.[109] Palaeoscientific Orientalism appealed to Blavatsky, who was particularly fond of 'the monstrous *Sivatherium*', an extinct mammal named by Falconer after the Hindu god Shiva.[110] In 1871, an author in the *Geological Magazine* had only with rhetorical difficulty restrained himself from

---

[103] O'Connor, *Earth on Show*, 360, 422. See also Eugen Georg, *The Adventure of Mankind*, trans. by Robert Bek-Gran (New York: E. P. Dutton, 1931), 169.
[104] Blavatsky, *Secret Doctrine*, II, 299.
[105] Charles Carter Blake, 'The Third Eye', *Lucifer*, 4 (1889), 341–45 (343–44).
[106] Blake, 345.
[107] Helena Petrovna Blavatsky, *Isis Unveiled: A Master-Key to the Mysteries of Ancient and Modern Science and Theology*, 2 vols (New York: J. W. Bouton, 1877), I, 603.
[108] Isidore Geoffroy Saint-Hilaire, 'Note on Some Bones and Eggs Found at Madagascar, in Recent Alluvia, Belonging to a Gigantic Bird', *Magazine of Natural History*, 7 (1851), 161–66 (166).
[109] Chakrabarti, *Inscriptions of Nature*, esp. chapter 3. For Theosophical Orientalism, see the essays in Tim Rudbøg and Erik Reenberg Sand, eds., *Imagining the East: The Early Theosophical Society* (Oxford: Oxford University Press, 2020).
[110] Blavatsky, *Secret Doctrine*, II, 218.

speculating about this mammal, fearing to 'conjure a picture rivalling modern Eastern tales'.[111] Theosophy's efforts to close the gap between 'Eastern tales' and scientific fact differed from the efforts of naturalistic science only in degree. Significantly, C. Mackenzie Brown suggests that Blavatsky, in *Isis Unveiled*, was one of 'the first to proclaim the avataric evolutionary theory'.[112] This theory ruthlessly naturalized and rationalized the story of another god, Vishnu—who progresses through various incarnations or avatars, including that of a fish, turtle, and boar, before becoming a man—by interpreting his transformations as a veiled allegory of evolutionary progress.

Detailed Investigation of the Astral Plane

The anthropocentric and Orientalist leanings that distinguished Theosophists' explorations of deep time from those of the Dentons were accompanied by differences in literary technology. These became especially apparent in the clairvoyant investigations published from the 1890s onwards, following Blavatsky's death. William Denton had used a scientific vocabulary to make sense of the visionary experiences of his family and friends, but Theosophists cultivated even more extensive scholarly stylings, presented in the distinctively occult mode that Alex Owen calls a 'matter-of-fact language of realism'.[113] This urbane manner reflected their assertive confidence: after all, in their clairvoyant practices Theosophists combined the originally distinct roles of the knowledgeable naturalist William and the psychically sensitive Elizabeth or Sherman into one authoritative figure.[114] Elizabeth, admittedly, had pushed back against conventional associations of mediumistic spirituality with feminine receptivity (353), but her husband's methods nonetheless relied on the alleged sketchiness of her scientific learning. After all, if Elizabeth understood the geological objects she examined, then this compromised the neutrality of her readings. In Cameron Strang's words, the gendered or infantilized authority of psychometers was more that of 'sensitive instruments than that of human experts'.[115] The result was that psychometers courted credibility—and the trust of virtual witnesses—through the tentative, amateurish, and experiential language I quoted from previously, punctuated with melodramatic exclamations

---

[111] James Murie, 'On the Systematic Position of the *Sivatherium giganteum* of Falconer and Cautley', *Geological Magazine*, 8 (1871), 438–48 (448).

[112] C. Mackenzie Brown, 'The Western Roots of Avataric Evolutionism in Colonial India', *Zygon*, 42 (2007), 425–49 (437).

[113] For Theosophical scientism, see Olav Hammer, *Claiming Knowledge: Strategies of Epistemology from Theosophy to the New Age* (Leiden: Brill, 2004), 218–29. For matter-of-factness, see Alex Owen, *The Place of Enchantment: British Occultism and the Culture of the Modern* (Chicago: University of Chicago Press, 2004), 158; and Hanegraaff, 4, 13–14.

[114] Hanegraaff, 17, 28–29, 31.

[115] Strang, 88.

linked to bodily sensations: 'Oh, what shells!' (37); 'Mercy!' (115); 'It thrills me, and I fairly tremble' (218).

The Dentons' style therefore contrasts significantly with that of Theosophical research, of which an influential 1895 paper on 'The Lunar Pitris', delivered before the Theosophical Society's London Lodge by elite occultists Patience Sinnett (A. P. Sinnett's wife) and William Scott-Elliot, may be taken as representative. The paper, referring in its title to one of the Theosophical cosmogony's various superhuman orders of being, communicated the results of secretive *akashic* explorations undertaken by Leadbeater and by Scott-Elliot's enigmatic wife Maude (*née* Boyle-Travers).[116] It was published in the *Transactions of the London Lodge of the Theosophical Society*, the title of which connoted the *Philosophical Transactions of the Royal Society* of London and thus the hoariest models of scientific publication (unlike more accessible Theosophical periodicals, such as *Lucifer*).[117] In line with the journal's august image, the named authors adopted the terse, passive language of judicious argumentation, announcing that 'it is proposed now to trace—of course in very rough outline—the process of evolution through the past ages on this earth'.[118] Sinnett and Scott-Elliot obliquely explained that they were 'given to understand' the *akashic* facts that, with understated scholarly distance, they related: 'it may be interesting to note'; '[f]or greater ease of reference'; 'as it were'.[119] This cool tone provoked one curious but exasperated Spiritualist to write that 'there is evidently not the slightest suspicion on the part of the authors that anyone could be so unreasonable as to ask them how they came to "know" all the wonderful things they narrate'.[120]

In 'The Lunar Pitris', humanity's prehistoric Root Races were described by Sinnett and Scott-Elliot with natural-historical precision. A clairvoyantly accessed 'astral image' of a human of the Second Race was described as 'jelly-like in substance', floating, with 'two centres of force', namely the 'pineal gland' and primitive 'spleen', and capable of conveying 'rudimentary feeling and expression'.[121] If this being resembled the seashore molluscs described by the likes of naturalist Philip Henry Gosse, a subsequent description of a Lemurian human of the Third Race was inspired, for all its supposedly Jurassic origins, by the racial categories of contemporary ethnology. Despite the Society's professed antipathy towards racism, deterministic ethnological classifications were becoming an obsession as its members mapped out the deep evolving history of humanity, in effect a story of the rise and fall of particular racial groups, in increasing detail. The Lemurian was

---

[116] Godwin, *Atlantis*, 88–90
[117] Ferguson, 'Luciferian Public Sphere', 92–93.
[118] Patience Sinnett and William Scott-Elliot, 'The Lunar Pitris', *Transactions of the London Lodge of the Theosophical Society*, 26 (1895), 3–30 (4).
[119] Sinnett and Scott-Elliot, 6, 8, 11.
[120] 'The Lunar Pitris', *Light*, 15 (1895), 430.
[121] Sinnett and Scott-Elliot, 16.

'a giant of about twelve to fifteen feet', in colour 'dark yellow-brown' with sloping forehead and short hair, holding a 'sharpened staff' that it used to lead 'a huge and hideous reptile, somewhat resembling the plesiosaurus'.[122] In this stark literalization of nineteenth-century science's routine characterization of people of colour as prehistoric survivals, the Lemurians, descendants of modern black people in Theosophical thought, coexist with malformed Mesozoic saurians.[123]

Having zoomed in to capture these glimpses of prehistoric life, the authors returned to a more abstract, textbook view. They appended 'a tabular summary of Ernest [sic] Haeckel's *History of Creation*', featuring added occult columns that boldly inserted humans into every geological period in the German naturalist's framework, while detailing the rise and fall of Lemuria, Atlantis, and the other occult continents.[124] This table stood as a succinct visual statement of the authors' anthropocentrism, anti-materialism, and confidence in appropriating modern science's visual language. The reference to Haeckel was no arbitrary choice. Although Theosophists had long enthused over suggestive Haeckelian notions such as the origin of humans on Lemuria, his name was also a byword for the disenchanted secularity espoused by elite evolutionists. Here, Sinnett and Scott-Elliot impishly repackaged his *Natürliche Schöpfungsgeschichte* (1868)—translated as *The History of Creation* in 1876—to enchanted ends, showing how geology's visual language could be adapted to map the esoteric concepts of Root Races and continental cycles on to the more widely accepted geological column.

When these papers were revised and elaborated upon by the Theosophical Publishing Society for individual publication and wider circulation, efforts were made to format the research in an even more academic manner. At first, when Scott-Elliot's 1896 London Lodge paper on 'Atlantis' was reprinted separately as *The Story of Atlantis*, nothing was altered beyond an added preface by A. P. Sinnett, explaining the principles of clairvoyance (which had needed no introduction to Lodge insiders). A sequel, *The Lost Lemuria* (1904), however, displayed more concerted attention to scholarly presentation: a running gloss now accompanied the text, and, whereas *Atlantis* had bracketed the sources of quotations, *Lemuria* presented them in the more exacting form of footnotes.[125] When *Atlantis* was revised in 1909, its text was reset to incorporate these more polished conventions.[126] The resulting publications made for visually sober and readable booklets on natural

---

[122] Sinnett and Scott-Elliot, 18–19.
[123] Ramaswamy, 67–69. For colonial subjects as prehistoric, see Chakrabarti, *Inscriptions of Nature*, esp. chapter 4.
[124] Sinnett and Scott-Elliot, 13.
[125] William Scott-Elliot, *The Story of Atlantis: A Geographical, Historical, and Ethnological Sketch* (London: Theosophical Publishing Society, 1896); William Scott-Elliot, *The Lost Lemuria with Two Maps Showing Distribution of Land Areas at Different Periods* (London: Theosophical Publishing Society, 1904).
[126] William Scott-Elliot, *The Story of Atlantis: A Geographical, Historical, and Ethnological Sketch*, 2nd rev. edn (London: Theosophical Publishing Society, 1909). When the two volumes were formally combined in 1925, however, the unrevised version of *Atlantis* was used, perhaps to dispose of excess stock.

history and anthropology, extensively citing authorities like Darwin, Alfred Russel Wallace, and T. H. Huxley, even while their contents described Atlantean flying machines and Lemurian triclopes. These works were part of an increasingly sophisticated publishing operation, intent on communicating insistently complicated Theosophical concepts to diverse audiences.[127]

By the early twentieth century, Besant and Leadbeater had become prominent authors of clairvoyant research in various scientific fields. Their *akashic* publications grew in ambition, evidenced by series like the 'Rents in the Veil of Time' published in the Adyar *Theosophist* from 1909, as well as by epic accounts like *Man: Whence, How and Whither* (1913). These titles shifted focus even more obsessively to the human civilizations of prehistory, with little interest in their palaeontological context, although the Root Races of early geological periods still merited brief mention. Clairvoyant investigators remained unruffled by these strange beings. Now personally recounting his clairvoyant experiences in deep time, rather than allowing them to be relayed by others, Leadbeater cultivated a down-to-earth style. As he was later known as a fan of H. G. Wells's scientific romances, it is notable that the paper on 'The Lunar Pitris', with its vast leaps through time, was read to the London Lodge just a few months after *The Time Machine* had completed its 1895 serialization.[128] Leadbeater's candid descriptions of the stranger inhabitants of deep time may also have been inspired by Wells's characteristically blunt descriptions of weird life forms, like the 'round thing, the size of a football' that the time traveller sees 'hopping fitfully about' on a beach in the distant future.[129] In Leadbeater and Besant's *Man*, we hear how primordial humans were 'pudding-bag creatures' emitting 'a cheerful kind of chirruping' to express a 'general sense of *bien-être*'.[130] Here and elsewhere, potentially disturbing aspects of this psychic encounter with gelatinous prehistoric monsters were defused by unflappable good humour.

Leadbeater's colleague, Jinarajadasa, later claimed that the *Transactions of the London Lodge* had pioneered a newly scientific approach to clairvoyant research. Unlike previous enlightened adepts, who had focused on the 'life side' of nature, Leadbeater, perhaps due to his origins in the materialistic West rather than the spiritual East, focused on its 'form side'.[131] As a result, his 'detailed investigation of the Astral Plane' resembled the work of 'a botanist in an Amazonian jungle'.[132] Leadbeater was framed as a globetrotting imperial naturalist, a Theosophical

---

[127] Morrison, 'The Periodical Culture of the Occult Revival', 4; Ferguson, 'Luciferian Public Sphere', 92.
[128] Gregory Tillett, *The Elder Brother: A Biography of Charles Webster Leadbeater* (London: Routledge and Kegan Paul, 1982), 115, 120.
[129] H. G. Wells, 'The Time Machine [5/5], *New Review*, 12 (1895), 577–88 (583).
[130] Annie Besant and Charles Webster Leadbeater, *Man: Whence, How and Whither: A Record of Clairvoyant Investigation* (Adyar: Theosophical Publishing House, 1913), 83.
[131] Curuppumullage Jinarajadasa, 'Introduction', in Charles Webster Leadbeater, *The Astral Plane*, rev. edn (Adyar: Theosophical Publishing House, 1973 [1933]), vii–xxi (xiv).
[132] Jinarajadasa, 'Introduction', xv.

Alexander von Humboldt, rationalizing and categorizing his exotic discoveries. Parallelling contemporary scientific practices, his approach combined (psychic) fieldwork with research undertaken in the museum, albeit a museum consulted remotely, so to speak. The mysterious 'museum of the Adept Brotherhood', Jinarajadasa explained, provided clairvoyants with unparalleled access to 'fossils and skeletons, maps, models and manuscripts, illustrative of the development of the earth'.[133] Superior Theosophical equivalents seemingly existed for all the methods, materials, and even publication venues of modern science.

Jinarajadasa himself was becoming skilled at repurposing materialistic science's tools. In his *First Principles of Theosophy* (1919–21), a serial which began life as a lecture series performed in Chicago in 1909, we again find a concerted attempt to graft occult concepts upon the visual language developed by scientific savants. Its lantern-slide diagrams were mostly drawn, to Jinarajadasa's specifications, by railway draftsman Ralph E. Packard, and they were left out of copyright to encourage dissemination.[134] The text on many of these diagrams boldly and misleadingly implied that they represented the thinking of the likes of some of the previous century's most famous scientific savants, including Herbert Spencer and Darwin. One table depicting 'THE PRINCIPLES OF EVOLUTION', subtitled 'SPENCER', featured the philosopher's famous dictum that life progresses from the 'Homogeneous' to the 'Heterogeneous', but his terms, creatively redefined as 'ADHARMA' and 'DHARMA', were given a transformative Indian gloss.[135] Another diagram expanded Charles Darwin's iconic evolutionary tree into non-material planes (Figure 2.5). Theosophists' scepticism of realistic imagery meant that their artists usually preferred to produce schematic diagrams like these, rather than the scenes from deep time illustrated by the likes of Tuttle and Sherman Denton.[136] In any case, the audacious implication of Jinarajadasa's images was that the *First Principles of Theosophy* (a title echoing that of Spencer's own 1862 book *First Principles*, among others) were more comprehensive than any 'principles' that had come before.

Infinite Visioning

Engaging accounts of occult palaeoscience, from the Dentons' immersive transcripts to the Theosophists' scholarly booklets, intrigued many readers who otherwise rejected the authors' claims. Those fiction writers who, in the 1920s and 1930s, contributed to the development of 'scientifiction', or rather science fiction,

---

[133] Jinarajadasa, *First Principles*, 30–31.
[134] Jinarajadasa, *First Principles*, vii–viii.
[135] Jinarajadasa, *First Principles*, 11.
[136] Massimo Introvigne, 'Paintings the Masters in Britain: From Schmiechen to Scott', in *The Occult Imagination in Britain, 1875–1947*, ed. by Christine Ferguson and Andrew Radford (London: Routledge, 2018), 206–26.

were particularly alert to the literary potential of these heterodox notions.[137] As Aren Roukema notes, early science fiction was, to a significant degree, a technoscientific repackaging of occult plots.[138] Deep clairvoyant narratives, in which occult concepts were already intertwined with palaeoscientific ideas, presented especially ideal fodder for repackaging. Leadbeater himself was alert to the literary possibilities of presenting his experiences in a more suspenseful manner. In a short story, 'A Test of Courage' (1911), avowedly based on real events, he provided a chilling depiction of how, under uncontrolled conditions, 'the gigantic monsters of the

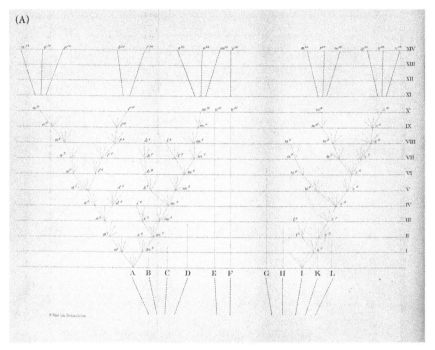

Figure 2.5 (A) Charles Darwin, *On the Origin of Species by Means of Natural Selection, or, The Preservation of Favoured Races in the Struggle for Life* (London: John Murray, 1859). (B) Curuppumullage Jinarajadasa, *First Principles of Theosophy*, 9th edn (Madras [Chennai]: Theosophical Publishing House, 1951), 116. Reproduced courtesy of the Main Library, University of Birmingham. Ralph E. Packard's 1909 diagram, based on Jinarajadasa's design, inverts Darwin's sketch of the tree of evolution, implying refining progress to higher planes rather than specializing descent.

---

[137] For the emergence of science fiction, see Will Tattersdill, *Science, Fiction, and the* Fin-de-Siècle *Periodical Press* (Cambridge: Cambridge University Press, 2016), 10–14.
[138] Roukema, 'The Esoteric Roots of Science Fiction', 23–35.

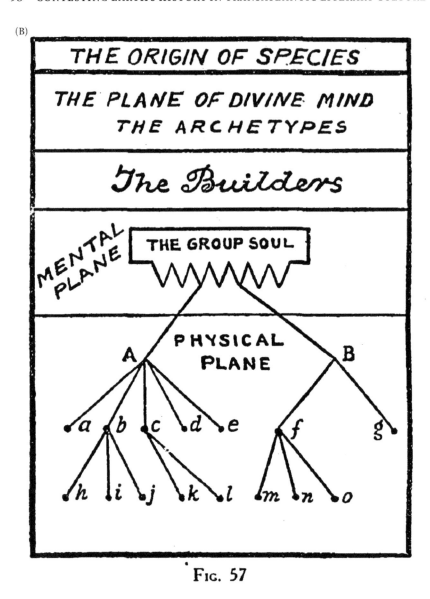

Figure 2.5 Continued

so-called antediluvian era' might be able to terrorize the psychic voyager.[139] However, it was his more nonchalant Theosophical style, and that of collaborators like Scott-Elliot, that was his most distinctive contribution.

One of the most skilled authors to translate these occult ideas to science fiction was E. T. Bell, a Scottish-born mathematician who spent most of his life working in

---

[139] Charles Webster Leadbeater, *The Perfume of Egypt, and Other Weird Stories* (Adyar: Theosophist Office, 1911), 103.

universities on the West Coast of the United States and writing novels under the name John Taine. His many interests included palaeontology: as his biographer notes, Bell repeatedly stressed his fondness for Hawkins's Crystal Palace models.[140] He also took his position as a scientist writing fiction very seriously. Bell told an interviewer in 1930 that scientifiction would 'disseminate science in popular fiction form', adding that the 'popularizing of learning' was 'just as important as its creation'.[141] Bell insisted, moreover, that his own scientifiction was immeasurably enhanced by his scientific expertise. He informed the president of Dutton, his primary publisher, that 'scientific imaginativeness is not possible to one who is not primarily a scientist', and that the 'logically selfconsistent [sic]' nature of his own stories could not have been achieved by 'a purely literary man'.[142] The exclusively scientific firm Williams & Wilkins was so taken with the plausibility of Bell's fictional technique that it took the unusual decision of publishing his novel *Before the Dawn* (1934), justifying the decision by inventing a new genre, '*fantascience*', to mediate between science fiction and science fact.[143]

Despite Bell's rhetoric and these canny marketing claims, the novel was a scarcely disguised narrativization of the principles of psychometry. The 'selfconsistent' invention at the heart of *Before the Dawn* was the televisor: a device that 'does for light what a phonograph does for sound' (5), projecting three-dimensional visuals of 'the record of the days and nights' stored up in an inanimate object (7), based on the order in which that object has lost electrons. Applying the televisor to fossils, a team of scientists replay entire generations of Mesozoic life. As Gary Lachman has observed, Bell's televisor simply instrumentalizes psychometry.[144] The narrator foregrounds rather than dispels this connection by unconvincingly insisting that 'the televisor has nothing whatever to do with spiritualism' as '[i]t does *not* invoke spirits' but rather is 'a purely physical apparatus' producing effects through 'well known physical laws' (21). Aside from the explicit mechanization of the psychometer, this was nothing new: over half a century earlier, the Dentons had equally claimed their research relied on purely physical laws, as did Spiritualists in general.

Others terms used by his narrator imply Bell was not overly embarrassed by the occult antecedents of his serious fantascience, published by a scientific imprint: the investigators' desire to uncover 'what language is nature's own, and in what script she conceals her secrets' (7), distinctly recalls *Nature's Secrets*, the British title of *The Soul of Things*, while the reference to 'unaging records of light itself

---

[140] Constance Reid, *The Search for E. T. Bell, Also Known as John Taine* (Washington, D.C.: Mathematical Association of America, 1993), 12, 65, 80. See also James L. Campbell, Sr., 'John Taine', in *Science Fiction Writers: Critical Studies of the Major Authors from the Early Nineteenth Century to the Present Day*, ed. by E. F. Bleiler (New York: Scribner, 1981), 75–82.

[141] Braxton Blades, 'Dr. Bell, alias John Taine', *Los Angeles Times*, 1 June 1930, J1.

[142] Quoted in Reid, *Search for E. T. Bell*, 240.

[143] Eric Temple Bell (John Taine), *Before the Dawn* (Baltimore, MD: Williams and Wilkins, 1934), v–vii. Subsequent references included in text.

[144] Gary Lachman, *Dreaming Ahead of Time: Experiences with Precognitive Dreams, Synchronicity and Coincidence* (Edinburgh: Floris Books, 2022), chapter 4, n.p.

stored up in the ultimate particles of matter' highlights the device's resemblance to Theosophists' *akashic* records (22). Thus nineteenth-century borderline palaeoscience was repurposed, with little difficulty, for a work of pioneering science fiction (or fantascience) in the 1930s. Indeed, the fact that the story's 'television in time' (vi) is persistently compared not to the more obviously topical technology of cinema but to two more hoary immersive media—the theatre (23, 25) and 'those old fashioned [*sic*] indoor panoramas' (42)—hints at Bell's familiarity with the history of these visual metaphors.

The fantascience of *Before the Dawn* was indebted to occult literature, but, ironically, Bell's dedication to verisimilitude impelled him to consider the logistics of psychometry more systematically than had the Dentons. Although a section of *The Soul of Things* had been dedicated pre-emptively to answering potential queries, Elizabeth made only tentative attempts to explain how she was sometimes able to move around in the psychometric scenes she witnessed, or how she could exercise senses other than vision, such as taste. Bell's world-building was more meticulous: petrified plants, due to their stationary nature, provide the best fossil fuel for the televisor's three-dimensional projections (21); unlike in Elizabeth's visions, no sound has been preserved (22); the projector must be carefully calibrated to project the images at life size (27); and walking inside opaque objects sends participants into an unrecorded void (44). By mastering these quirks, the protagonists are able to document the life of a single *Tyrannosaurus*. Much like Elizabeth's visions, the projections are so convincing that intangible dinosaurs seem to threaten the lives of their human observers (38).

If Bell made deep clairvoyance fictionally viable, another, more famous writer involved in the interwar birth of science fiction made it weird. The interest expressed in occultism by the American literary circle gathered around H. P. Lovecraft is reasonably well known.[145] Although he was, like Bell, no believer in clairvoyance, Lovecraft found substantial inspiration in the imaginative pretence. The rationalistic Rhode Islander dealt with the perceived disenchantments of modernity—disenchantments not felt by contemporaries like the Theosophists—by employing what we have, in my introduction, already heard Michael Saler refer to as 'ironic imagination', taking pleasure in the realistic construction of fantastical worlds and scenarios.[146] Lovecraft's letters, as much as his stories, evidence this pleasure in rigorous make-believe, sometimes in language that borders on the psychometric. When, in 1930, fellow author Clark Ashton Smith mailed his friend what he claimed was a dinosaur bone, Lovecraft reported a pseudo-Dentonian

---

[145] Robert M. Price, 'HPL and HPB: Lovecraft's Use of Theosophy', *Crypt of Cthulhu*, 5 (1982), 3–9.
[146] Michael Saler, *As If: Modern Enchantment and the Literary Prehistory of Virtual Reality* (Oxford: Oxford University Press, 2012), chapter 2.

reading of the 'dinosaurian reliquiae': 'I can close my eyes & behold the steaming fungoid morass in Lemuria through which It once floundered'.[147]

Lovecraft's clairvoyant predilections were likely prompted by the 1925 compilation and reissue of Scott-Elliot's *The Story of Atlantis and the Lost Lemuria*. He referred to this book in *The Call of Cthulhu* (written 1926, published 1928), a story which also alluded to the 'strange survivals' described by 'Theosophists' with 'bland optimism'.[148] This latter comment presumably referred not just to the religious 'optimism' of the Theosophical worldview, but also to the characteristic Theosophical literary register: the ingenuous manner in which Scott-Elliot and others described the outlandish anatomies of early Root Races. When encountering cosmic horrors, characters in *The Call of Cthulhu* lack this Theosophical sangfroid. Outside of the bleak world of his fiction, however, Lovecraft's personal imaginative fantasies closely resembled the coupling of visionary power and cosy domesticity practised by Leadbeater. Ideally, he told Smith, he wished to possess '*infinite visioning & voyaging power*, yet without loss of the familiar background which gives all things significance'.[149] Lovecraft's interest in the subject's utility grew when, in early 1933, he expanded his acquaintance with Theosophical thought. As he remarked to Smith, '[t]hese *Akashic records* tickle my imagination'.[150]

A significant product of this interest in Theosophy was *The Shadow Out of Time*, a novella that appeared in the pulp science fiction magazine *Astounding Stories* in June 1936, (although it was, again, written several years earlier). The story concerns a university professor, Nathaniel Wingate Peaslee, whose mind has been swapped with that of a time-travelling alien. This alien belongs to the Great Race, an advanced species that appropriates the bodies of other beings to learn about their societies (Figure 2.6). Unbound by time, the Great Race has taken up residence on Earth 150 million years ago, 'when the Paleozoic Age was giving place to the Mesozoic', on what is implied to be the continent of Lemuria—associated here, as it often was in the twentieth century, with Australia.[151] While the alien explores Peaslee's present day using Peaslee's body, the man is free to roam the alien's Lemurian city in the alien's body, although he tries to convince himself that these out-of-body experiences are dreams. Robert M. Price notes that this plot draws upon a pivotal event in the Theosophical cosmogony: the time when the majestic Lords of Venus implant consciousness into the Lemurian humans of the

---

[147] David E. Schultz and S. T. Joshi, eds., *Dawnward Spire, Lonely, Hill: The Letters of H. P. Lovecraft and Clark Ashton Smith: 1922–1937*, 2 vols continuously paginated (New York: Hippocampus Press, 2017), I, 209.
[148] H. P. Lovecraft, 'The Call of Cthulhu', *Weird Tales*, 11 (1928) 159–78 (159, 161).
[149] *Dawnward Spire*, I, 263.
[150] *Dawnward Spire*, II, 414.
[151] H. P. Lovecraft, 'The Shadow Out of Time', *Astounding Stories*, 17 (1936), 110–54 (131). Subsequent references included in text.

Jurassic period.[152] What has not received attention, however, is the way Lovecraft's story subverts the aforementioned 'bland optimism' of Theosophical writing. This subversion reflected Lovecraft's stylistic evaluation not just of Theosophical writing but also of much science fiction, in which he felt protagonists responded to events with unrealistic nonchalance.[153]

In *The Shadow Out of Time*, Lovecraft subverted the godlike power usually granted to the Theosophical clairvoyant by splitting this figure back in two—although not in the manner of the Dentons' collaboration. One half is represented by the alien of the Great Race who takes over Peaslee's body. The Race's mental forays into other times are depicted as 'sinister experiments' by beings with little regard for mortal lives (117), 'blasphemous reachings and seizures' (153). These invasive creatures, who have 'conquered the secret of time' to 'study the lore of every age' (122), are amoral versions of Theosophists: their cyclopean 'central archives' (123) of 'terrible books' on 'endless shelves' (128) represent a perversion of the encyclopaedic *akashic* records and the Museum of the Adept Brotherhood. Here, the more sinister implications of omnipotent observation are stressed, suggesting that near-infinite access to knowledge is by no means as benign as Theosophical seers implied.

The other clairvoyant figure in the novella is the human, Peaslee. When his body is stolen by the Great Race for study, Peaslee, in turn, begins to study the Race and their Permian-Triassic lair. In the process, he sounds just like Leadbeater or Scott-Elliot in their mild and industrious surveys of the natural history and culture of other ages. From his reading of Scott-Elliot's book, Lovecraft would have been familiar with that author's tendency to discuss not only the occult wonders of prehistoric Atlantis but also mundane matters like its 'council of agricultural advisers and coadjutors' and 'systems of land tenure'.[154] Peaslee's discussion of Great Race sociology, given his disturbing and *outré* situation, takes on a subtly parodic Theosophical blandness: 'Crime was surprisingly scant' in the Great Race's prehistoric city, he observes, 'and was dealt with through highly efficient policing' (132). Scott-Elliot also depicted Lemurians as coexistent with 'Plesiosauri and Ichthyosauri', 'Dinosauria', and 'Pterodactyls', and Peaslee, too, is attentive to palaeontology.[155] 'Of the animals I saw, I could write volumes' (131), Peaslee notes, describing 'lesser, archaic prototypes of many forms—Dinosauria, Pterodactyls, Ichthyosauria, Labyrinthodonta, Plesiosauri, and the like' (132). When he comes to doubt his memories of these experiences, Peaslee even echoes a criticism levelled against clairvoyants, wondering if he is merely unconsciously regurgitating 'text-book knowledge of the plants and other conditions of the primitive world' (120).

[152] Price, 'HPL and HPB'.
[153] *Dawnward Spire*, I, 194.
[154] Scott-Elliot, *The Story of Atlantis* (1896), 59, 61.
[155] Scott-Elliot, *The Lost Lemuria* (1904), 17.

Figure 2.6 Howard V. Brown, front cover of *Astounding Stories*, 17 (1936). The cover, based on H. P. Lovecraft's *The Shadow Out of Time*, depicts the alien Great Race whose minds leap across the universe to collect information about other worlds. Brown depicts the repositories of this information as gigantic folios.

This rational exploration for what Peaslee calls his 'pseudomemories' (117) was precisely Lovecraft's explanation for his own pseudo-clairvoyant experiences. These included the pleasurable but illusive notions that he had, for example, been alive in Roman times, which he attributed to his unconscious mind's production of 'a composite of places I have visited, pictures I have seen, & things I have read'.[156] At the end of the novella, as Peaslee finds the ruins of the Great Race's archives in present-day Australia, he realizes that what he thought were dreams and pseudomemories were, in fact, real happenings. Here, his mental travels disturbingly intrude upon his sense of reality in a manner that never happens to the detached Theosophical clairvoyant, nor to Lovecraft himself, coolly aware of the fictionality of his own time-hopping fantasies. In the characteristic fashion of a Lovecraftian narrator, Peaslee responds to these mind-bending revelations by concluding that 'time and space had become a mockery' (150) in which 'there is no hope' (154). In a weird, atheistic universe, the clairvoyant is no longer a blithe voyager through deep time, but rather is crushed into insignificance by its vast inhumanity.

Conclusion

For thinkers at the radical edges of science, bursting the limits of time legitimized the bursting of other, weirder limits. Clairvoyants' direct access to the deep past solved otherwise unanswerable scientific questions and inserted humanity back into the foreground of prehistory. It also allowed participants excluded from the highest echelons of the scientific community, including women, to attempt high-level contributions to knowledge that were, on their own terms, founded on empirical research. The price of this unrestrained agency was credibility outside occult and radical circles, but it was a price that these clairvoyants were willing to pay. After all, for those not invoking personal occult experience as the source of their knowledge, recruiting virtual witnesses for claims about Earth's deep history was a messy, compromised process. We have previously seen that day-agers, addressing broadly Christian publics, strained simply to make their speculations valid matter for debate within what remained of the scientific public sphere. Practitioners of deep clairvoyance bypassed much of this negotiative process and did not even sanction experimental tests of their abilities, such as those applied to sufficiently game Spiritualist mediums by believers like Alfred Russel Wallace.[157] Although they were open to evaluation by participants, sympathetic observers, and contributors to counter-public spheres, these were not scientific claims to be vulgarly tested, as their primary importance lay in the practitioners' sociocultural—and temporal—empowerment.

---

[156] *Dawnward Spire*, I, 249.
[157] Sera-Shriar, esp. chapter 1.

This did not, however, mean, that the clairvoyants I have discussed made no exertions to recruit virtual witnesses to their side. Even sceptics like Bell and Lovecraft looked to their writings for inspiration precisely due to their immersive verisimilitude. It is well known that practitioners of fringe science commonly adopt forbidding technical jargon and formatting akin to that of respected disciplines, and this was certainly the case for Theosophical clairvoyants.[158] Significantly, I have argued that the boldest clairvoyants also emulated, or rather actualized, science's imaginative literary technologies, namely those compellingly visual—even when textual—reconstructions of deep history pioneered by authors like Miller and artists like Riou.

The idea that geologists' and palaeontologists' resplendent deductions and word-paintings represented deeper powers of mind than the authors realized was a Romantic one, a fact hinted at in the works I have discussed through routine allusions to the intuitive genius of poets like Wordsworth, Goethe, and Keats.[159] This late Romanticism was not merely skin-deep: the Romantic notion that the human imagination could intuit truth had become a core tenet of nineteenth-century esotericism. In Olav Hammer's terms, occultists in this period radicalized Romantic-era science's interest in the anti-materialistic powers of the mind.[160] Palaeoscience, as we have seen, was interfused with transcendental *Naturphilosophie*, Orientalism, and visionary self-fashioning in the works of some of its most respected practitioners. By 'radicalizing' susceptibly Romantic attributes like these, paranormal researchers simply took an extra step. In so doing, they aimed to transcend the provisional access to reality claimed by authoritative scientific predecessors and see the deep past with their own eyes. In the next chapter, we will encounter authors whose radical claims about geology required no such psychic assistance.

---

[158] Michael D. Gordin, *On the Fringe: Where Science Meets Pseudoscience* (Oxford: Oxford University Press, 2021), 43.
[159] See, for instance, Jinarajadasa, *First Principles*, 39, 191, 201, 203.
[160] Hammer, *Claiming Knowledge*, 329. See also Wouter J. Hanegraaff, 'Romanticism and the Esoteric Connection', in *Gnosis and Hermeticism from Antiquity to Modern Times*, ed. by Roelof van den Broek and Wouter J. Hanegraaff (Albany: State University of New York Press, 1998), 237–68; and Alex Owen, 128–32.

3
Hollow Ground

On 8 August 1892, Charles Lapworth delivered the Presidential Address to the Geological Section of the British Association for the Advancement of Science in Edinburgh. His colleagues, the geologist observed, 'regret that there are, nowadays, no new geological worlds to conquer, no new systems to discover and name'.[1] Lapworth, however, saw this apparent closure of the geological frontier, the end of geology's heroic age of mapping, as only apparent. The final frontier was not more mapping, but rather the creation of a global geophysical synthesis, through which such mysteries as the origin of mountain chains and the nature of continents and oceans could be connected and explained. Concluding an address in which he proposed that Earth's surface was an interconnected structure of wave-like folds, Lapworth considered what this meant for its interior structure. The consideration led him to a provocative question: 'Is it not just possible, after all, that, as others have suggested, our earth is such a hollow shell, or series of concentric shells, on the surface of which gravity is at a maximum, and in whose interior it is practically non-existent?'[2]

His suggestion was a radical one. Various figures might be numbered among the 'others' obliquely referred to as having conceived Earth as a 'hollow shell' or a 'series of concentric shells', but foremost among such heterodox thinkers would almost certainly have been found the name of John Cleves Symmes Jr. In 1818, Symmes, a former United States army officer and trader, had informed the world that 'the earth is hollow, and habitable within; containing a number of solid concentrick spheres, one within the other, and that it is open at the poles' (Figure 3.1). By descending into these polar depressions, which became known as 'Symmes holes', one would encounter hidden interior territories beyond all previous imaginings. Symmes's hollow-world system, which received minimal endorsement from the nascent geological establishment when it was proposed, has come to represent one of the United States' premier examples of fringe science.[3]

---

[1] Charles Lapworth, 'Presidential Address to the Geological Section', *Report of the Sixty-Second Meeting of the British Association for the Advancement of Science held at Edinburgh in August 1892* (London: John Murray, 1893), 695–707 (697).

[2] Lapworth, 707.

[3] Peter W. Sinnema, '10 April 1818: John Cleves Symmes's "No. 1 Circular"', in *BRANCH: Britain, Representation and Nineteenth-Century History*, ed. by Dino Franco Felluga, Extension of *Romanticism and Victorianism on the Net* (2012) [accessed 19 November 2023].

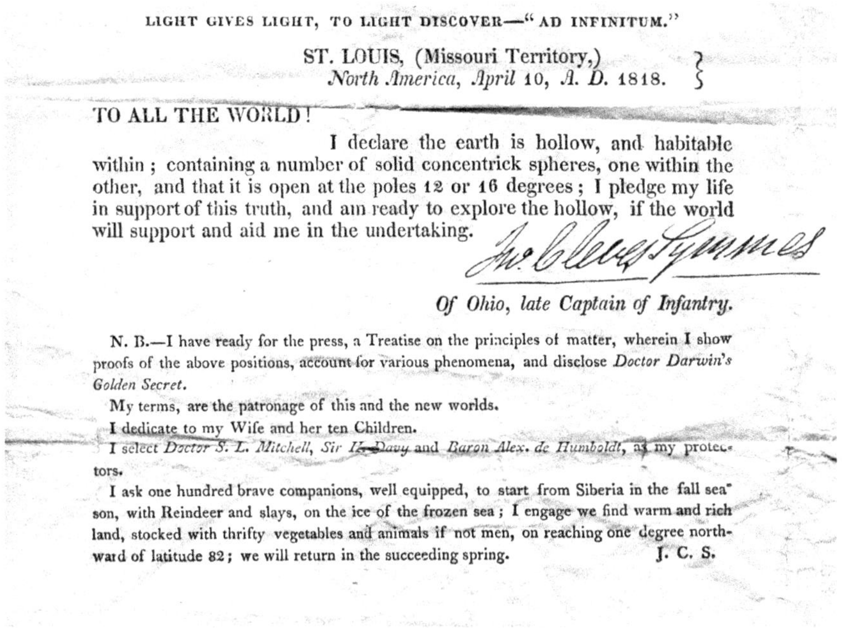

> LIGHT GIVES LIGHT, TO LIGHT DISCOVER—"AD INFINITUM."
>
> ST. LOUIS, (Missouri Territory,)
> North America, April 10, A. D. 1818.
>
> TO ALL THE WORLD!
> I declare the earth is hollow, and habitable within; containing a number of solid concentrick spheres, one within the other, and that it is open at the poles 12 or 16 degrees; I pledge my life in support of this truth, and am ready to explore the hollow, if the world will support and aid me in the undertaking.
>
> *Of Ohio, late Captain of Infantry.*
>
> N. B.—I have ready for the press, a Treatise on the principles of matter, wherein I show proofs of the above positions, account for various phenomena, and disclose *Doctor Darwin's Golden Secret*.
> My terms, are the patronage of this and the new worlds.
> I dedicate to my Wife and her ten Children.
> I select *Doctor S. L. Mitchell*, Sir *H. Davy* and *Baron Alex. de Humboldt*, as my protectors.
> I ask one hundred brave companions, well equipped, to start from Siberia in the fall season, with Reindeer and slays, on the ice of the frozen sea; I engage we find warm and rich land, stocked with thrifty vegetables and animals if not men, on reaching one degree northward of latitude 82; we will return in the succeeding spring. J. C. S.

Figure 3.1 John Cleves Symmes Jr's signed circular declaring the Earth to be hollow and filled with concentric spheres, distributed across the world in 1818. Note the appeal to famous savants Humphry Davy and Alexander von Humboldt.

Lapworth's foray into borderline realms of science, a friend later claimed, was inspired by a transcendent mental experience. In the words of his Birmingham physician Theodore Stacey Wilson, when Lapworth began writing his address,

> the whole subject of wave movement in connection with geology opened out before him, and he could see that continents with their mountain ranges and ocean basins were but the expression of wave movements in what we call solid land. He had a vision of the laws of movement which showed him that not only light and sound but chemical affinity and the structure of matter was explainable by wave movement, and that atoms might only be internodal points of such waves.[4]

Wilson recalled that Lapworth had been 'overbalanced by the greatness of the revelation' and appears to have seen it as one cause of his persistent ill health.[5] Descriptions of his patient's all-synthesizing geophysical insights as a divine

---

[4] Theodore Stacey Wilson, untitled and undated note, CL/2/4/1 (in former catalogue L2B.14), Charles Lapworth Archive Collection, Lapworth Museum of Geology, University of Birmingham.

[5] Theodore Stacey Wilson, 'Plotting out the Action of Wave Movements on Paper', CL/2/4/1 (in former catalogue L2B.17), Charles Lapworth Archive Collection.

'revelation' or occult 'vision', akin to those discussed in the previous two chapters, should not immediately be dismissed as figurative, given that Wilson was a student of psychic phenomena and author of a volume on *Thought Transference* (1925). Lapworth did not share these predilections but, like several famous geologists, he described efforts at correlating heterogeneous geological phenomena as rovings of the mind's eye.[6] At the close of his British Association address, Icarus-like speculation had carried Lapworth, '*in imagination*, to the fiery eddies of the sun' itself.[7] He does not appear to have taken the hollow-earth component of his thinking any further, nor do his auditors seem to have drawn attention to its discreditable connotations (even though the stimulating address itself lingered in geologists' minds for decades).[8] Nonetheless, Lapworth's speculations indicate the sublime scope of the uncertainty concerning the nature of Earth's interior during the late nineteenth century. While Symmes's idea that Earth's centre was 'habitable' was likely too much even for Lapworth, these regions remained the subject of debate.[9] Symmes's ideas about the North and Soul Poles' 'open' nature were more pertinent still in an age of intense polar exploration.[10] Maverick thinkers planted their flag at the crossroads of this establishment uncertainty, making the long-dead Symmes's theory more influential than ever before.

Depictions of Symmesian lands on the inverse side of Earth's thin crust were just one subcategory, albeit a major one, of the nineteenth and early twentieth century's vast literature on the exploration of underground worlds. Michele Kathryn Yost groups this wider category of writing under the umbrella of *terra cava* literature, and it has been the subject of an immense quantity of scholarship.[11] Rather than retread all the steps of this work, I focus in this chapter on a traceable phenomenon that has only intermittently received serious scholarly attention and which is highly pertinent to the themes of *Contesting Earth's History*: the revival of Symmes's ideas in the Gilded-Age United States. Indeed, my focus is yet more precise. I am interested in an oft-noted but rarely interrogated aspect of this literature: its tendency to textualize truth claims in ways that complicate, rather than clarify, authorial intent—blurring the difference between fact and fiction. The generic

[6] James A. Secord, 'Global Geology and the Tectonics of Empire', in *Worlds of Natural History*, ed. by H. A. Curry, N. Jardine, J. A. Secord, and E. C. Spary (Cambridge: Cambridge University Press, 2018), 401–17 (404–5). For Lapworth as writer, see Adelene Buckland, *Novel Science: Fiction and the Invention of Nineteenth-Century Geology* (Chicago: University of Chicago Press, 2013), 88–93.
[7] Lapworth, 707.
[8] For example, see Edward Hubert Cunningham Craig to William Whitehead Watts, 27 July 1921, L2B.21a, Charles Lapworth Archive Collection.
[9] Martin J. S. Rudwick, *Earth's Deep History: How It Was Discovered and Why It Matters* (Chicago: University of Chicago Press, 2014), 241–42.
[10] Sarah Moss, *Scott's Last Biscuit* (Oxford: Signal Books, 2006); Will Tattersdill, *Science, Fiction, and the Fin-de-Siècle Periodical Press* (Cambridge: Cambridge University Press, 2016), chapter 4.
[11] Michelle Kathryn Yost, 'American Hollow Earth Narratives from the 1820s to 1920', unpublished PhD thesis, University of Liverpool (2014), 6. For key general works, see Walter Kafton-Minkel, *Subterranean Worlds: 100,000 Years of Dragons, Dwarfs, the Dead, Lost Races & UFOs from Inside the Earth* (Port Townsend, WA: Loompanics Unlimited, 1989); and Peter Fitting, *Subterranean Worlds: A Critical Anthology* (Middletown, CT: Wesleyan University Press, 2004).

spectrum of these hollow-earth writings shades from earnest technical monographs to works of pure fantastic fiction, but, in practice, many sit uneasily at either end. Hollow-earth non-fiction often featured elaborate fictionalized thought experiments, for instance, while authors of fiction regularly used bizarre framing devices to imply, sometimes seriously, that the story was contiguous with fact. The most complex in this line, John Uri Lloyd's novel *Etidorhpa* (1895), will be one of my main subjects, but even the pulp 'Pellucidar' romances of Edgar Rice Burroughs, clearly marketed as fiction, thematized the thin line separating romance from reality.

In part, these aspects of the first hollow-earth revival—which ran from the 1870s to the 1910s—are explained by its contemporaneity with two rising genres that foregrounded the relationship between fantasy and truthfulness. Starting in the mid-1880s, a new kind of adventure fiction was emerging; usually depicting the exploration of lost worlds, its authors employed what Michael Saler calls 'ironic imagination' in the self-conscious construction of verisimilitude for even the most fabulous stories (an approach which, as discussed in chapter 2, later fed into H. P. Lovecraft's writing).[12] This being said, understanding hollow-earth texts simply as contributions to the lost world genre is insufficient. Yost's perceptive thesis sees most narrative hollow-earth texts less as 'works of fiction' than 'didactic dialogues between science and religion, author and reader, told through allegorical narratives'.[13] In this respect, the hollow-earth revival also reflected the era's growing taste for utopias. Utopian texts, in Fátima Vieira's words, narrativize an unsolvable 'tension between the affirmation of a possibility and the negation of its fulfilment', a tension also extant in many of the works I discuss.[14] If hollow-earthers drew from both lost worlds and utopias, however, their most original productions also brought more critical interest in how scientific claims are made.

To understand the subjects of this chapter, then, a particular vocabulary is required. Speaking of the early modern context in which recognizable notions of literary fiction and scientific non-fiction gestated, Frédérique Aït-Touati plots imaginative scientific texts on a spectrum. Its two poles are '*fictionalizing*' and '*factualizing narratives*', the narrative techniques of the latter tendency, represented by the likes of Johannes Kepler's *Somnium* (written 1608), being carefully anchored and 'localized' for 'heuristic purposes', in contrast with less disciplined, fictionalizing forms of imaginative science.[15] Attending, as Aït-Touati proposes, to the localization of fiction techniques helps to clarify matters that have confused previous critics about many hollow-earth texts, but it does not explain everything.

---

[12] Michael Saler, *As If: Modern Enchantment and the Literary Prehistory of Virtual Reality* (Oxford: Oxford University Press, 2012). See also Anna Vaninskaya, 'The Late Victorian Romance Revival: A Generic Excursus', *English Literature in Transition, 1880–1920*, 51 (2008), 57–79.

[13] Yost, 234–35.

[14] Fátima Vieira, 'The Concept of Utopia', in *The Cambridge Companion to Utopian Literature*, ed. by Gregory Claeys (Cambridge: Cambridge University Press, 2010), 3–27 (6).

[15] Frédérique Aït-Touati, *Fictions of the Cosmos: Science and Literature in the Seventeenth Century*, trans. by Susan Emanuel (Chicago: University of Chicago Press, 2011), 194, 196.

While thinking of some of these texts as factualizing narratives reveals that their use of fiction is relatively instrumentalized, it does not explain away the frequency with which ambiguities of intent and interpretation are placed before the reader in serious-minded hollow-earth writings.

Various scholars have seen unruly and confounding literary forms as tools for challenges to the intellectual status quo. This tradition goes back at least as far as the classical genre that has been dubbed Menippean satire, authors of which combine 'genres' and 'tones . . . to combat a false and threatening orthodoxy', '[h]eterogeneous form' being this genre's 'hallmark'.[16] Labyrinthine literary form is here conceived as a way of unsettling prior beliefs, including scientific ones, by tangling readers in an unfamiliar web from which they must escape. Outlandish works like Thomas Carlyle's *Sartor Resartus* (1833–34), itself a Romantic update of Menippean satire, packaged robust criticism of modern science in just such a bewildering manner.[17] Literary scholars of esotericism and conspiracy theories are well versed in parsing baroque ambiguities in works apparently written to put forth truth claims, but students of geology are far less used to encountering such apparently counterintuitive ingredients.[18] As I shall demonstrate, hollow-earthers regularly challenged not just the fundamentals of modern geology and its hierarchies of specialist expertise, but also the careful distinctions between fact and fiction, and sincerity and irony, we have seen maintained even in such heterodox works as the Denton family's *The Soul of Things*. In so doing, they were not postmodernists *avant la lettre*. Rather, they hoped to reveal brave new scientific worlds.

Enterprising Spirit

Symmes's theory of the Earth as hollow, and containing multiple concentric spheres, had its origin in the seventeenth- and eighteenth-century speculations of the savant Edmond Halley in Britain—likely one of the 'others' alluded to by Lapworth—and clergyman-philosopher Cotton Mather in colonial America.[19] Indeed, even in its later incarnations, the theory bore the impress of early modern geotheory. Geotheoretical works like René Descartes' *Principia Philosophiae*

---

[16] Howard D. Weinbrot, *Menippean Satire Reconsidered: From Antiquity to the Eighteenth Century* (Baltimore, MD: Johns Hopkins University Press, 2005), xi ('orthodoxy'); David Musgrave, *Grotesque Anatomies: Menippean Satire since the Renaissance* (Newcastle upon Tyne: Cambridge Scholars Publishing, 2014), 21 ('hallmark').

[17] For a relevant reading of *Sartor Resartus*, see James A. Secord, *Visions of Science: Books and Readers at the Dawn of the Victorian Age* (Oxford: Oxford University Press, 2014), chapter 6.

[18] For example, see Tanner F. Boyle, *The Fortean Influence on Science Fiction* (Jefferson, NC: McFarland, 2021), 24–27; and Christine Ferguson, 'Beyond Belief: Literature, Esotericism Studies, and the Challenges of Biographical Reading in Arthur Conan Doyle's *The Land of Mist*', *Aries*, 22 (2022), 205–30.

[19] Conway Zirkle, 'The Theory of Concentric Spheres: Edmond Halley, Cotton Mather, & John Cleves Symmes', *Isis*, 37 (1947), 155–59; Peter W. Sinnema, '"We Have Adventured To Make the Earth Hollow": Edmond Halley's Extravagant Hypothesis', *Perspectives in Science*, 22 (2014), 423–38.

(1644; 1647) and Thomas Burnet's *Sacred Theory of the Earth* (1681; 1690) had presented elegant, directional models of the formation of the Earth, rationally determined from iron physical laws, the likes of which allowed Halley and Mather to reach their conclusions about Earth's hollowness. By keeping alive this manner of thinking in subsequent centuries, hollow-earthers rejected the idea—so integral to our understanding of the development of geology thanks to the work of Martin J. S. Rudwick—that Earth has a complex, contingent history.[20] Planetary history, rather than a convoluted tale painstakingly reconstructed through fieldwork analysis of rocks and fossils, was, instead, so logical that it could be deduced wholesale. Geotheories had been unfashionable in the early nineteenth century, but, by the century's end, when the Symmes revival flourished, even elite geologists like Lapworth were searching for the kind of generalizing approach to global geology that Symmes had proposed, albeit one allied to a contingent conception of geohistory.

Varieties of hollow-earth theory have existed, but most followed similar principles. As Elizabeth Hope Chang observes, the focus of hollow-earth literature was 'self-sustaining ecology'.[21] Proponents believed that Earth is a steady-state system in which seawater flows into a giant hole at one pole and flows out a hole at the other, this detail being Symmes's addition to Halley's concentric hypothesis (Figure 3.2). The gradients of Symmes holes are subtle, so that polar explorers would not immediately notice that they were descending into the earth. To explain how this descent was feasible, hollow-earthers revised Newtonian conceptions of gravity, insisting that the force lies in the exact centre of the Earth's crust. This, crucially, means that the underside of the Earth's crust can be walked upon. The interior world is lit by sunlight shining into the poles, or sometimes by an interior sun, and usually imagined as a place of abundant natural resources. Among the mysteries hollow-earth theory was purported to explain were odd animal migration patterns, the rings of Saturn, the erratic behaviour of compass needles near the North Pole, the aurora borealis, and the existence of elephantine remains in cold climes. Although it was one of the most striking aspects of the theory, the idea that multiple concentric spheres existed inside the hollow earth was rarely explored in detail, ultimately being abandoned even by Symmes.

Symmes's campaign began with his 1818 'circular', reproduced above, which called for an expert team to join him in an expedition to the North Pole. A captain in the War of 1812, Symmes sent the circular from his home in St. Louis, where he was working as a trader. Moving to Newport, Kentucky, in 1819, he subsequently spent much of his time petitioning Congress to enact his proposal, writing to newspapers, and lecturing across the Ohio River in neighbouring Cincinnati (his uncle, John Cleves Symmes, had bought much of the territory of Ohio in the so-called 'Symmes Purchase' of 1788).[22] Being '[r]eared at the plough' and

---

[20] Martin J. S. Rudwick, *Bursting the Limits of Time: The Reconstruction of Geohistory in the Age of Revolution* (Chicago: University of Chicago Press, 2005), 193.

[21] Elizabeth Hope Chang, 'Hollow Earth Fiction and Environmental Form in the Late Nineteenth Century', *Nineteenth-Century Contexts*, 38 (2016), 387–97 (389).

[22] Sinnema, '10 April 1818', n.p.

thus, he stated, 'better fitted for thinking than writing', Symmes never published his beliefs in book form.[23] While this was not unknown among the savants of the early nineteenth century, his reluctance to publish will have contributed to his comparative isolation from the scientific communities of the young United States. The most detailed exposition of his thinking is found in a book by his disciple James McBride, *Symmes's Theory of Concentric Spheres; Demonstrating that the Earth is Hollow, Habitable Within, and Widely Open About the Poles* (1826), and a lecture transcription in McBride's personal scrapbook. The lecture, delivered in Cincinnati on 25 March 1820, establishes themes that would evolve in the later hollow-earth revival, as do numerous newspaper items pasted in the scrapbook.

The lecture's text shows that Symmes was at least passingly familiar with the newly christened science of geology, including Georges Cuvier's *Theory of the Earth* (1813)—the English title of which misleadingly characterized the French savant's geohistorical work as geotheoretical.[24] Significantly, but appropriately, Symmes's concept was dubbed 'the new Theory of the Earth' (C1–2). His preference for cosmogonical theory over more fashionably empirical approaches was indicated by the fact that his discussions of 'Geology' in fact concerned astronomy (C7–8), while his nonchalant proposal that 'a planet made by a miracle' would look 'precisely' like one 'made by the slowest possible process' even allowed for a young-earth creationist interpretation of geological phenomena (C38). For the time, his levelling attitude to geological authority was unusual only in degree, rather than in kind, and in the *Western Spy* he called for 'the Geologists of the day' to 'declare publicly for or against my new theory' (C83).[25] The *North American Review* grudgingly lauded his commitment to falsifiability, unlike savants like Cuvier, who, the *Review* joked, shunned 'the test of an actual expedition' into the Earth to prove the truth of their claims.[26] Symmes was an exemplary proponent of the public sphere of knowledge: he wished his work to circulate 'as free as air' (C11) and '[e]very community and individual throughout the world' was 'invited, by the author, to investigate' his proposals (C120).[27] The most significant, and symbolic, literary manifestation of this communal investigation was his dialogic manuscript journal. This journal was co-written by Symmes and scientifically inclined 'visitors of the Reading-Room' at Cincinnati (C60).

---

[23] 'James McBride's Scrapbook of Articles on the Hollow Earth Theory Lectures of John Symmes', James McBride Collection of John Symmes' Hollow Earth Theory, 1819–1859, Academy of Natural Sciences of Drexel University, Library and Archives, Philadelphia, 17. Subsequent references to separately numbered lecture (L) and clippings (C) included in text.
[24] Rudwick, *Bursting the Limits of Time*, 510.
[25] For a comparable attitude to geological authority, see Ralph O'Connor, *The Earth on Show: Fossils and the Poetics of Popular Science, 1802–1856* (Chicago: University of Chicago Press, 2007), 204–5.
[26] Untitled review of *Symzonia*, by 'Adam Seaborn', *North American Review*, 13 (1821), 134–43 (142).
[27] For the public sphere of science, see Thomas Broman, 'The Habermasian Public Sphere and "Science *in* the Enlightenment"', *History of Science*, 36 (1998), 123–49.

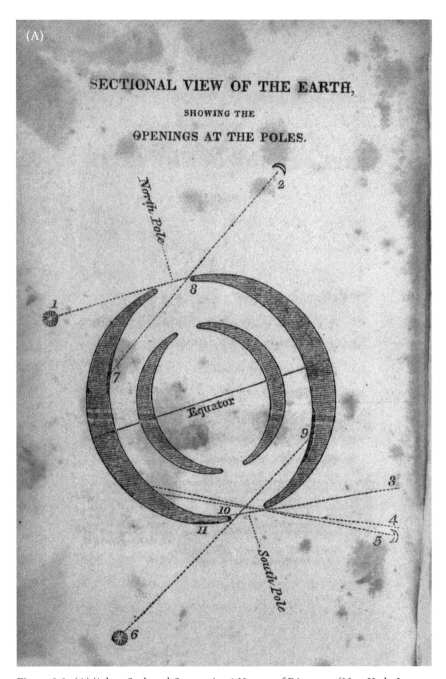

Figure 3.2 (A) 'Adam Seaborn', *Symzonia: A Voyage of Discovery* (New York: J. Seymour, 1820), frontispiece. (B) William R. Bradshaw, *The Goddess of Atvatabar; Being the History of the Discovery of the Interior World and Conquest of Atvatabar* (New York: J. F. Douthitt, 1892), frontispiece. Typical schematic representations of the Earth's hollow interior.

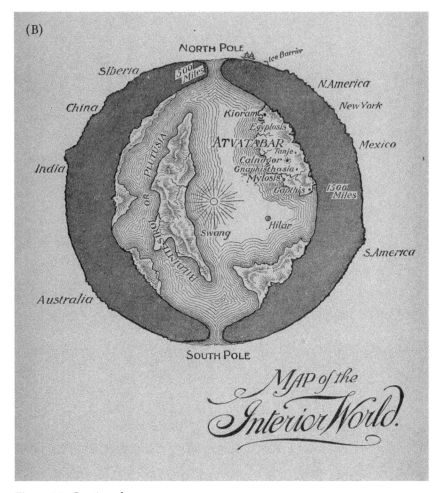

Figure 3.2 Continued

Key to this participatory mentality was a confidence in the democratic deducibility of phenomena from physical laws. Symmes told his Cincinnati auditors that his theory was 'founded on geometric and self-evident principles' that were irrefutable until the poles were reached (L4–5). The theory was not, however, purely theoretical. After all, Symmes was a practical man, even cutting his own crude 'diagrams' with a 'pen-knife' to use as woodcut blocks in newspapers (C21). In this style, he offered 'strong experimental proof' for the theory in the behaviour of magnetized iron filings as well as providing 'undisputed ocular demonstration' for nature's concentric tendencies by pointing to the rings of Saturn (L5). Symmes uneasily combined his passion for dialogue with a desire to be recognized as an original thinker. In a rhetorical trope we have seen elsewhere, he denied having read Halley's work before positing his own theory (C33). Heredity explained how a man 'without a liberal education' could generate such insights: Symmes pointed to canny ancestors, including a minister who was one of 'the first emigrants, who settled at Plymouth' (L25).

This latter point hinted at the most tangible attraction of a habitable inner earth: new colonial territory. Shortly after the Louisiana Purchase of 1803, France's sale of what would become the Great Plains and Midwestern United States, Thomas Jefferson commissioned the Lewis and Clark Expedition to survey this vast area of new Western territory.[28] Symmes was living at the febrile edge of this huge mass of newly part-colonized land, and expansionism across and beyond the continent must have seemed, to him, inevitable. In his lecture he expressed his hope that the United States would adopt the 'enterprising spirit' that had led 'the Russian and British nations' to examine 'the unexplored parts of the earth' (L21). McBride's point of comparison for the maverick Symmes, tellingly, was Christopher Columbus.[29] As Hester Blum has pointed out, this nationalistic framing was largely the work of Symmes's backers. His own rhetoric, she points out, was chiefly 'supranational' and 'planetary in scope', calling to his aid not just Americans but the global scientific community.[30] Journalists attending his lectures pointed to his humble manner, his rhetorical 'inarticulateness' evidencing 'sincerity'.[31] Symmes's self-fashioning thus balanced megalomania and modesty.

One thing Symmes lacked was a sense of irony. His theory's ironic associations began with the Swiftian mock-travelogue *Symzonia: A Voyage of Discovery* (1820). This book was published under the name of 'Captain Adam Seaborn', an adventurous unreliable narrator who stumbles upon a utopian white supremacist community within the hollow earth. Once these pale 'Internals' learn of the imperfect and bellicose state of Seaborn's own American society they eject him, and, upon his return to the surface, his stories are not believed. A 1965 facsimile attributed the work to Symmes himself, who had indeed referred to a theorized interior continent as 'SYMMESONIA' (L96). This attribution has been maintained in some recent scholarly works, although Everett Bleiler called it 'not tenable, since, among other reasons, the work obviously satirizes Symmes'.[32] Peter Fitting points

---

[28] Kris Fresonke and Mark Spence, eds., *Lewis & Clark: Legacies, Memories, and New Perspectives* (Berkeley: University of California Press, 2004).

[29] [James McBride], *Symmes's Theory of Concentric Spheres; Demonstrating that the Earth is Hollow, Habitable Within, and Widely Open About the Poles* (Cincinnati, OH: Morgan, Lodge and Fisher, 1826), 20, 136.

[30] Hester Blum, 'John Cleves Symmes and the Planetary Reach of Polar Exploration', *American Literature*, 84 (2012), 243–71 (249, 251).

[31] Blum, 256.

[32] Everett F. Bleiler, *Science-Fiction: The Early Years* (Kent, OH: Kent State University Press, 1990), 662. For modern works attributing *Symzonia* to Symmes, see David Seed, ed., 'Breaking the Bounds: The Rhetoric of Limits in the Works of Edgar Allan Poe, his Contemporaries and Adaptors', in *Anticipations: Essays on Early Science Fiction and Its Precursors* (Liverpool: Liverpool University Press, 1995), 75–97 (77); and Peter W. Sinnema, 'Gender Trouble in the Hollow Earth: *Pantaletta, Mizora*, and

out that an least one early critic thought Symmes the author and concludes that the work is a satire on United States politics that 'includes a defence of Symmes's theory'.[33] This balanced judgement appeared as early as 1866, when one New York litterateur called *Symzonia* 'semi-ridicule, semi-illustration'.[34] Yost, who sees the pseudonymous 'Seaborn' as 'affiliated in some way with Symmes', likewise stresses that most hollow-earth works should not be characterized as 'parodies' of hollow-earth thinking, whatever satire they may employ.[35] Certainly, even the po-faced Symmes would have sympathized with *Symzonia*'s apparently face-value condemnation of 'abstract theories' and its praise of 'practical knowledge' and 'common sense'.[36] The evidence against Symmes's authorship is strong but the case is symbolic: writing a book suspended between seriousness and parody would become characteristic of the hollow-earth tradition. This tradition also retained Symmes's enthusiasm for participatory approaches to science.

Resolved to Revive the Theory

After Symmes's death in 1829, attempts to promote his theory dried up, but Symmes holes remained familiar landmarks in American culture. Most allusions to these polar openings were facetious, but they were occasionally associated with introspection and even insight. Henry David Thoreau's *Walden* (1854), for instance, proposed that readers 'find some "Symmes' Hole" by which to get inside' their own 'being' to attain self-knowledge.[37] More literally, a vast treatise called *The Hollow Globe; or The World's Agitator and Reconciler* (1871) argued for a Symmesian hollow earth on the basis of the clairvoyant insights of a Spiritualist medium called Manly L. Sherman. A more mainstream catalyst of the hollow-earth revival was a sympathetic 1873 article in the widely circulated *Atlantic Monthly*. The author, 'P. Clark' of Brooklyn, had attended a lecture by Symmes at Union College at Schenectady, New York, in 1827. The article called him 'a bold and original thinker' whose 'popular' theory, Clark claimed, with some exaggeration, had 'commanded the attention of the learned and scientific men of the day'.[38]

---

the American Antifeminist Romance', *ESQ: A Journal of Nineteenth-Century American Literature and Culture*, 69 (2023), 201–34 (209). For an eighteenth-century precursor to the hollow-earth satire of *Symzonia*, see the discussion of Ludvig Holberg's *The Journey of Niels Klim to the World Underground* (1741) in Fitting, 37–40.

[33] Fitting, 103, 106. See also Gretchen Murphy, '*Symzonia*, *Typee*, and the Dream of U.S. Global Isolation', *ESQ: A Journal of the American Renaissance*, 49 (2003), 249–84.

[34] [William Henry Bogart], *Who Goes There? Or, Men and Events* (New York: Carleton, 1866), 111

[35] Yost, 11, 26.

[36] 'Adam Seaborn', *Symzonia: A Voyage of Discovery* (New York: J. Seymour, 1820), 157, 161.

[37] Henry David Thoreau, *Walden; or, Life in the Woods* (Boston, MA: Ticknor and Fields, 1854), 344.

[38] P. Clark, 'The Symmes Theory of the Earth', *Atlantic Monthly*, 31 (1873), 471–80 (471).

Growing interest in Symmes would have been buoyed by recent interest in (non-Symmesian) subterranean worlds by several of the era's most popular novelists. Jules Verne's *Voyage au centre de la Terre* was translated into English in the early 1870s, while Edward Bulwer-Lytton's *The Coming Race* (1871) depicted an American's encounter with a quasi-utopian underground society.[39] The underground was in the air, so to speak.

Not long after Clark's article appeared, one of Symmes's sons, Americus Vespucius Symmes (named in an apparent lapse of his father's aforementioned cosmopolitanism and modesty), took up the baton. Americus, a farmer living near Louisville, had erected a monument to his father's theory some time before 1871, possibly as early as the 1840s.[40] Early in 1875, Americus began lecturing in Louisville, 'reviving the theory of his father', one newspaper reported, 'as a filial duty'.[41] On his deathbed, reported the *Louisville Courier-Journal*, the elder Symmes had beseeched Americus, then seventeen, 'to investigate the theory', and now 'at this advanced period of his life' Americus had 'resolved to revive the theory'.[42] Heterodox geoscience ran in the family: Americus's younger brother, also named John Cleves, wrote to Charles Darwin in 1874, informing the British naturalist that he was developing a theory 'as weighty, *phy*sically, as yours is, *psychi*cally', namely the theory 'that "all atoms revolve"'. He believed the rolling atom hypothesis would 'show the truth or falsity' of his father's hollow-earth speculations.[43]

The main product of this filial piety was a short book by Americus, *The Symmes Theory of Concentric Spheres* (1878). Like his father, Americus was no author. That the book is largely reprinted but entirely unattributed material is often noted, although the sheer variety of its sources has not previously been recognized. The book, supposedly 'compiled' by Americus '[f]rom the writings of his father', begins with a life of Symmes taken from James McBride's posthumous *Pioneer Biography: Sketches of the Lives of Some of the Early Settlers of Butler County, Ohio* (1871).[44] The first part of the exposition of the theory derives from Clark's article (13–24), interlaced with a paragraph from an *American Quarterly Review* article on McBride's 1826 book (14–15), before turning to a lightly edited resetting of that

---

[39] For a helpful contextualization of these works, see Fitting, *Subterranean Worlds*, chapters 12 and 13.
[40] James McBride, *Pioneer Biography: Sketches of the Lives of Some of the Early Settlers of Butler County, Ohio*, 2 vols (Cincinnati, OH: Robert Clarke, 1869–1871), II, 248.
[41] 'Polar Voids', *Chicago Daily Tribune*, 22 February 1875, 3. Reprint from the *Louisville Ledger*.
[42] 'The Undiscovered World', *Helena Weekly Herald* [MT], 2 December 1875, 3. Reprint from the *Louisville Courier-Journal*.
[43] 'Letter no. 9269', *Darwin Correspondence Project* <https://www.darwinproject.ac.uk/letter/?docId=letters/DCP-LETT-9269.xml> [accessed on 6 January 2023].
[44] Americus Symmes, *The Symmes Theory of Concentric Spheres, Demonstrating that the Earth Is Hollow, Habitable Within, and Widely Open about the Poles* (Louisville, KY: Bardley & Gilbert, 1878), title page. Subsequent references included in text.

book itself (24–49).⁴⁵ Nine pages of quotations from the Arctic explorer William Morton follow (up to 58). The only apparently original, although perhaps simply unidentified, additions are a discussion of the aurora borealis (58–60) and an account of an 1871 Arctic expedition by Charles Francis Hall of Cincinnati (60–64). Towards the end, a soaring peroration is inserted:

> Reason, common sense, and all the analogies in the natural universe, conspire to support and establish the theory, and an examination will prove that it is the most natural and harmonious view of this subject that has ever been presented for the consideration of the human mind. (60)

These lines were taken almost verbatim from Lyon and Sherman's Spiritualist *Hollow Globe*.⁴⁶ The closing words return to Clark's article (65–66). *The Symmes Theory* was a collage.

Ida Elmore Symmes, who dated her father Americus's conversion to the theory to 1874, suggested that the book's 'hasty publication' was an attempt to exploit the 'fabulous prices' for which McBride's by-then antiquarian volume was sold.⁴⁷ Its patchwork composition, aside from dating most of its scientific sources to the dawn of the century, gives it a heteroglossic character. McBride's biography of Symmes had been edited by its publisher, no true believer. As a result, Americus's preface, ingenuously reprinting the publisher's deflating conclusion, was made to declare that the evidence favoured not Symmes holes but 'the more plausible theory of *open polar seas*' (xii). Clearly, despite references to heroic iconoclasts like 'Franklin or Newton' (66), Symmes's family did not believe his theory required the textual integrity and coherence demanded of a Romantic genius, or indeed of a modern scientific author. Rather, the theory had become a communal effort, just like the public journal in the Cincinnati reading room. While Sinnema considers Americus's book 'plagiarized' from McBride, the latter apparently was happy with unattributed reuse of his work (hardly a novelty in the nineteenth-century United States), and the same may have applied to the sympathetic Clark.⁴⁸ All data, once public, even if contradictory, could be subsumed into the rhizomic Symmes project.

The theory received further exposure in a *Harper's Magazine* article of 1882. The author was 'indebted to Mr. Americus Symmes . . . for much interesting information', although Ida Elmore Symmes, who later explained that the article

---

⁴⁵ Untitled review of *Symmes's Theory of Concentric Spheres*, *American Quarterly Review*, 1 (1827), 235–53 (239).
⁴⁶ Manly L. Sherman and William F. Lyon, *The Hollow Globe; or The World's Agitator and Reconciler* (Chicago: Religio-Philosophical Publishing House, 1871), 802.
⁴⁷ Ida Elmore Symmes, 'John Cleves Symmes, The Theorist', *Southern Bivouac*, 2 (1886–86), 555–66, 621–31, 682–93 (690–91).
⁴⁸ Sinnema, '10 April 1818', n.p.; Robert Clarke, 'Publisher's Notice', in McBride, *Pioneer Biography*, I, i–iii (iii).

was written around 1878, complained that it contained 'numerous' inaccuracies and she suspected 'satire' at points.[49] Ida's own detailed articles in the *Southern Bivouac* demonstrated a more rigorous approach to writing the history of her much-mythologized grandfather. Despite Ida's reservations about 'satire', the *Harper's* journalist had painted an endearing picture of the elder Symmes. The writer quoted McBride on Symmes's 'bright blue eye[s] that often seem fixed on something beyond immediate surrounding objects', before delving into speculative terrain, imagining 'splendid visions of untold wonders' floating 'through his nightly and daily thoughts'.[50] This Gilded-Age version of Symmes, remembering 'neglected Columbus' at every setback, was 'human, faulty enough, but still a self-denying, steadfast man—a man with a purpose'.[51]

Presented to the World as Romance

Copycat thinkers emerged in the decades after Americus and Ida Symmes recirculated hollow-earth theory. Among the confident new breed of Symmes revivalists can be found Frederick Culmer of Salt Lake City, an English-born Mormon who penned *The Inner World* (1886); Franklin Titus Ives, retired merchant of Meriden, Connecticut, and author of *The Hollow Earth* (1904); and New-York-based William Reed, not merely author of *The Phantom of the Poles* (1906) but also 'one of the best known among the fire insurance men of the country'.[52] Although biographical details are sparse, these were mostly businessmen—eminent men, albeit not in the scientific world—for whom the hollow earth held tangible material prospects in addition to the kudos of geological iconoclasm. Ives's interest in the water-circulating Symmes holes appears to have been partly agricultural in nature, and his volume was advertised in *Green's Fruit Grower*.[53] Reed's monograph led to the establishment of a 'William Reed Hollow Earth Exploring Club' in 1908, its membership consisting of prominent New York figures 'prepared to spend $1,500,000' in search of the polar openings.[54] Although it led to nothing in the face of almost immediate claims that the (solid) poles had been reached—the North by Robert Peary in 1909 and the South by Roald Amundsen in 1911—this was the closest thing to a hollow-earth expedition plan since Symmes's day.

These texts were almost all prefaced with photographic portrait frontispieces, demonstrating the authors' sober appearances and unflinching willingness to be

---

[49] 'Symmes and His Theory', *Harper's New Monthly Magazine*, 65 (1882), 740–44 (740); Ida Elmore Symmes, 693.
[50] 'Symmes and His Theory', 741, 743.
[51] 'Symmes and His Theory', 744.
[52] 'Franklin Titus Ives', *New York Times*, 31 January 1910, 7; 'Going to Look for a Big Hole at the Top of the World', *Marion Daily Mirror* [OH], 25 April 1908, 9.
[53] 'Book Notices', *Green's Fruit Grower*, 30 (1910), 2.
[54] 'Going To Look for a Big Hole', *Marion Daily Mirror*, 9.

associated with a maverick geological theory (Figure 3.3). All professed originality, having lit upon the theory without the help of Symmes. They explained the hollow-earth world-system in judicial language liberally peppered with propositions, lists of facts, pointed rhetorical questions, and suggestive quotations cherry-picked from travelogues and periodicals. Hierarchies of expertise had hardened since the time of Symmes and many of these hollow-earthers recognized, as put most bluntly by Ives, their perceived status as 'Cranks', albeit ones ranked alongside 'such Cranks as Copernicus, Galileo, Columbus'.[55] The appeal of colonizing a resource-rich interior world was now more apparent than ever. While Symmes's fantasy was born in the continental expansionist age of Lewis and Clark, this renewed interest emerged when the closure of the American frontier ignited reflections on the colonial future of the United States.[56] Few authors failed to stress the economic value of the concave land within the Earth, whether annexed by the US or magnanimously donated to the civilized world at large.

Despite the authors' exploitation of genuine gaps in geological and geographical knowledge, their defiance of established scholarly apparatus, not to mention their anti-Newtonianism, indicated their preference for addressing the public sphere rather than scientific specialists. In response, these books were mostly ignored by the major scientific periodicals. Demeaning acknowledgements of the kind made in the *Bulletin of the American Geographical Society*, which declared one such book '[m]ore pathetic than amusing', were the alternative.[57] Many of these publications were widely reviewed in magazines for general readers, however, rarely enjoying endorsement from bemused critics but often praised for their thought-provoking individualism.[58] For scientifically inclined members of the public, journalists' lack of interest in boundary policing was frustrating. In August 1906, teenage amateur astronomer H. P. Lovecraft wrote to the *Providence Sunday Journal* to refute a sympathetic notice of Reed's book, asserting that his 'novel' theory was 'certainly not possible'.[59] For purveyors of popular newspapers, though, there was little reason to discredit hollow-earthers. During the 1890s, the sensationalized New Journalism, also known as 'Yellow Journalism', had achieved unprecedented sales by making news as exciting as fiction, so it is unsurprising to see hollow-earth subject matter receive uncritical and lavishly illustrated spreads in popular newspapers.[60]

---

[55] Franklin Titus Ives, *The Hollow Earth* (New York: Broadway Publishing, 1904), 2.
[56] Yost, 69.
[57] Untitled review of *A Journey to the Earth's Interior*, by Marshall B. Gardner, *Bulletin of the American Geographical Society*, 46 (1914), 543.
[58] For instance, see 'The Phantom of the Poles', *The Four-Track News*, 11 (1906), 172.
[59] H. P. Lovecraft, *Collected Essays Volume 3: Science* (New York: Hippocampus Press, 2004), 21.
[60] 'The Phantom of the Poles', *Newtown Bee* [CT], 21 February 1908, 5, 12; 'Is There a World Inside of the World?', *The Age-Herald* [Birmingham, AL], 3 August 1913, n.p. For the New Journalism, see Karen Roggenkamp, *Narrating the News: New Journalism and Literary Genre in Late Nineteenth-Century American Newspapers and Fiction* (Kent, OH: Kent State University, 2005), chapter 2.

HOLLOW GROUND 121

Figure 3.3 (A) Americus Symmes, *The Symmes Theory of Concentric Spheres, Demonstrating that the Earth Is Hollow, Habitable Within, and Widely Open about the Poles* (Louisville, KY: Bardley & Gilbert, 1878), frontispiece © The British Library Board, 8561.cc.5. (B) William Reed, *The Phantom of the Poles* (New York: Walter S. Rockey, 1906), frontispiece. (C) Marshall B. Gardner, *A Journey to the Earth's Interior; or, Have the Poles Really Been Discovered* (Aurora, IL: Marshall B. Gardner, 1913), frontispiece. Frontispieces depicting the horizon-scanning resolve of hollow-earthers, starting with an engraving of John Cleves Symmes Jr himself.

Figure 3.3 Continued

HOLLOW GROUND 123

Figure 3.3 Continued

However sensational their claims, Culmer, Ives, and Reed themselves wrote in the no-nonsense tradition of Symmes rather than that of the author of the slyly evasive *Symzonia*. Nonetheless, reviewers of these books consistently suspected irony, even if they could not pinpoint it. Discussing *The Phantom of the Poles*, an intrigued critic in the *American Journal of Clinical Medicine* failed to determine whether 'Mr. Reed is in real earnest, or only fictitiously so'.[61] Although Reed may, perhaps, have found this sceptical reading of his ideas irritating, other neo-Symmesians found that the ambiguity between fiction and non-fiction, and even earnestness and irony, could be a feature, rather than a bug.

An early example of this approach is encapsulated in the title of Washington L. Tower's *Interior World: A Romance Illustrating a New Hypothesis of Terrestrial Organization, with an Appendix Setting Forth an Original Theory of Gravitation* (1885). Tower's preface announced two goals: highlighting the 'original theory' of gravity that made hollow-earth theory work and presenting a wholesome 'narrative entertaining to boys'.[62] The main body of the text is a didactic boy's own adventure, interspersed with pious poems and featuring conversions to Christianity, damsels in distress, and enterprising settler colonialism. The plot follows a sailor, Anian Belivast, and a cabin boy, Rolin Widmore. In the wilds of Oregon, also the location of the publisher (based in the city of Oakland, rather than the wilderness), the two discover a chasm leading to the reverse surface of the Earth. This aurorally lit world contains animals thought extinct, including 'gigantic birds' (56) and 'a mastodon or mammoth' (71). Tower's theorization of the hollow earth is incorporated into the dialogue and the story's final sentence contains a twist: Belivast has 'permitted a sketch of his marvelous history to be published, but insists that it shall be presented to the world as romance' (172).

Belivast's open secret was surely not a serious claim, although this conclusion did foreground the notion that Tower's plot was not strictly impossible. An appendix followed, however, recognizing that the story's contradictions of 'received teachings of science' needed to be justified by outlining 'proof' (174) for Tower's theory of 'negative gravitation' (181). Inverted gravity explained why the Earth had formed in a hollow manner, and why walking on the concave interior of the outer crust was physically possible. One of the book's few reviewers, writing in the Boston *Literary World*, found the juxtaposition of content downright dangerous: while the romance 'may give harmless amusement' to children, the appendix 'would be mischievous to young readers, who might mistake it for earnest opinion, and be perplexed and confused by it'.[63] The critic, then, saw the appendix as a piece of too-convincing literary verisimilitude, but Tower seems actually to have been professing his 'earnest opinion'. The adventures of Belivast and Widmore should

[61] 'Reed's "Phantom of the Poles"', *American Journal of Clinical Medicine*, 14 (1907), 931.
[62] Washington L. Tower, *Interior World: A Romance Illustrating a New Hypothesis of Terrestrial Organization, with an Appendix Setting Forth an Original Theory of Gravitation* (Oakland, OR: Milton H. Tower, 1885), 3. Subsequent references included in text.
[63] 'Fiction', *Literary World*, 16 (1885), 405.

be understood, in Aït-Touati's terms, as a factualizing rather than fictionalizing narrative: a thought experiment intended to cultivate suspension of disbelief and pave the way for acceptance of the author's anti-establishment science. That the romance was intended to be returned to with an open mind was indicated by the alphabetization of its events in the book's index, a rare feature in a work of fiction.

A similar interdependence of theoretical treatise and imaginative thought experiment was produced by Marshall B. Gardner, a sewing-machine entrepreneur of Aurora, Illinois.[64] His slim but elaborate neo-Symmesian treatise, *A Journey to the Earth's Interior; or Have the Poles Really Been Discovered*, twenty years in the making, was self-published in 1913. Gardner demanded 'a fair hearing' for his theory in the face of the 'conservatism of scientists' towards 'discoveries . . . made independently of the great universities'.[65] At the end of his brief non-fiction exposition, with the usual talk of polar curvatures and mammoth migrations, Gardner proposed to turn full 'attention to the interior of the earth and ascertain what may be found there . . . by taking the following imaginary journey'.[66] He then guided readers in first-person plural through the probable sights of the polar holes and the interior world. The book's factualizing link to imaginative fiction was hinted at in the title, which closely echoed *A Journey to the Interior of the Earth*, one of the early translated titles of Verne's famous romance. Admitting the need to focus on 'probable facts', even in an imagined voyage, the author allowed himself leeway to 'speculate' that the interior may contain 'strange animals' perhaps even larger 'than the prehistoric mammoth or mastodon', as well as 'a race of people whose existence is entirely unknown'.[67] This moment of 'idle speculation' over, Gardner continued his equally imaginary, but more evidence-based, 'journey southward'.[68]

When expanding *A Journey to the Earth's Interior* for a new edition, Gardner proudly reprinted a letter from the first edition's most high-profile reader. Gardner explained that Arthur Conan Doyle, recipient of a gift copy of the 1913 edition, was 'one of the most scientifically-minded men in England', as demonstrated in his 'Sherlock Holmes detective stories'.[69] Doyle's letter, dated 31 January 1914, expressed polite incredulity, claiming that the theory's capacity to 'explain so many facts' would have made him 'a convert' were it not for Peary and Amundsen's recent discoveries of the poles (399). Given that Gardner believed the discoverers to be mistaken, he presented the novelist's letter as an indirect endorsement. Significantly, his book had followed hot on the heels of *The Lost World* (1912), Doyle's tale of adventure on an isolated South American plateau populated with

---

[64] Kafton-Minkel, 82.
[65] Marshall B. Gardner, *A Journey to the Earth's Interior or Have the Poles Really Been Discovered* (Aurora, IL: Marshall B. Gardner, 1913), 10.
[66] Gardner (1913), 53.
[67] Gardner (1913), 62.
[68] Gardner (1913), 63.
[69] Marshall B. Gardner, *A Journey to the Earth's Interior; or Have the Poles Really Been Discovered*, rev. edn (Aurora, IL: Marshall B. Gardner, 1920), 398. Subsequent references included in text.

prehistoric animals. As their exchange antedated Doyle's public conversion to Spiritualism, and thus his reputation for openness to fringe science, it was *The Lost World* that seems to have convinced Gardner he would be a potential convert: Gardner stressed that Doyle 'has read widely along geological lines in his search for material for some of his romances' (398). As readers of *The Lost World* will recall, its spectacular events refute the declaration, made by a weary newsman at the book's opening, that the 'blank spaces in the map are all being filled in, and there's no room for romance anywhere'.[70] Gardner's own geological reading had convinced him that Earth held even more 'room for romance' than Doyle's fictional plateau.

Rather than sobering in response to the increasing scrutiny of the poles, the imaginary subterranean voyage in Gardner's expanded 1920 edition grew in proportion to over twenty pages. He again declared that this voyage would not 'invent any new facts', but Gardner's ambitions were now slipping into fictionalizing territory (321). Slipping from modal to present to future tense, the virtual voyage resembled those employed in popular geological writing: the sea inside the Earth 'is fairly alive with organisms' (325) and 'we shall certainly see species of shell fish that we have previously only seen in their fossilized condition' (326). Gardner, in the manner of Hugh Miller, John William Dawson, and Elizabeth Denton, took his readers on a virtual tour of a hollow ecosystem: 'Occasionally under the dense undergrowth we should espy a serpent or serpent-like creature wending its silent way' (330). Attempting to 'penetrate our new world a little more systematically' (335), and straying further from his 'facts', Gardner even encountered dinosaurs, those 'cretaceous' reptiles resembling 'the kangaroo' (336). While mammoths were a staple of hollow-earth monographs, such a diverse range of survivals was unheard of outside contemporary novels, and bore even closer resemblance to those of Edgar Rice Burroughs (to whom I will later turn) than to Doyle's *The Lost World*.

The close relationship between hollow-earth monographs and lost-world romances, whether on the fictionalizing or factualizing end of the spectrum, is suggestive. For authors wishing to replicate the success of the first lost worlds proper, H. Rider Haggard's *King Solomon's Mines* (1885) and *She* (1886), it had become de rigueur to frame fantastic events with hoax-like prefaces, realistic images, and mock-scholarly footnotes.[71] *The Lost World* (a late arrival, despite retrospectively giving the genre its name), for example, was presented by Doyle as a mock-journalistic exposé. Doyle's exchange with Gardner suggests that the latter saw these works of verisimilar fantastic fiction as the thin end of the wedge when

---

[70] Arthur Conan Doyle, *The Lost World* (London: Hodder and Stoughton, 1912), 19. For fact and fiction in *The Lost World*, see Richard Fallon, *Reimagining Dinosaurs in Late Victorian and Edwardian Literature: How the 'Terrible Lizard' Became a Transatlantic Cultural Icon* (Cambridge: Cambridge University Press, 2021), chapter 4.

[71] Saler, chapter 2.

it came to opening readers' minds. Hollow-earth theory's proximity to romance is complicated further by the fact that, during this period, a flood of derivative lost-world fiction emerged in which the hollow earth acted as a fantastic setting akin to Africa, South America, or the poles. William Jenkins Shaw even republished his first attempt at a hollow-earth romance, *Under the Auroras* (1888), as *Cresten, Queen of the Toltus* (1892), hammering home its potentially saleable similarities with Haggard's sensational *She*. Meanwhile, other authors took the *Symzonia* route, using the hollow earth's interior, with its symbolic connotations of inversion and depth, as a novel setting for utopian speculations—but without demonstrating much investment in vindicating Symmes. Likely the first in the vein was Mary E. Bradley Lane's feminist utopia *Mizora*, serialized in the *Cincinnati Commercial* between 1880 and 1881.[72]

Although they did not typically contain endorsements of Symmes's theory, hollow-earth utopias and lost-world romances, framed as rediscovered manuscripts and testimonials, contributed to the challenging complexities of tone that had been introduced to Symmesian literature by 'Adam Seaborn'. Nineteenth-century readers were attuned to parsing fiendishly knotty narratives like these. For instance, John De Mille's *A Strange Manuscript Found in a Copper Cylinder*, serialized in 1888, dramatizes an unresolved debate over the interpretation of the titular manuscript, the possibly satirical story of an adventurer's encounter with a dystopian society living in a deep (but not technically Symmesian) polar valley.[73] However, nobody ever believed that books like this were intended as making scientific claims. The same cannot be said for hollow-earth texts like *The Smoky God*, a lost-world romance by Los Angeles banker and real estate broker Willis George Emerson, serialized in the *National Magazine* (1907–8).[74] Emerson's story of the discovery of the Garden of Eden inside a Symmes hole, presented as a posthumous manuscript, is accompanied by so many technical footnotes that modern scholars have disagreed on the author's level of investment in the theory.[75] While Emerson's contemporaries seem to have read it simply as employing the 'vraisemblance of vivid reality', later hollow-earthers saw it as prophetic.[76] But this is not just a case of modern readers misunderstanding earlier literary stylings. After all, we have already seen evidence of contemporaries' confusion over *Symzonia*, Tower's *Interior World*, and Reed's *Phantom of the Poles*. 'How much of the book is to be taken seriously', remarked a reviewer of Ives's *The Hollow Earth*, 'Mr. Ives does not say'.[77] Therein lies what made hollow-earth writing a unique contribution to the

---

[72] Christine Mahady, 'No World of Difference: Examining the Significance of Women's Relationships to Nature in Mary Bradley Lane's *Mizora*', *Utopian Studies*, 15 (2004), 93–115.
[73] Crawford Killian, 'The Cheerful Inferno of James De Mille', *Journal of Canadian Fiction*, 1 (1972), 61–67.
[74] Kafton-Minkel, 247.
[75] Kafton-Minkel, 208, 247; Yost, 245.
[76] William Jackson Armstrong, 'The Smoky God', *Los Angeles Times*, 30 August 1908, III13.
[77] Untitled review of *The Hollow Earth*, by Franklin Titus Ives, *Current Literature*, 37 (1904), 476.

culture of borderline palaeoscience: its authors challenged readers to re-examine their assumptions about how writing can communicate scientific truth.

Interior World of the Soul

So far, we have not seen religious belief take the central motivating role it had for the protagonists of my previous chapters, but this should not be taken as a sign of absence. While it was often peripheral in technical hollow-earth writings, an early modern Christian or deist conception of God's providential use of space and natural resources typically underlay heterodox geotheoretical convictions.[78] In the religious ferment of the *fin de siècle*, the hollow earth also became attached to more novel religious questionings. This, too, was shared with the lost-world tradition: many of the figures Saler cites as proponents of the 'ironic imagination', including Haggard and Doyle, were, in their personal lives, by no means sceptics or agnostics, despite the appeal Saler argues their works held for disenchanted modern mass readerships.[79] Even *The Lost World*, for instance, intertwined Doyle's ludic humour with his desire to promote fringe and anti-materialistic science.[80] Likewise, authors of hollow-earth romances, rather than constructing imaginary worlds to distract from the godless disenchantment of reality, were more likely to use them for exploring religious belief in the modern world.

The trope of spiritually enlightening underground journeys, *katabasis*, is an ancient one that received regular scientific updates in the nineteenth century.[81] The most influential jumping-off point was Dante Alighieri's fourteenth-century *Inferno*, a poem with considerable popularity in nineteenth-century America.[82] Updating the Catholic moral confidence of Dante's pilgrimage through hell using modern religious syncretism, the likes of which we saw in chapter 2, authors of hollow-earth stories worked through the challenges science presented to orthodox belief. For instance, Mark Twain's brother, Orion Clemens, an independent thinker who was excommunicated from his local Presbyterian branch, began but never finished an adventure set inside a Symmes hole. Here, the Symmesian cosmogony acted as the setting for what Nathaniel Williams calls 'a strident

---

[78] For the origins of this belief, see Sinnema, 'We Have Adventured To Make the Earth Hollow', 426-27. See also McBride's scrapbook, ANSDU, C12; and Ives, 59, 63.

[79] Saler, chapters 2 and 3.

[80] Fallon, *Reimagining Dinosaurs*, chapter 4.

[81] Rachel Eames, 'Geological *Katabasis*: Geology and the Christian Underworld in Kingsley's *The Water-Babies*', *Victoriographies*, 7 (2017), 195-209.

[82] Dennis Looney, 'Dante Alighieri and the Divine Comedy in Nineteenth-Century America', in *The Routledge History of Italian Americans* (New York: Routledge, 2018), 91-104.

articulation of religious doubt'.[83] Even Lane (another Cincinnati native), in *Mizora*, gradually shifts focus from the novel's now better-known feminist argumentation to the crisis of faith that drives the novel's final chapters, when the Christian protagonist struggles to accept the atheism of the otherwise idealized Mizoran society.[84]

Others, such as William Richard Bradshaw in *The Goddess of Atvatabar* (1892), used the lost-world genre to adapt more conservative religious conclusions to the tastes of urbane readers. A preface by novelist Julian Hawthorne, son of the more famous Nathaniel, situated Bradshaw's book in the Haggard tradition, presenting it as a masculine romance exploring 'an interior world of the soul', inspired by 'magical achievements of theosophy and occultism' but ultimately revealing their superficiality.[85] The novel's hero is Lexington White, whose name encapsulates his role as Nordic inventor-adventurer—an all-American stock character of the period. White finds a race of spiritually advanced beings inside the hollow earth, whose worship of 'the aggregated universal human soul' is initially presented as seductively attractive (84). White, however, soon comes to see Atvatabarerse society as excessively effeminate. In subduing it, he is assisted by the British and American navies, who have arrived at the centre of the Earth to claim its incredible mineral wealth. White hopes soon 'to see the Christian faith rule the souls of those who had so recently worshipped themselves under the guise of . . . the universal human soul' (316). He predicts that Christianity, backed up by Anglo-American military technology, will ultimately absorb the weaker Atvatabarese religion. If White's faith is tested by Atvatabar's stand-ins for the attractive new religious systems circulating during the *fin de siècle*, including Theosophy, it is refined in the purgatorial flames. Uncoincidentally, Dantesque allusions permeate Bradshaw's narrative (e.g. 114, 132, 209).

Other Christians tapped into the vogue for hollow lost worlds to provide scientific support for specific interpretations of the Bible. This approach was particularly attractive to members of that quintessentially American branch of Christianity, the Church of Jesus Christ of Latter-Day Saints, founded in the early nineteenth century by Joseph Smith. Culmer, the British-born Mormon whose 1886 *Inner World* I mentioned above, used the hollow-earth framework to explain a long-standing biblical mystery: the location of the Ten Lost Tribes of Israel. This subject, introduced by a cryptic passage of the Book of Kings, was of particular significance to Mormons, whose traditions suggested that these dispersed tribes

---

[83] Nathaniel Williams, *Gears and God: Technocratic Fiction, Faith, and Empire in Mark Twain's America* (Tuscaloosa: University of Alabama Press, 2018), 108–9.
[84] Mahady, 108. For a valuable overview of religious hollow-earth narratives, see Yost, chapter 3.
[85] Julian Hawthorne, 'Introduction', in William R. Bradshaw, *The Goddess of Atvatabar; Being the History of the Discovery of the Interior World and Conquest of Atvatabar* (New York: J. F. Douthitt, 1892), 9–12 (9, 11–12). Subsequent references included in text refer to the body of Bradshaw's novel.

survived in the icy north.[86] For Culmer, the existence of habitable land beyond the North Pole (inside a Symmes hole, that is) explained how these tribes had managed comfortably to remain lost.

Ten years after Culmer's investigations, another Mormon writer, De Witt C. Chipman, elaborated upon this syncretism of Symmes and Joseph Smith in a romance called *Beyond the Verge: Home of Ten Lost Tribes of Israel* (1896). Like Tower's *Interior World*, Chipman's book leant far to the factualizing end of the hollow-earth fiction spectrum, accompanied by a non-fictional appendix. Following a frontispiece that depicted the author in horizon-scanning pose, Chipman's introduction offered up to 'the considerate judgment of all fair-minded and intelligent investigators' his theory for the survival, inside the hollow earth, of the titular lost tribes.[87] The main body of Chipman's text is a pseudo-historical version of lost-world romance set two thousand years ago, following a Native American called Nardo, who, through an improbable conversion, has become a Christian. Nardo encounters the tribes of Israel as they migrate northwards through America towards the pole and joins them on their quest. Recalling the Spiritualist origins of *The Hollow Globe*, an Israelite priest learns through visionary communication with God that Earth is 'a hollow sphere' accessible by 'open' poles (66). Nardo and the tribes subsequently enter the open pole to reach an inner-earthly paradise. Chipman's theological conservatism, including belief in the Mosaic authorship of Genesis, is flavoured with syncretic elements, not least acknowledgement of the historicity of the 'Atlantis mentioned by Plato' (57). More radically, it is revealed that the Himalayan 'Mahatmas' so integral to the lore of the Theosophical Society are, in fact, Israelites from inside Earth (157).

Although superficially adopting Haggardian subject matter and style, Chipman, a patent attorney living in Anderson, Indiana, differed from fellow lost-world romancers in his minimal commitment to verisimilar narrative absorption.[88] Furthermore, rather than complicating factual exposition with fiction, in the manner of Marshall, he complicated his fiction through recourse to fact. An omniscient narrator interrupts the story without compunction, as when he announces that 'the natural thread of our story is here broken to introduce scientific and historic evidence' on Symmesian theory (72). The digression that follows cites Americus's book as well as articles in *Popular Science Monthly* and local newspapers. Chipman never expects his readers to suspend their disbelief, a privilege one might expect to be enjoyed by a novelist, ensuring that, whenever 'credulity is severely tested' (95), secondary sources are at hand. Indeed, despite his media-savvy recognition

---

[86] Elizabeth Fenton, *Old Canaan in a New World: Native Americans and Ten Lost Tribes of Israel* (New York: New York University Press, 2020), 125–26, 133–40.

[87] De Witt C. Chipman, *Beyond the Verge: Home of Ten Lost Tribes of Israel* (Boston: James H. Earle, 1896), 6. Subsequent references included in text. For Chipman and American theories about the lost tribes, see Fenton, *Old Canaan*, chapter 6.

[88] Fenton, 171.

that the tale could be marketed as a lost-world romance, Chipman wielded an artfully artless style. His apparently blasé attitude towards his own plot is demonstrated not just in scientific digressions, but also when the narrator ventures 'to quote some poetry about the stars', admitting that this has 'no relation to our story' (90).

In addition to engaging the general reader's ability to evaluate Symmes's theory, and testing their tolerance for poetic non sequiturs, Chipman promotes local science, citing, for example, one 'McKee' from his own home city of Anderson, who makes 'photographs by electricity, and after night' (238). This McKee, of course, is far removed from the two thousand-year-old story of Nardo's adventures. The heterogeneous content and sparsity of literary illusionism in *Beyond the Verge* cannot simply be attributed to the author's naivety, although he was not a professional writer. Rather, Chipman's presentation embodies his fusion of practical, accessible science and a 'simple faith' in the Bible's literal accuracy (5). The book, employing fiction but never deception, is a testament to the marriage of participatory palaeoscience and a defence of the Bible's historicity, but it was the hollow-earth literary tradition, in which truth claims were bolstered by infusions of fiction, that facilitated this union.

Sarcasm Deep

Chipman's ingenuous style makes for an interesting contrast with John Uri Lloyd's labyrinthine *Etidorhpa; or The End of Earth: The Strange History of a Mysterious Being and The Account of a Remarkable Journey as Communicated in Manuscript to Lllewellyn Drury Who Promised to Print the Same, but Finally Evaded the Responsibility which Was Assumed by John Uri Lloyd* (1895). Lloyd was a successful pharmacist, inventor, and, from the 1870s on, a leading figure in Eclectic medicine (Figure 3.4).[89] This populist branch of American herbal medicine stemmed from the same anti-establishment culture that had produced the psychometrist Joseph Rodes Buchanan. Indeed, the two men knew each other (although Buchanan's career at the Eclectic Medical Institute in Cincinnati, Lloyd's home city, concluded before the latter's time). At the end of his life, Buchanan was still contacting his 'Friend' Lloyd to boast of his continuing ambition 'to master all science'.[90] *Etidorhpa*, Lloyd's story of an occult journey inside the hollow earth, was a sensation in its day, challenging readers to decipher its obscure comments about the nature of science and religion.

---

[89] Michael A. Flannery, *John Uri Lloyd: The Great American Eclectic* (Carbondale: Southern Illinois University Press, 1998), xiii, 27–31.
[90] Joseph Rodes Buchanan to John Uri Lloyd, 18 April [1898 or 1899] and 21 February 1899, John Uri Lloyd Papers, 1849–1936, series VII.6, folder 80, Lloyd Library and Museum, Cincinnati, OH.

Figure 3.4 John Uri Lloyd, *c.*1915–1920, George Grantham Bain Collection, LC-B2-5190-5, Library of Congress Prints and Photographs Division, Washington, D.C. This photograph shows Lloyd at the height of his career, a leader in Eclectic Medicine.

Its narrative, set in the 1850s, recounts how Llewyllyn Drury, a pseudo-version of Lloyd, is forced to hear the story of a supernatural stranger known as I-Am-the-Man-Who-Did-It. Thirty years prior, the Man betrayed the secrets of a

Rosicrucian-masonic society, and as punishment is sent down a Kentucky cavern towards the centre of the hollow earth, where he traverses surreal underground realms of crystals, monsters, and giant mushrooms. During the adventure, the Man converses with various enlightened guides on questions of science and religion, in the process shedding his former agnosticism. In the narrative's 1850s present, the Man uses his hard-won discoveries to challenge Drury's commonplace conceptions of what is scientifically possible. In its fictional aspects, *Etidorhpa* combines with Dante's *Inferno* elements of *A Christmas Carol*, *Alice's Adventures in Wonderland*, *The Pilgrim's Progress*, and *The Rime of the Ancient Mariner*. In its non-fictional aspects, the book belongs with the philosophical works of science dubbed 'reflective treatises' by James A. Secord, especially Carlyle's aforementioned *Sartor Resartus*, which also takes the form of a discussion of the writings of a fictional character.[91]

The book was lavishly illustrated, purportedly to the fictive Man's specifications, by J. Augustus Knapp, later known for his work on Manly P. Hall's occult classic *The Secret Teachings of All Ages* (1928). Saler, noting Lloyd's nested structure and playful paratexts, frames it as a work of ironic imagination, intended to challenge 'modern disenchantment without relinquishing modern rationality'; Lloyd's biographer, Michael A. Flannery, who observes that 'consensus on *Etidorhpa* remains elusive', considers it an attack on 'scientism'.[92] Both readings—ironic enchantment and sceptical satire on modern science—are, in part, convincing, but there is far more to say about their uneasy coexistence in one visionary book.

The work's heterogeneity was a consequence both of circumstance and design. As Lloyd later explained, it was begun, around 1880, not as a novel but as multiple 'studies of different subjects in which I allowed my imagination to run in connection with such scientific problems as were before me'.[93] In the process of synthesizing this material, Lloyd reached out to an acquaintance, Cincinnati litterateur William Henry Venable. Venable recalled that Lloyd's 'experience in writing had been limited to purely practical subjects, mainly pertaining to chemistry and pharmacy', and that he had asked the more experienced man to edit his 'manuscript of a semi-scientific and very discursive story'.[94] Although Venable polished Lloyd's rough material, some of *Etidorhpa*'s most salient attributes were afterthoughts. Lloyd's wife Emma complained that the story included no women, so Lloyd 'threw in' several Bunyanesque scenes in which the Man's temperance is tested by the beautiful Etidorhpa (Aphrodite spelled backwards), an idealized representation of love, leading him to name the book after this character 'as a final thought'.[95] Reflecting the book's somewhat torturous origins, one of Lloyd's several prefaces admitted sole responsibility for '[s]tructural imperfections as well as word

---

[91] Secord, *Visions of Science*, chapter 7.
[92] Saler, 34; Flannery, 119, 124.
[93] John Uri Lloyd, 'A Statement of Fact' (1906), John Uri Lloyd Papers, series XVII.A.28, folder 403.
[94] William Henry Venable to Emerson Venable, John Uri Lloyd Papers, series XVII.A.28 folder 404.
[95] Lloyd, 'A Statement of Fact'.

selections and phrases that break all rules in composition'.[96] Given that he immediately went on to claim responsibility for employing 'ideas nearer to empiricism than to science', however, the admission of literary rule-breaking should be read as a boast and a compositional parallel with the book's scientific iconoclasm (ii).

That Lloyd (in probable contrast with Chipman) could also pen more generically stable novels is indicated by the success of his later series of nostalgic Kentucky romances, including *Stringtown on the Pike* (1900). *Etidorhpa*'s maverick features, therefore, represent a blend of amateurism and strategy. The first edition, published via subscription, was carefully produced to the then-fashionable aestheticist precepts of the Arts and Crafts movement, conspicuously crediting the companies responsible for its 'exquisite presswork and binding' (n.p.), but the book's mystifying mode is introduced right from the start. The halftone facsimile of a letter from Lloyd to his subscribers, given authenticity by crossings-out, addresses those who question 'if a line can be drawn between fact and fiction' and if 'enthusiasm intense is far separated from sarcasm deep', calling them to avoid categorizing the book's contents into either 'speculation and science' or 'sarcasm, ignorance and irony' (n.p.). Contemporaries may have been primed for such paradoxes by the subversive maxims prefacing another recent aestheticist novel, Oscar Wilde's *The Picture of Dorian Gray* (1891). Wilde, however, never met his creation, Dorian: in contrast, Lloyd's main preface, set in his own library, weaves the author himself into the story as the recipient of Drury's manuscript, which, in turn, details Drury's conversations with the Man. The preface's Lloyd-narrator reflects on how books outlive their owners and maintain a ghostlike influence in the world, foreshadowing the authorial spectrality of *Etidorhpa*, over which no figure, fictional or even non-fictional (given Venable's substantial editorial work), could claim sole authorship.

Lloyd's prefatory obfuscations dovetailed with the book's empirical, open-minded attitude towards scientific knowledge. The extent to which the Ohioan pharmacist used hollow-earth theory purely symbolically, and to which he believed in its accuracy, is debatable, but his philosophy of science harkened back to the original Symmes debates. Lloyd called on subscribers to share their 'impressions', including 'adverse criticism' (n.p.), imitating the story's dialectic debates between Drury and the Man, and between the Man and his guides. When Dury first meets the Man, the latter chides him for making indefensibly 'positive statements' on subjects beyond his experience (9). Throughout the novel, after making radical scientific claims, he provides intervals for Drury to go away and 'search the records, question authorities, and note such objections as rise therefrom' (149). In each case, Lloyd provides—and the Man shortly debunks—all materialistic counterarguments to the novel's enchanted science. The most extensive passage of this

---

[96] John Uri Lloyd, *Etidorhpa; or The End of Earth* (Cincinnati, OH: John Uri Lloyd, 1895), ii. Subsequent references included in text.

kind appeared in the eleventh edition of 1901, where Lloyd restored previously deleted chapters in which an 'aged scientific friend' of Drury appears to demolish the Man's claims.[97] When the scientific friend confronts the Man, however, the latter easily gains the upper hand, demonstrating the frailty of establishment science.

The book's scientific claims are multifarious. Peppered throughout are practical, replicable experiments, another classic trait of hollow-earth argumentation. Drury, for instance, mistrusts the Man's description of 'the dried bed of an underground lake' filled with giant 'cubical crystals' of salt (130), the water's evaporation seemingly defying the everyday principle 'that liquids seek a common level'. In response to his scepticism, the Man provides 'a few glass tubes and some blotting or bibulous paper' (135) to refute this principle. A textbook-style annotated diagram illustrates the process of this simple experiment, clearly intended to be testable. Some of the book's most careful readers apparently did just that: T. H. Norton, a University of Cincinnati chemist, wrote that his recreations of the Man's experiments 'showed conclusively' that Lloyd 'introduced into his narrative actual facts'.[98] The submersion of Drury's use of characterization in these expository sections may reflect their pre-novelistic origins as thought experiments intended to challenge chemical commonplaces. These passages function most clearly as truth claims and even attestations of Lloyd's scientific priority. Below a conversation about the substances found in Earth's 'upper atmosphere', 'J. U. L.' [John Uri Lloyd] wrote that the book's theory 'has since been partly supported by the discovery of the element Argon' (in 1894). To refute charges of 'plagiarism' on this score, he supplied a list of witnesses who could attest to having read the manuscript of *Etidorhpa* as far back as 1887 (221).

Etidorhpa is thus interwoven with truth claims regarding physics and chemistry, but Lloyd himself took an active role in heterodox palaeoscience too. Having grown up near Kentucky's fossiliferous Big Bone Lick, he felt confident to propose, controversially, that mammoths had gone extinct only a century before Europeans landed on American shores. He later became president of the Big Bone Lick Association, formed to apply 'objective methods' to the excavation of new specimens, presumably seeking further proof of their recent extinction.[99] Palaeontology, admittedly, takes a back seat in *Etidorhpa* itself, although geology plays both thematic and material roles. These are unfavourable: the Man's superhuman

---

[97] John Uri Lloyd, *Etidorhpa; or The End of Earth*, 11th rev. edn (New York: Dodd, Mead, 1901), 360.

[98] John Uri Lloyd, *Etidorhpa, or The End of Earth*, 2nd edn (Cincinnati, OH: Robert Clarke Company, [1896?]), 378.

[99] John Uri Lloyd, 'When Did the American Mammoth and Mastodon Became Extinct', *Records of the Past*, 3 (1904), clipping, John Uri Lloyd Papers, series IX.54, folder 757; Grace Haller, 'Man and Mammoth at Big Bone Lick, Ky.', *Enquirer Sunday Magazine*, 21 July 1935, clipping, John Uri Lloyd Papers, series IX.43, folder 760.

guide uses geological science as a symbol of shallow learning. Geologists grasp 'feebly the lessons left in the superficial fragments of earth strata' in their 'endeavors to formulate a story of the world's life' (109). By travelling down into Kentucky's subterranean system, the Man takes a more empirical approach to deepening knowledge. His sagely guide, for example, explains that the traditional explanation of volcanoes as channels to the 'molten interior of the earth' was made 'in ignorance' (211–12). A deeper understanding—actually witnessed by the Man, but, the book suggests, deducible from simple physical principles—shows that volcanoes are 'local' conflagrations caused by chemical reaction. This proves that 'the interior of the earth is not a molten mass' (214). The claim is justified on the same non-Newtonian principles as the hollowness of Earth itself, described as a 'sphere of energy' covered in a 'thin' crust, like 'dust on a bubble' (216). The book's characters have no time for geophysicists' untested speculations about Earth's unseen core. As a result, Lloyd, ever the empiricist, presents a Symmesian interpretation as no less probable than more reputable alternatives.

Etidorhpa's reception was lively. Lloyd was especially proud of praise from the actor Henry Irving, who visited Cincinnati on tour with his manager, Bram Stoker, in March 1896. *Etidorhpa* 'pleased' Irving 'to such an extent', Lloyd later claimed, 'as to lead him to purchase many copies from time to time and present them to his friends'.[100] Some less worldly readers took the book at its word. Kentucky writer William Courtney Watts informed Lloyd that a 'friend . . . tells me *he is certain* that he knew I-Am-the-Man and that his name was Waller'.[101] Lloyd's fiction even gained retrospective tangibility through inscription upon the landscape itself: Watts added that a landowner near Bizzell's Mount, a landmark featured in the story, 'had named his cave the "Cave of Etidorhpa"' to exploit a growing tourist trade.[102] In follow-up editions Lloyd reprinted nearly twenty pages of excerpts from reviewers. Most of these were taken from amenable wings of the pharmaceutical and occult press and from Midwestern newspapers. Frames of reference varied from Dante to Verne, but most critics agreed with the *American Druggist and Pharmaceutical Record*, which observed that 'Lloyd has availed himself of the romance form merely as a means of exploiting tentatively philosophical theories and speculation, which, as yet, he is not prepared to openly indorse [sic]'.[103]

Some attempted to diagnose the book's formal and tonal idiosyncrasies more precisely. The Cincinnati *Eclectic Medical Gleaner* declared that the realms of 'uncanny mystery' into which *Etidorhpa* strayed were a trial for the reader, calling 'up a mental tension, and awesome uncertainty, which is not easily compatible

---

[100] John Uri Lloyd, 'Reminiscences of Sir Henry Irving', John Uri Lloyd Papers, series II.2, folder 22.
[101] William Courtney Watts to John Uri Lloyd, 8 May 1897, John Uri Lloyd Papers, series XVII.A.28, folder 404.
[102] William Courtney Watts to John Uri Lloyd, 25 May 1897, John Uri Lloyd Papers, series XVII.A.28, folder 404.
[103] Lloyd, *Etidorhpa*, 2nd edn, 370.

with sanity'.[104] Its form thus enacted a dark night of the scientific soul, withdrawing stable waypoints not just of materialistic knowledge but also of literary genre in order to encourage absolute empiricism. Straying into journals not excerpted by Lloyd, we find that the book's textual presentation could backfire. The British *Saturday Review* complained that Lloyd could 'hardly expect his readers to follow' his experiments and sarcastically implied that the flattering review reprinted from the *Eclectic Medical Gleaner* was, like the book's other mock-verisimilar touches, 'the offspring of his imagination'.[105]

Over subsequent years, Lloyd simplified *Etidorhpa*. Drastic changes appeared in the eleventh edition, published not, like most earlier versions, by the Cincinnati-based Robert Clarke, but by the mainstream New York publisher Dodd, Mead, also responsible for Lloyd's Kentucky romances. In addition to cutting Lloyd's confusing preface, this version relegated the dialogic interludes in the Man's story to the end of the book. These alterations regularized *Etidorhpa*'s various nested narrative levels. Lloyd evidently wished to provide retrospective clarity on his book's 'sarcasm deep' (n.p.): in an unpublished 1906 'Statement of Fact', he confirmed that the 'manuscript left in the author's hands' by Drury 'is altogether a fiction'. Predictably, however, even in explaining his motives he refused to homogenize *Etidorhpa* in the direction of either fictionalization or factualization. 'At the time the book was written', he wrote, 'a number of my friends were wrapped up in what some people choose to call occult studies, and whether "Etidorhpa" may be taken as a bit of sarcasm, or as an attempt at anything else than playfulness, I will leave to the reader'.[106] That Lloyd maintained this ambiguity even when attempting to shed light on his intentions is further evidence for the book's experimental rejection of utilitarian models of scientific communication. Rather than attempting to recruit virtual witnesses to his various claims with maximum textual efficiency, to return to the language of Shapin and Schaffer, Lloyd threw down the virtual gauntlet and challenged readers to pan for the gold in his murky silt. This was a literary technology of provocation.

True to the Imagination

By April 1914, when Edgar Rice Burroughs' lost-world romance *At the Earth's Core* began serialization in Frank Munsey's pulp *All-Story Weekly*, the Symmesian craze was dying down. Peary and Amundsen's widely circulated accounts showed that the poles were not, as Symmes has contended, open channels to Earth's inverted interior. While we have already heard Marshall express scepticism about these

---

[104] Lloyd, *Etidorhpa*, 2nd edn, 380.
[105] Untitled review of *Etidorhpa*, by John Uri Lloyd, *Saturday Review*, 82 (1896), 271–72 (272).
[106] Lloyd, 'A Statement of Fact'.

claims, hollow-earth theory, at least in its Symmesian form, was losing its currency for most heterodox thinkers. Its popularity nonetheless improved, from the point of view of book sales, as it became the imaginative framework for Burroughs' romances about Pellucidar: a land on the underside of the hollow earth, lit by a central sun, filled with dinosaurs, and initially reached by the protagonists with a powerful vehicular drill. Muscular heroes meander through the Pellucidar stories, protecting cavewomen like Dian the Beautiful and uncovering uncanny concave geographies. Born in Chicago, Burroughs undertook numerous financially unsatisfying careers before shooting to literary stardom late in life, following the publication of *Under the Moons of Mars* and *Tarzan of the Apes* in 1912, and he continued to produce Pellucidar novels until his death in 1950.

Although one fan, submitting a letter to *All-Story Weekly*, wondered if Burroughs 'injects enough cocaine to make himself believe he is living in the prehistoric ages', the author's imaginative verisimilitude was the product of his career-long habit of producing glossaries, diagrams, and maps to ensure continuity.[107] He was even reasonably knowledgeable about geology, having, somewhat implausibly, taught it alongside Gatling gun instruction at Michigan Military Academy in 1895. He later claimed that his pupils 'knew that I knew nothing about geology', but he 'studied geology harder than they did and always kept about one jump ahead of them'.[108] One of his later sources of geological details—perhaps even his cram-book at the Academy—was Thomas George Bonney's *The Story of Our Planet* (1893).[109] Working notes for *At the Earth's Core*, dated 14 January 1913, affirm his commitment to precision. The Pellucidarian genus called the 'sithic', for instance, is glossed as 'labyrinthodon—amphibian, toad like body, alligator jaws, weighed tons, Triassic', while the distance to the centre of the Earth is calculated and the geological periods accurately listed.[110] So careful was Burroughs with these details that his secretary, Cyril Ralph Rothmund, was able to send out to inquirers complete lists of the prehistoric animals mentioned in the series.[111] Burroughs was no Symmes convert, but, as we shall see, some of the preoccupations of true believers that I have discussed in this chapter survived in his work in surprising ways.

Burroughs had been ten when *King Solomon's Mines* was published. Growing up during the height of the lost-world romance's popularity likely inspired the ironic framing devices characteristic of his novels, surrounded in their pulp venues by

---

[107] H. H. R., 'How Does He Do It?', *All-Story Weekly*, 30 (1914), 671. For examples of diagrams and notes, see Edgar Rice Burroughs, quoted in Irwin Porges, *Edgar Rice Burroughs: The Man Who Created Tarzan* (Provo, UT: Brigham Young University Press, 1975), 171, 471, 499, 532–33.

[108] Edgar Rice Burroughs, quoted in Porges, 49.

[109] Phillip R. Burger, 'Afterword', in Edgar Rice Burroughs, *At the Earth's Core* (Lincoln: University of Nebraska Press, 2000), 279–90 (280).

[110] 'At the Earth's Core: The Inner World' notes, 14 January 1913, Irwin Porges Papers, Edgar Rice Burroughs Memorial Collection, Archives and Special Collections, University of Louisville, KY. For Burroughs' detailed notes for the book's sequel, see Porges, 735.

[111] Cyril Ralph Rothmund to Jerry Cady, 27 May 1936, Irwin Porges Papers.

the more unselfconscious third-person narratives coming into fashion in the early twentieth century. In *At the Earth's Core*, a prologue recounts how the author himself met the book's protagonist, David Innes, while hunting lions in the Sahara, and wrote down Innes' story of incredible underground adventures.[112] The prologue of its sequel, *Pellucidar* (1915), also serialized in *All-Story Weekly*, elaborated on this conceit. Now, Burroughs has received a letter from a man, Cogdon Nestor, who explains that he once believed the novel *At the Earth's Core* 'impossible trash'.[113] While travelling in the Sahara himself, however, Nestor has stumbled upon a buried 'telegraph instrument' communicating genuine messages from Innes at the Earth's core. He begs Burroughs for clarity:

> There is no David Innes.
> There is no Dian the Beautiful.
> There is no world within a world.
> Pellucidar is but a realm of your imagination—nothing more.
> But—[114]

The *All-Story* played along with the conceit. The editor's preview of *Pellucidar* explained that Innes 'told his story to Burroughs, and Burroughs, being a good-natured author, fixed up the English, had the story copied, and sent it to us'.[115] Readers of a version serialized in the British boys' magazine *Pluck* in 1923 saw the boundary between text and paratext diminish further. The editor's removal of chapter numbers in subsequent issues, and addition of bold subheadings like 'A Joyful Meeting—Armed Gorillas!' to the three-columned text, gave the novel a veneer of journalistic reportage.[116]

These novels received belated book versions in 1922 and 1923, but, kept busy by Tarzan, Burroughs did not add to the series until *Tanar of Pellucidar*, serialized in another pulp heavy-hitter, *Blue Book Magazine*, in 1929. Fourteen years on, Burroughs' framing became more playful than ever: a man called Jason Gridley, living on Burroughs' own Tarzana ranch in the San Fernando Valley, has invented a radio wave that picks up another communication from Pellucidar. Readers of *Tanar* were reminded in each instalment that the story came 'to Edgar Rice Burroughs unexpectedly one night by means of the far-reaching waves of the radio'.[117] *Tanar* ends on a cliff-hanger, with its protagonists calling upon Tarzan for help, and the crossover sequel, *Tarzan at the Earth's Core*, began serialization the following month. This story finally demonstrates that Pellucidar is connected to the outside world through Symmes holes. By flying an airship into the hole at the

---

[112] Edgar Rice Burroughs, 'At the Earth's Core [1/4]', *All-Story Weekly*, 30 (1914), 1–21 (1–2).
[113] Edgar Rice Burroughs, 'Pellucidar [1/5]', *All-Story Weekly*, 44 (1915), 385–411 (386).
[114] Burroughs, 'Pellucidar [1/5]', 387.
[115] 'Heart to Heart Talks', *All-Story Weekly*, 44 (1915), 322–25 (322).
[116] Edgar Rice Burroughs, 'At the Earth's Core [2/11]', *Pluck*, 1 (1923), 651–53 (652).
[117] Edgar Rice Burroughs, 'Tanar of Pellucidar [2/6]', *Blue Book Magazine*, 48.6 (1929), 76–97 (76).

North Pole, a notion probably inspired by the 1920 edition of Gardner's *Journey to the Earth's Interior* (which, as we have seen, itself alluded to lost-world fiction), Tarzan enters Pellucidar—and Pellucidar promptly ejects Edgar Rice Burroughs.[118] For the first time in the series, *Tarzan at the Earth's Core* exclusively uses a third-person narrator, eschewing Pellucidar's trademark framing devices. Hollow-earthers had long played with notions of authorship, but never before had an author performed such a sudden vanishing act. Burroughs, it must be admitted, was not always careful with literary craft. Known to change tense mid-story, he had almost as fraught a relationship with grammar and structure as Lloyd.[119] That said, his meticulousness when it came to worldbuilding has already been noted. Rather than carelessness, Burroughs' decision was possibly motivated by desire to remove himself from Tarzan's fictional plane. Whatever his reasoning, magazine paratexts immediately contradicted his efforts. Despite Burroughs' abandonment here of the fourth-wall-breaking devices popularized by *fin-de-siècle* lost-world romances, the *Blue Book*'s monthly recaps insisted that Burroughs had, again, received the story via 'his friend' Gridley's radio.[120]

It was not only mischievous and confusing framing that the Pellucidar series shared with many hollow-earth sources. These novels were, magazine editors claimed, to be understood as having some claims upon truth. *Blue Book*'s editors persistently suggested that they communicated eternal verities about the masculine desire to encroach upon new frontiers—new lost worlds. In an editorial of February 1929, the editors lamented the 'loss of the mystery, romance and fascination' that had been available to Europeans in the age of Columbus.[121] Columbus was, of course, the very name to which Symmes himself had so regularly drawn comparison. The antidote to this familiar cliché of modern disenchantment and closed frontiers was Burroughs, who 'restores to us', in *Tanar* (soon to begin serialization), the 'ancient world of glamour'.[122] The next month, the editors ventriloquized the 'inaccurate' view of a sceptic, who complained that the 'frontier has been pushed far back', rendering the world 'relatively humdrum'.[123] The editors admitted that, while Burroughs' romances dealt not with 'actual frontiers', their characters were 'true to the imagination', reviving in men the 'boyhood hours' spent reading romances like *King Solomon's Mines*.[124] Aptly, opposite the first page of *Tanar*, readers' eyes met an illustrated feature on 'Men Who Won the West'.[125]

---

[118] For Gardner's influence, see John Taliaferro, *Tarzan Forever: The Life of Edgar Rice Burroughs, Creator of Tarzan* (New York: Scribner, 2002), 236.
[119] Taliaferro, 96–97, 110.
[120] Edgar Rice Burroughs, 'Tarzan at the Earth's Core [6/7]', *Blue Book Magazine*, 50.4 (1930), 28–44 (28).
[121] 'Tanar of Pellucidar', *Blue Book Magazine*, 48.4 (1929), 5.
[122] 'Tanar of Pellucidar', 5.
[123] 'New Worlds for Old', *Blue Book Magazine*, 48.5 (1929), 5.
[124] 'New Worlds for Old', 5.
[125] 'Men Who Won the West: John C. Frémont—Explorer', *Blue Book Magazine*, 48.5 (1929), 6.

During this period, the magazine was a lucrative vehicle for Burroughs. In Mike Ashley's words, *Blue Book*, formerly a family publication, now 'made no pretence to being anything other than a men's adventure magazine'.[126] As such, editors continued to insist that Pellucidar romances held intuited insights into masculinity. In a piece titled 'To Understand the World', they suggested that *Tanar* 'most of all illuminates and interprets life through the force of its brilliant contrasts', presumably referring to 'contrasts' between Earth's savage interior and its overcivilized urban surface.[127] They contended that Tarzan makes 'real for us our instinctive desire for tremendous experience', awakening 'the hereditary thrill' of the 'savage' in the 'primeval' jungle.[128] Thus, while Symmesians had preached the material advantages of extending Manifest Destiny underground, in the *Blue Book* the sentiment took on a psychological colouring. For their young, broadly male readerships, Pellucidar stories offered a proxy frontier, legitimized by the claim that they dramatized truths about the masculine desire to test oneself in a rugged, prehistoric environment. Burroughs even sacrificed his own long-term plotting to push this angle, briskly revealing early in the 1929 *Tanar* that the colonization and industrialization of the interior world that began in the 1914 *Pellucidar* had utterly failed.

Tarzan at the Earth's Core did not end the series. Tarzan had left a member of his airship crew, Wilhelm von Horst, stranded in Pellucidar and, five years later, in 1935, Burroughs returned to this lost thread. The result, *Back to the Stone Age*, traversed a rocky road to publication, exacerbated but not uniquely caused by the Great Depression. Continued omission of framing devices may have contributed to editors' impressions that this work was fatally disjointed. In January 1936, Donald Kennicott, veteran editor of the *Blue Book*, told Burroughs' secretary that the manuscript 'starts in the middle of things, with no attempt to explain Pellucidar'.[129] In June, Leo Margulies of the publisher Standard Magazine advised that, as 'it hasn't got a plot', the best option was to 'cut up' the episodic novel 'into a series of short novelettes' for *Thrilling Adventures*.[130] Ultimately, it was serialized, heterogeneous but entire, by the Munsey *Argosy* in 1937 with the arbitrary, although distinctly colonialist, new title of *Seven Worlds to Conquer*. A prologue, written by *Argosy* staff, reintroduced Pellucidar, that land 'scorned and derided by timid savants', although this was removed in the book, which also restored the original title.[131]

In the most curious episode of this patchwork novel, the series unexpectedly turned to the spiritual speculations sometimes found in the *fin-de-siècle* Symmes

---

[126] Mike Ashley, 'Blue Book—The Slick in Pulp Clothing', *Pulp Vault*, 14 (2011), 210–53 (228).
[127] 'To Understand the World', *Blue Book Magazine*, 49.3 (1929), 5.
[128] 'Men Worth Knowing', *Blue Book Magazine*, 50.3 (1930), 5.
[129] Donald Kennicott to Cyril Ralph Rothmund, 24 January 1936, Irwin Porges Papers.
[130] Leo Margulies to Cyril Ralph Rothmund, 11 June 1936, Irwin Porges Papers.
[131] Edgar Rice Burroughs, 'Seven Worlds to Conquer [1/6]', *Argosy Weekly*, 270 (1937), 4–27 (4).

revival. Generally a religious sceptic, Burroughs had no faith in occult phenomena, declaring that many, 'perhaps all, so-called supernatural phenomena are the result of injured or diseased brains'.[132] A library press release for *Back to the Stone Age* appealed to readers with healthy brains only, declaring that it provided 'a diverting interlude for your customers who may have had too much psychiatry, stream of consciousness or history in their recent fiction'.[133] Despite this assurance of level-headed adventure fare, Burroughs' latest Pellucidar entry introduced baffling new metaphysical implications.

Halfway through the novel, von Horst is captured by a tribe of albino cannibals called Gorbuses (Figure 3.5). Speaking with a Gorbus named Durg, von Horst notices that it used 'two English words—cleaver and dagger'.[134] Durg grasps that von Horst comes 'from that other world' from which Gorbuses receive 'fleeting glimpses ... of almost forgotten memories' (141). In what Irwin Porges, Burroughs' biographer, calls an 'odd, undeveloped section' of the book (which he notes, Burroughs wrote during a period of 'guilt and mental conflict' regarding his divorce and remarriage), it emerges that the Gorbuses possess the memories of human murderers and sinners from the surface world.[135] They are, effectively, the ghosts of these criminals, living underground in punishment for their crimes. Von Horst remarks that the Gorbuses' predicament 'answers a question that has been bothering generations of men of the outer crust' (145), presumably referring either to the location of hell or the authenticity of reincarnation and afterlife. Here Burroughs unexpectedly drew on the hollow earth's associations with Dante's *Inferno* and its penal system of *contrapasso*: the fitting punishment of sins. Ejected from the surface world, Gorbuses must live in the caverns of the concave interior, feeding on human flesh as they once fed on others' lives.

This idea, of a more spiritual character than anything else in the Pellucidar stories, is never returned to after this chapter, but it hints at a vast metaphysical system that dwarfs the inconsequential events of the novel itself. The Darwinian world of Pellucidar suddenly gains a moral economy, while the surface and interior worlds are reconnected far more intimately than they were by the Symmes holes into which Tarzan's airship flew. Again, it should not be thought that Burroughs had given up on consistent worldbuilding. His notes to the illustrator of the book, his son John Coleman Burroughs, insisted that the map of Pellucidar 'has to be accurate', however 'silly' it may seem, 'because readers have a way of discovering

---

[132] Edgar Rice Burroughs to Boston Society for Psychic Research, 18 February 1929, Irwin Porges Papers.
[133] Cyril Ralph Rothmund, 'New Bonanza for Rental Libraries', 10 September 1937, Irwin Porges Papers.
[134] Edgar Rice Burroughs, *Back to the Stone Age* (Tarzana, CA: Edgar Rice Burroughs Incorporated, 1937), 138. Subsequent references included in text.
[135] Porges, 576, 763.

Behind him, hissing and roaring, galloped a small dinosaur

Figure 3.5 Edgar Rice Burroughs, *Back to the Stone Age* (Tarzana, CA: Edgar Rice Burroughs Incorporated, 1937), facing p. 114. An illustration of a dinosaur chasing a 'Gorbus', painted by the author's son. John Coleman Burroughs Pellucidar art copyright © 1937 Edgar Rice Burroughs, Inc. All Rights Reserved. Trademarks Tarzan®, Pellucidar®, At the Earth's Core™, Edgar Rice Burroughs® Owned by Edgar Rice Burroughs, Inc. Used by Permission.

errors'.[136] Using the shifting forms of early twentieth-century romance, Burroughs instilled in readers the same urge to push back frontiers, challenge orthodoxy, question the binary between fact and fiction, and expect the unexpected, that had motivated hollow-earth true believers.

Conclusion

There were many routes to discovering the secrets of the hollow earth, but they were unified by scepticism about top-down authority in the domain of knowledge. The hollow-earthers' contemporaries, the day-agers, were driven by commitment to the authority of Genesis, correctly understood; occultists, meanwhile, claimed personal authority through unique insight into the deep past. With partial exceptions, as in the case of Mormonism, most hollow-earthers rejected claims to privileged knowledge of important matters of science and religion, on their own part and especially on the part of anyone else. These iconoclasts had hit, they believed, at the true constitution of Earth, but they refused to let this go to their heads, remaining willing to answer criticisms and invite modifications to their ideas on an open playing field, ultimately bringing the public sphere closer to enlightenment. In this democratic vision of scientific practice, one man's sheer dedication to the truth could bring order to the messy geohistorical narrative pieced together by geologists, revealing Earth's formation and constitution to be a neat, satisfying matter of basic (and somewhat early modern) physics, while opening up a brand-new space for colonization, resource extraction, and manly endeavour.

In pursuing these goals, we have seen hollow-earthers work through literary forms like dialogic manuscripts, ventriloquial monographs, appendices in adventure fiction, and virtual voyages to the Earth's core. Their efforts mingled with those of a more nebulous array of fellow travellers, who, working hollow-earth ideas into utopias and lost-world romances, usually displayed minimal dedication to the Symmesian cause. The results moved along a spectrum of factualization and fictionalization, irony and candour—indeed, as we have seen, even in works using the hollow earth in a (mostly) symbolic manner, like *Etidorhpa*, truth claims came thick and fast. These models challenged readers to put aside their preconceptions about the nature of the globe and to treat the author not as an authority, but as a provocateur. The project of establishing that the earth was hollow, and that the science promulgated in the metropole was flawed, required a more participatory approach not just to geology, but also to writing about it. That approach did not always mean giving the reader an easy ride.

---

[136] Edgar Rice Burroughs to John Coleman Burroughs, undated *Back to the Stone Age* notes [1937], Irwin Porges Papers.

Although the hollow interior of Earth remained unseen by human eyes, its proponents, from the start, asserted that it had no less tangible reality than other equally untested hypotheses about the contents and history of the planet. The anti-authoritarian ethos of this system involved a scepticism about anything other than personal eye-witnessing in establishing scientific claims—a scepticism we will see return, in different form, in chapter 5's discussion of young-earth creationism. As long as neither Symmes nor anyone else had explored the poles, or Earth's core, all bets were off. This time, virtual witnessing was, in the end, simply not enough. Burroughs hinted at this fact in *At the Earth's Core*, when Abner Perry—the geologist friend of David Innes, Pellucidaran hero—responds to the discovery of surviving prehistoric animals:

'David, I used to teach geology, and I thought that I believed what I taught; but now I see that I did not believe it—that it is impossible for man to believe such things as these unless he sees them with his own eyes.

'We take things for granted, perhaps, because we are told them over and over again, and have no way of disproving them—like religions, for example; but we don't believe them, we only think we do. If you ever get back to the outer world you will find that the geologists and paleontologists will be the first to set you down a liar, for they know that no such creatures as they restore ever existed. It is all right to *imagine* them as existing in an equally imaginary epoch—but now? poof!'[137]

This kind of distrustful, hyper-empirical rhetoric, when espoused by proponents of borderline and fringe theories, was not calculated to appease elite practitioners. Outbursts against the establishment had to be carefully rationed by those authors discussed in my next chapter, who used all the respectable scientific paraphernalia they could muster to argue that a colossal deluge had ravaged early human culture.

---

[137] Burroughs, 'At the Earth's Core' (1/4), 19–20.

4
Submerged in the Public Sphere

In 1920, during a Spiritualist tour of Australasia, Arthur Conan Doyle conversed with Auckland occultists about the lost continent of Atlantis. Writing up his travels in *The Wanderings of a Spiritualist* (1921), Doyle, who deemed himself 'a bit of an Atlantean', related his personal theory that the lost continent had been destroyed in the same ancient flood recounted in Genesis.[1] His explanation, which cited the evidence of the 'mammoth remains which strew the Tundras of Siberia', implied familiarity with Atlantean scholarship, although he considered his own theory 'original and valid'.[2] A fellow novelist of lost worlds, Edgar Rice Burroughs, was also ruminating on this legendary sunken civilization. Responding, in May 1922, to the suggestion of an English Theosophist that he pen a series of Atlantis stories, Burroughs admitted to having had a 'similar idea' and asked for book recommendations providing 'whatever information science has been able to deduce', in order to 'make such a series of stories as approximately true to life as possible'.[3] His correspondent eagerly responded with a shortlist, including the works of William Scott-Elliot, which Burroughs promised to purchase via his London agent.[4] He added, however, that 'before attempting to enter a field that has for so long attracted the interest of scientific men' he needed carefully to gather 'all obtainable facts'.[5] Writing on Atlantis was not to be undertaken lightly.

Burroughs was ultimately discouraged from writing the Atlantean series by magazine editor Bob Davis, who complained that the fashionable lost continent 'has already been hit so damned hard that nobody cares whether it ever reappears again'.[6] We saw in chapter 3 that neither Doyle nor Burroughs were willing seriously to countenance hollow-earth theory, but Atlantis, whatever a canny editor's reservations about its cultural oversaturation, was a different story. The myth is known from Plato's fourth-century BCE dialogue, *Timaeus*: this text recounts how the Atlantean continent, overseen by Poseidon and inhabited by a sophisticated civilization, was hit by an immense catastrophe and sank into the Atlantic around nine thousand years prior to Plato's time. The dialogue

---

[1] Arthur Conan Doyle, *The Wanderings of a Spiritualist* (London: Hodder & Stoughton, 1921), 190.
[2] Doyle, *Wanderings*, 190.
[3] Edgar Rice Burroughs to William Gifford Hale, 6 May 1922, Irwin Porges Papers, Archives and Special Collections, University of Louisville, KY.
[4] Burroughs to Hale, 20 June 1922, Irwin Porges Papers.
[5] Burroughs to Hale, 20 June 1922.
[6] Quoted in Irwin Porges, *Edgar Rice Burroughs: The Man Who Created Tarzan* (Provo, UT: Brigham Young University Press, 1975), 366.

Contesting Earth's History in Transatlantic Literary Culture, 1860–1935. Richard Fallon, Oxford University Press.
© Richard Fallon (2025). DOI: 10.1093/9780198926191.003.0005

attributes knowledge of Atlantis to Egyptian records and scholars have long argued over whether Plato's account was merely allegorical, those in favour of the continent's historicity locating traces of it across the world.[7] Although Doyle was proud of his ingenuity, the theory that the biblical Deluge and the flood that destroyed Atlantis were identical was conventional among mythographers and syncretists.[8]

Speculations about continents sunk in primordial catastrophes, such as Atlantis, enjoyed cultural and at times scientific respectability during the late nineteenth and early twentieth centuries.[9] Indeed, shorn of the ancient super-city described by Plato, soberly conceived lost continents, even versions of Atlantis, belonged to comparatively mainstream science.[10] These bodies were instrumental for explaining the global distribution of plants and animals. They came into especial vogue as theoretical centres from which primitive humans spread across the world, submerged lands proving attractive to those who wished to situate a migratory white or Aryan race as the origin of world culture.[11] In basic outline, this 'hyper-diffusionist' understanding of culture dated back at least to Genesis's story of the post-diluvial dispersal of Noah's sons.[12] For those on palaeoscience's borderlines, lost-continent hypotheses were a means of pushing civilization deeper into the past than otherwise permitted, clawing back some of the human exceptionalism dented by palaeoanthropology, and proving that ancient texts like *Timaeus* and the Bible could be taken seriously. We have already heard about these matters in earlier chapters, but this chapter approaches them from a different direction by focusing on authors who attempted to stay, albeit reluctantly, within scientific naturalism's humbler methodological bounds. While Atlantis forms the chapter's core, I have, due to the overlapping nature of research on flood-catastrophism, found it necessary to refer to other lost lands (like Eden, Lemuria, and 'Mu') and to diluvial speculations more generally.

Proponents of flood-catastrophism were intellectually wide-ranging, but their knowledge was typically the product of time spent in libraries. This was a problem in the eyes of nineteenth-century reformers of science, who asserted that polymaths whose knowledge came purely from reading, however voracious, lacked

---

[7] L. Sprague de Camp, *Lost Continents: The Atlantis Theme in History, Science, and Literature*, rev. edn (New York: Dover Publications, 1970); Joscelyn Godwin, *Atlantis and the Cycles of Time: Prophecies, Traditions, and Occult Revelations* (Rochester, VT: Inner Traditions, 2011).

[8] Colin Kidd, *The World of Mr Casaubon: Britain's Wars of Mythography, 1700–1870* (Cambridge: Cambridge University Press, 2016), 187.

[9] Sumathi Ramaswamy, *The Lost Land of Lemuria: Fabulous Geographies, Catastrophic Histories* (Berkeley: University of California Press 2004).

[10] Peter J. Bowler, *Life's Splendid Drama: Evolutionary Biology and the Reconstruction of Life's Ancestry, 1860–1940* (Chicago: University of Chicago Press, 1996), 396–98.

[11] Dan Edelstein, 'Hyperborean Atlantis: Jean-Sylvain Bailly, Madame Blavatsky, and the Nazi Myth', *Studies in Eighteenth-Century Culture*, 25 (2006), 267–91.

[12] George W. Stocking, Jr, *After Tylor: British Social Anthropology 1888–1951* (London: Athlone Press, 1996), 180–81.

true expertise.[13] If trends in palaeoscience were moving in the reformers' favour in the late nineteenth century, however, this development was less clear in the professionally heterogeneous human sciences, and, when talking about lost continents and deluge myths, geology shaded imperceptibly into archaeology, ethnology, and anthropology.[14] Roving between permeable disciplinary silos, well-read theorists of Atlantis had little reason to see their research as less legitimate than respectable if contentious books like James George Frazer's massive, mythography-synthesizing *The Golden Bough* (1890; 1906–15). The Cambridge-based Frazer, after all, distinguished the field-working 'descriptive ethnologist' from 'comparative' ethnologists like himself, who were, in effect, prolific readers.[15] After all, interpreting flood myths seemed to demand familiarity with several thousand years' worth of textual matter more than it required geological fieldwork. What did mark out the authors I analyse in this chapter from academics like Frazer was that, rather than establishing abstractions about the evolution of human culture, they invariably argued, more controversially, for the historicity of ancient legends. Their work thus built upon traditions of Christian mythography, practitioners of which sought proofs of Genesis events preserved in pagan culture, but the most scholarly flood-catastrophists, in line with their attempts to toe the line of scientific naturalism, downplayed apologetic intent.

In thinking about these writers, I have found useful David Morris's analysis of 'suppositional' scientific texts which refuse to fit 'within the fact-fiction polarity', a tradition he traces through Victorian Theosophy to paranoid investigators of UFOs and ancient astronauts.[16] Evading the 'fact-fiction polarity' would be an apt description of the proudly heterodox practices of chapter 3's hollow-earthers, but it interests me that Morris cites nineteenth-century US politician Ignatius Donnelly, one of the most influential theorists of Atlantis, as a pioneer in this tradition.[17] While linking Donnelly to post–World War II countercultures makes sense in Morris's diachronic analysis, and Donnelly's later career turned in conspiratorial directions, focusing upon these aspects of his work occludes the more complex location of his Atlantean scholarship in the scientific culture of his own day.[18] In fact, rather than raging against the scientific establishment, my

---

[13] Gowan Dawson, *Show Me the Bone: Reconstructing Prehistoric Monsters in Nineteenth-Century Britain and America* (Chicago: University of Chicago Press, 2016), 245–47; Melinda Baldwin, *Making Nature: The History of a Scientific Journal* (Chicago: University of Chicago Press, 2015), 77, 83.

[14] Julia Reid, 'Archaeology and Anthropology', in *The Routledge Research Companion to Nineteenth-Century British Literature and Science*, ed. by John Holmes and Sharon Ruston (Abingdon: Routledge, 2017), 357–71.

[15] Quoted in Robert Fraser, ed., 'Introduction', in James George Frazer, *The Golden Bough* (Oxford: Oxford University Press, 1994), ix–xliii (xxi).

[16] David Morris, *The Masks of Lucifer: Technology and the Occult in Twentieth-Century Popular Literature* (London: B. T. Batsford, 1992), 44.

[17] Morris, *Masks of Lucifer*, 7, 44.

[18] For Donnelly and conspiracy theories, see Alex Beringer, '"Some Unsuspected Author": Ignatius Donnelly and the Conspiracy Novel', *Arizona Quarterly*, 68 (2012), 35–60.

subjects—Donnelly included—described catastrophe-wracked prehistoric worlds in comparatively sober terms, intending to win the sympathy of educated general readerships but also to court the grudging tolerance of elite naturalists. While naturalists typically signalled that these texts were problematic, generally educated publics often could not, or, more to the point, would not do so.

To understand why this was the case, my chapter explores the literary strategies through which ideas deemed beyond the palaeoscientific pale when penned by Theosophists and conservative evangelicals were made palatable to readers with neither presumed religious commitments nor antagonism towards the fundamentals of established science. My focus is a genre that I, adapting Morris's term, call the *suppositional synthesis*: a mannered approach to writing controversial science, of which works on diluvial catastrophes and sunken continents form one of the major thematic concerns. The coinage is necessary, as, perhaps uniquely in *Contesting Earth's History*, we can here talk about a consistent union between form and content. This is because the flood-catastrophist wing of the suppositional synthesis genre more often than not directly modelled itself upon Donnelly's lawyerly and systematic book *Atlantis: The Antediluvian World* (1882). Flooding readers with an arrangement of hard facts in support of spectacular reconfigurations of prehistory, these texts courted intellectual respectability by halting at the brink of imaginative incaution overleapt by the likes of hollow-earthers.

However, I will also show that suppositional synthesizers strained against the urge to envision more concretely the sunken worlds they gestured towards. In particular, they struggled to resist the siren call—and confident appeal to virtual witnesses—of occultism. Theosophical contemporaries were satisfied only with direct, clairvoyant access to truths about Atlantis, Lemuria, and the archaic catastrophes that had destroyed them, exhibiting an extreme version of what Jeb J. Card describes as the common fringe 'preference for the literal certainty of historical or sacred texts over the uncertainty and questioning inherent in scientific investigation'.[19] This preference contrasts with what Simon Goldhill identifies, in more mainstream forms of nineteenth-century research into ancient human history, as an acknowledgement of 'the impossibility of reaching the goal of absolute authenticity', aiming 'rather at fictionalized, idealized reconstruction'.[20] In a distinctively borderline manoeuvre, suppositional synthesizers precariously mated these two apparently contradictory methods, recognizing that intellectual credibility compelled muted ambitions regarding authentic accounts of ancient catastrophes, but shying away, in their desire for 'literal certainty', from 'fictionalized' reconstructions. The results were Janus-faced texts, exhibiting reams of circumstantial evidence while crying out for certain knowledge of deep human history.

---

[19] Jeb J. Card, *Spooky Archaeology: Myth and the Science of the Past* (Albuquerque: University of New Mexico Press, 2018), 114.
[20] Simon Goldhill, 'Ad Fontes', in *Time Travelers: Victorian Encounters with Time and History*, ed. by Adelene Buckland and Sadiah Qureshi (Chicago: University of Chicago Press, 2020), 67–85 (81).

A Thousand Converging Lines of Light

During the late nineteenth century, there was good reason to believe that prominent myths would be traced to historic and prehistoric actualities. George Smith's 1872 translations of what was later called the *Epic of Gilgamesh* from Assyrian tablets, for instance, provided a powerful corroborating source for Genesis. Thanks to the archaic flood story contained in *Gilgamesh*, the Noachian Deluge, which had lost its last remaining credibility among leading geologists back in the 1830s, again became a subject of (qualified) scientific investigation.[21] During the 1880s, textual accounts of deluge myths were being taken seriously by Eduard Suess, arguably the world's leading geologist.[22] While diluvial catastrophes were being rehabilitated, lost continents, geologically speaking, were already routine. In an age of disagreement about the nature of crustal movements, sunken land bridges made for valuable speculative tools. We have already heard how Lemuria was posited to explain biogeography. The Austrian palaeobotanist Franz Unger, in an 1860 lecture translated into English in 1865, declared it excessively 'bold', given the 'sound' evidence of similarities between American and European fauna, to dismiss Plato's account of Atlantis as 'imaginary'.[23] This encouragement from on high gave borderline thinkers scientific leeway to draw together evidence, magpie-like, from diverse fields. Notably, Clare Pettitt positions syncretism as the main 'viewfinder' of knowledge-making in the mid-nineteenth century, Atlantis being the syncretic theory *par excellence*.[24]

Emerging in this amenable climate, Ignatius Donnelly became Atlantis's most influential theorist since Plato.[25] Soon after starting out on a legal career on the East coast of the United States, Donnelly moved to Minnesota, where he became a Republican congressman and then a senator, pursuing, with limited success, populist and agrarian policies. Writing from his isolated farm, Donnelly composed *Atlantis* in just two months during early 1881, synthesizing several decades of his own research.[26] The breathless work was snapped up for publication by the prestigious firm Harper & Brothers, and by Sampson Low, the leading British publisher of American works. A lengthy and well-illustrated treatise, *Atlantis* argued that the abyssal Mid-Atlantic Ridge, uncovered during the oceanographic voyage

---

[21] For earlier research on the biblical Flood, see Lydia Barnett, *After the Flood: Imagining the Global Environment in Early Modern Europe* (Baltimore, MD: Johns Hopkins University Press, 2022).

[22] James A. Secord, 'Global Geology and the Tectonics of Empire', in *Worlds of Natural History*, ed. by H. A. Curry, N. Jardine, J. A. Secord, and E. C. Spary (Cambridge: Cambridge University Press, 2018), 401–17 (404). See also Vybarr Cregan-Reid, *Discovering Gilgamesh: Geology, Narrative and the Historical Sublime in Victorian Culture* (Manchester: Manchester University Press, 2013), 203.

[23] Franz Unger, 'The Sunken Island of Atlantis', *Journal of Botany*, 25 (1865), 12–26 (25).

[24] Clare Pettitt, 'At Sea', in *Time Travelers: Victorian Encounters with Time and History*, ed. by Adelene Buckland and Sadiah Qureshi (Chicago: University of Chicago Press, 2020), 196–220 (209).

[25] De Camp, *Lost Continents*, 37–43.

[26] Martin Ridge, *Ignatius Donnelly: The Portrait of a Politician* (Chicago: University of Chicago Press, 1962), 197–98.

of the HMS *Challenger* between 1872 and 1876, was a relic of Atlantis. While, for most geologists, the *Challenger* had stripped Atlantis of its remaining hiding places, Donnelly located the remains of the sunken country where the Ridge pokes above the water as the Azores archipelago. He marshalled support for this hypothesis from geology, palaeontology, biogeography, archaeology, ethnology, anthropology, palaeoanthropology, and comparative religion.

Smith's translations of the *Gilgamesh* tablets provided evidence for Donnelly, as did a host of mysteries, including palaeontologist O. C. Marsh's work on the horse's evolution in, and then disappearance from, America, although his most substantial testimony came from purported similarities between Egyptian and Latin American antiquity—namely their shared cultures of pyramid-building and mummification.[27] Egyptology had a long history, but now Maya, Aztec, and Inca ruins were increasingly fascinating to researchers based in the United States and the European empires, providing valuable raw material for syncretists like Donnelly. All this evidence legitimized his euhemerist, hyper-diffusionist argument that all world mythologies represented distorted accounts of Atlantis, an advanced monotheistic civilization that 'was the original seat of the Aryan or Indo-European family of nations, as well as of the Semitic peoples, and possibly also of the Turanian races' (2).

While Atlantis had been highly civilized, it was not advanced beyond credibility, nor was it a land of magic, as it became in the hands of Theosophists and pulp romancers. Donnelly sternly rejected 'tales of gods, gorgons, hobgoblins, or giants' (22) and presented his dense argument as relying solely on hard, rationally arranged evidence. Formally, *Atlantis* gives the impression of generating proofs through accretive induction. The act of rhetorically overwhelming readers with immense quantities of discrete, neutral facts, justified as obeying the inductive precepts attributed to the natural philosopher Francis Bacon, was common across the nineteenth-century sciences, but it was especially associated with precarious disciplines seeking legitimacy.[28] Harper & Brothers proudly vaunted their author's 'admirable facility of induction', and it is no coincidence, given his reverence for induction, that Donnelly would later commit himself to proving Bacon's authorship of Shakespeare's plays in *The Great Cryptogram* (1887).[29] *Atlantis* starts out with a series of thirteen 'novel propositions' to be proven (1), while the remainder of the book is divided into thematic parts, and subdivided into disciplinary or problem-solving chapters with judicial titles like 'The Probabilities

---

[27] Ignatius Donnelly, *Atlantis: The Antediluvian World* (New York: Harper & Brothers, 1882), 54, 76–82. Subsequent references provided in text.

[28] For example, see Efram Sera-Shriar, *Psychic Investigators: Anthropology, Modern Spiritualism, & Credible Witnessing in the Late Victorian Age* (Pittsburgh, PA: University of Pittsburgh Press, 2022), 29. For the rise and fall of nineteenth-century Baconianism, see Richard Yeo, 'An Idol of the Market-Place: Baconianism in Nineteenth-Century Britain', *History of Science*, 23 (1985), 251–98.

[29] 'New Publications', *New York Times*, 18 February 1882, 5.

of Plato's Story' and 'The Testimony of the Flora and Fauna'. These bring together, in quasi-geometric manner, 'a thousand converging lines of light from a multitude of researches' (4). The contents of these chapters are overwhelming accumulations of anecdotes drawn entirely, often in large quotations, from other scholars. As Donnelly's hyper-diffusionism was predicated on the unlikelihood of similar developments arising on opposite sides of the world, he stressed the statistical improbability of cumulative coincidences, comparing doubters to those blinkered early modern thinkers who had dismissed the organic appearance of fossils as a superficial result of nature's 'plastic power' (419).

Donnelly's hypothesis was thus presented as if irresistibly drawn from neutral facts, allowing readers to glimpse vanished Atlantean life chiefly through a patchwork of quotations. This was a strategy familiar from the early days of palaeoscientific writing, some fifty years prior, when, in a Baconian climate hostile to speculation, most geological authors had preferred to allow a small number of memorably pictorial quotations from other savants, or from poets, to do the imaginative work of depicting prehistoric worlds for them.[30] Even Donnelly's tasteful poetry quotations, including lines by his own sister, Eleanor, did instrumentalized double-duty as textual survivals of Atlantean culture. Robert Burns's jocular 'Address to the Deil' (1786), for instance, preserved in the Scots word 'Clootie' (cloven) an association of the devil with the horned god Baal that could be traced to Atlantean bull-worship (428).[31] The politician was also attuned to the suggestive value of modern fiction, declaring that

> even the wild imagination of Jules Verne, when he described Captain Nemo, in his diving-armor, looking down upon the temples and towers of the lost island, lit by the fires of submarine volcanoes, had some groundwork of possibility to build upon. (44)

His allusion expressly conjured up one of the most compelling illustrations from Verne's *Vingt mille lieues sous les mers* (*Twenty Thousand Leagues under the Seas*) (1869–70) (Figure 4.1). Drawing on the imaginative credit of Verne and others, Donnelly thus enabled readers virtually to witness the wonders of lost Atlantis without being seen personally to step outside the bounds of induction.

The book enjoyed impressive sales, reaching twenty-three editions in the United States and twenty-six in Britain by 1890, accompanied by numerous translations and extensive reviews.[32] However, most scientific critics in esteemed journals demarcated *Atlantis* from works of true science. *Nature* insisted that its 'only

---

[30] Ralph O'Connor, *The Earth on Show: Fossils and the Poetics of Popular Science, 1802–1856* (Chicago: University of Chicago Press, 2007), 329.
[31] For folk songs as vestiges of primitive cultures in Burns's day, see Noah Heringman, *Deep Time: A Literary History* (Princeton, NJ: Princeton University Press, 2023), 136.
[32] Ridge, *Donnelly*, 202.

SUBMERGED IN THE PUBLIC SPHERE 153

Là, sous mes yeux, apparaissait une ville détruite (p. 297).

Figure 4.1 Jules Verne, *Vingt mille lieues sous les mers* (Paris: Hetzel, [1872]), 297. Reproduced courtesy of the Main Library, University of Birmingham. Alphonse de Neuville's depiction of Atlantean ruins, engraved by Henri Théophile Hildibrand in Verne's *Twenty-Thousand Leagues under the Seas*.

reason for noticing' the book was because Donnelly's abundant references to 'writers of authority', including Charles Darwin and Richard Owen, might 'lead some readers astray'.[33] The *American Naturalist* expressed similar impatience with the 'leveling democratic use or misuse of authors, which is characteristic of works' like Donnelly's *Atlantis*. Donnelly, the author implied, did not or could not evaluate the comparative value of his sources. Tellingly, however, the anonymous reviewer (possibly the co-editor, palaeontologist Edward Drinker Cope) suspected that not all readers would detect this problem. He directed those who felt 'that our criticisms are unjust' to authoritative sources on prehistory, like the work of palaeoanthropologist William Boyd Dawkins.[34] Donnelly's biographer attributes his subject's 'legal rather than scientific' approach to his judicial training, which encouraged the discarding, rather than transparent acknowledgement, of 'contradictory evidence'.[35] Only readers familiar with the modern scientific methodologies preferred by *Nature* and the *American Naturalist* would have been able confidently to characterize forensic methods as illegitimate. As a result, Donnelly's presentation was perceived by these critics as meriting discrediting, unlike more obviously heterodox work. For journals like *Popular Science Monthly*, mediators of science for wider publics, *Atlantis*, with its 'lawyer's brief' approach and 'array of circumstantial evidence', required even more detailed refutation.[36]

Nonetheless, Donnelly's courting of famous savants in his citations occasionally paid off. His ethnology had heavily relied on the work of University of Michigan geologist and polygenist Alexander Winchell, who, writing in the sophisticated literary *Dial*, stated that Donnelly's 'highly attractive book' possessed 'a marked degree of plausibility', concluding that there was 'no longer any formidable doubt' that the Atlantis myth contained a kernel of truth.[37] Those receptive to Donnelly included W. E. Gladstone, who provided an eminent model for the polymathic statesman-scholar that Donnelly wished to emulate.[38] He sent Gladstone a copy of the book, noting that it touched upon the latter's own Greek mythological researches. When Gladstone wrote back in encouraging detail, Donnelly ventured to suggest that the British prime minister send 'idle war vessels to complete the work of the H. M. S. *Challenger*', sounding the Azores region 'by many supposed to be the Lost Atlantis' and rewriting the 'prehistoric past of the human race'.[39]

Although many critics objected to Donnelly's haphazard evaluation of sources, and few expressed entirely unqualified assent with his propositions, almost none questioned the right of a legally trained politician to contribute to contested

---

[33] Untitled review of *Atlantis: The Antediluvian World*, by Ignatius Donnelly, *Nature*, 26 (1882), 341.
[34] 'Donnelly's Atlantis', *American Naturalist*, 16 (1882), 729–31 (730–31).
[35] Ridge, *Donnelly*, 198.
[36] Untitled review of *Atlantis: The Antediluvian World*, by Ignatius Donnelly, *Popular Science Monthly*, 22 (1883), 131–32.
[37] Alexander Winchell, 'Ancient Myth and Modern Fact', *Dial*, 2 (1882), 284–86 (286).
[38] Ridge, *Donnelly*, 202.
[39] Ignatius Donnelly to W. E. Gladstone, 25 March 1882, Add MS 44474, British Library, London.

questions of palaeogeography and ethnology, nor did they deny their own right, whatever their qualifications, to evaluate a work on these topics. Major American papers like the *Chicago Tribune* expressed awareness of the blow that the *Challenger* had dealt to Atlantis, all while conceding that 'Mr. Donnelly deserves credit for the ingenious contribution he has prepared on the subject'.[40] The inconclusive characterization of Donnelly's book as bold, 'ingenious', even admirable, but not fully persuasive, was a common one that we will continue to encounter in the reception of flood-catastrophist suppositional syntheses. Indeed, this was a category in which non-specialist reviewers regularly took refuge, especially when their democratic stance made them instinctively sympathetic to scientific outsiders.[41] Suggestively, unqualified rejection was rare.

The question of Donnelly's right to contribute to science in the public sphere was, in part, related to profit. Having been rejected as excessively unscientific by his original publisher, Harper, and by the similarly respected Scribner, Donnelly's sequel, *Ragnarok: The Age of Fire and Gravel* (1883), was taken on by Appleton, a publisher of respected scientific works who insisted that they had faith in the new book's marketability rather than its science.[42] As Sylvia Nickerson has shown, even outright detestation of a book's scientific thesis was not necessarily an obstacle to a respectable publisher when financial gain was likely to be significant.[43] When his convoluted *Ragnarok* failed to bring these projected gains, Donnelly's attempt to boost the flailing book's visibility was turned down even by Appleton's own *Popular Science Monthly*. Although this periodical had previously reviewed *Atlantis* expressly to discredit its simulation of scientific argumentation, the editor, Edward L. Youmans, joked that criticizing its patently absurd successor would be akin to 'gravely' replying to a novel by 'the romancer Jules Verne'.[44] This time, Youmans felt that no reader could reasonably be fooled by Donnelly's presentation. In contrast, Winchell, perhaps regretting his gentle treatment of *Atlantis*, was less sanguine about the general reader's ability to demarcate *Ragnarok*. Donnelly's 'feint of argumentation' meant that 'the unsuspecting reader absorbs the conclusions with avidity', but his 'garb of genuine science' was far from 'harmless'.[45] He thus explained to the educated literary and political readers of the *Forum* why *Ragnarok* should be seen as a spurious 'scientific romance'.[46] Despite an intervening forty years of scientific reforms, bringing with them much clearer markers

---

[40] 'Atlantis', *Chicago Tribune*, 25 March 1882, 9.
[41] For example, see untitled review of *Atlantis: The Antediluvian World*, by Ignatius Donnelly, *Western Christian Advocate* (Cincinnati, OH), 22 March 1882, 95.
[42] Ridge, *Donnelly*, 209.
[43] Sylvia Nickerson, 'Darwin's Publisher: John Murray III at the Intersection of Science and Religion', in *Rethinking History, Science, and Religion: An Exploration of Conflict and the Complexity Principle*, ed. by Bernard Lightman (Pittsburgh, PA: University of Pittsburgh Press, 2019), 110–28.
[44] Quoted in Ridge, *Donnelly*, 209.
[45] Alexander Winchell, 'Ignatius Donnelly's Comet', *Forum*, 4 (1887), 105–15 (106).
[46] Winchell, 'Ignatius Donnelly's Comet', 106.

of expertise, this was still the brittle, boundary-policing language of the attacks on the *Vestiges of the Natural History of Creation*, alluded to in my introduction.[47]

Every One Must Respect His Learning

Donnelly's suppositional approach thus garnered a mixed reaction, but the reach of *Atlantis* encouraged other scholars intrigued by flood legends. Whether arguing for Atlantis or the biblical Deluge, they built up a shared pool of evidence, making much of factors like the comparatively recent, rapid—and perhaps catastrophic—elevation of mountain chains, the existence of elephant remains in polar regions, and signs of high levels of civilization in the distant past. These authors presented themselves as eagle-eyed generalists, gathering vivid facts during snatches of free time away from their more practical daytime vocations. Their prose was that of logicians and men of common sense, denouncing imaginative speculation and overly elegant rhetoric. They chastised widely accepted scientific theories as blinkered, while also taking care to avoid being liable to dismissal as cranks, seers, or religious zealots. Moreover, just as Gladstone's status had made his otherwise isolated views on Genesis a topic hotly debated in heavyweight journals, so too the political and social eminence of these writers opened doors to publishing venues closed to more poorly connected mavericks.

The similarity of the evidence base used in these suppositional syntheses, a toolbox of heterodox facts circulated through literary replication, was such that a treatise arguing for the location of Eden could strongly resemble one arguing for the existence of Atlantis.[48] William Fairfield Warren's *Paradise Found* (1885) argued that 'THE CRADLE OF THE HUMAN RACE, THE EDEN OF PRIMITIVE TRADITION, WAS SITUATED AT THE NORTH POLE, IN A COUNTRY SUBMERGED AT THE TIME OF THE DELUGE'.[49] An evangelical Methodist and scholar of comparative religion, Warren was long-standing President of Boston University. That he worked in the apologetic tradition of 'collateral evidences' was apparent from his frontispiece, which flaunted a master-system of mythographic correlations previously outlined in his *True Key to Ancient Cosmology and Mythical Geography* (1882). However, unlike Mr Casaubon, the dusty pedant of George Eliot's *Middlemarch* (1871–72), who also develops a 'Key to all Mythologies' (and Cuthbert Collingwood, whose Swedenborgian 'Key' was discussed in chapter 1), Warren was attentive to modern

---

[47] James A. Secord, *Victorian Sensation: The Extraordinary Publication, Reception, and Secret Authorship of* Vestiges of the Natural History of Creation (Chicago: University of Chicago Press, 2000), 40.

[48] For literary replication, see Secord, *Victorian Sensation*, 126. For the context of a polar Eden, see Alessandro Scafi, *Mapping Paradise: A History of Heaven on Earth* (Chicago: University of Chicago Press, 2006), 352–59.

[49] William Fairfield Warren, *Paradise Found: The Cradle of the Human Race at the North Pole: A Study of the Prehistoric World* (Boston, MA: Houghton, Mifflin, 1885), 47.

biblical scholarship, having worked in Germany for much of the 1850s and 1860s. In the words of one of few scholars to examine his thinking, Warren was a mediating figure in Methodist theology, adopting a 'liberal' approach that stressed 'rigorous exercise of the human mental capacities' while rejecting a 'rationalistic standpoint' that failed to acknowledge 'revealed Christianity'.[50]

Warren presented *Paradise Found* as the testing of a hypothesis, in contrast with the unsystematic 'fancies' posited regarding the North Pole by thinkers like 'Symmes' (84). He legitimized the use of this form of reasoning, often denigrated by conservative scientists in favour of Baconian induction, with an epigraph taken from John Stuart Mill's defence of scientific hypotheses in the *System of Logic* (1843) (46), one of several invocations of the impeccably rational British logician (see also 50, 316). The structure of the synthesis was split into six parts: the first set out the 'STATE OF THE QUESTION' (viii); the second offered 'A FRESH HYPOTHESIS'; the third 'TESTED AND CONFIRMED' the hypothesis by calling upon the 'TESTIMONY' of various sciences (xiv); the fourth did the same using 'ETHNIC TRADITION' (xvii); the fifth offered 'FURTHER VERIFICATIONS' (xix); and the sixth explained 'THE SIGNIFICANCE OF OUR RESULTS' (xxii). These sections often ended in a summative conclusion. Building his argument in this almost mathematical manner, Warren requested that undecided but active readers 'turn back' and 'carefully collate' the facts 'enumerated' (301). *Paradise Found* moved from supposition towards near certainty over its 500 pages, from the tentative claim that nowhere was Warren's 'hypothesis inadmissible' (187), to his 'firm-grounded conviction' that 'LOST EDEN IS FOUND' (301).

Although he endeavoured to locate the historical Eden using evidence valid even to readers sceptical about revealed knowledge, Warren was by no means convinced that a 'narrow naturalism' was the only valid scientific methodology (ix). As he had preached at the university two years prior, 'purblind' positivistic 'scientists' who refused to allow 'questions as to the whence, whither, and wherefore of nature' reduced science to an arid study 'of *that which is*'.[51] If Warren chafed against the limits which he had imposed upon his own study of Eden, he ultimately recognized that his book's success depended on toeing this line. Like Donnelly, he refrained from reconstructing prelapsarian life on the evidence remaining, leaving the true nature of Eden and its natural history sublimely shadowy and vague. Even an illustrated plate tantalizingly labelled 'NIGHT SKIES OF EDEN' was, as the full caption explained, simply a modern 'Aurora Borealis' (facing 68), standing in for the cradle of life's unfallen beauty. Warren's accompanying 'prosaic' description of an Arctic dawn, moreover, was qualified with the assertion that even this phenomenon was 'indescribable' and 'unpicturable' (70), a qualification that implicitly applied in a

---

[50] Howard Eugene Hunter, 'William Fairfield Warren: Methodist Theologian', unpublished PhD thesis, Boston University (1957), 302, 330.

[51] William Fairfield Warren, *President's First Baccalaureate Sermon and Tenth Annual Report* (Boston, MA: University Offices, 1884), 14.

more extreme manner to any attempt to depict Eden itself. Warren concluded that, destroyed by the Flood, Eden's nature could never be known: 'Long-lost Eden is found; but its gates are barred against us' (432). This union of triumphant scientific discovery with absence and deferral was central to suppositional syntheses.

The book's publication by the Riverside Press of Cambridge, Massachusetts, and Houghton, Mifflin of Boston—the former, in John Tebbel's words, 'a guarantee of excellence' in 'printing' quality and 'literary superiority', and the latter among 'the most formidable in American publishing' during the late nineteenth century, joined in Britain by an edition from Sampson Low—meant that *Paradise Found* reached readerships unavailable to most attempts at reconciling Genesis with science.[52] Its reception resembled that of Donnelly's *Atlantis*, which Warren repeatedly cited (180–81, 186, 306)—to the distaste of the *New York Times*, whose critic saw both books as 'rammed with undigested facts'.[53] For Winchell, the book, like *Ragnarok*, was 'a phenomenal aggregation of varied learning sundered from its conclusions'.[54] Those critics concerned about readers' inability to recognize Warren's limitations, such as the *Atlantic Monthly*, continued to explain that 'men of general literary training' like Warren 'make avail of the scientific method in part alone', employing 'hypotheses' without applying 'the checks which the well-trained student of nature puts upon the use of the imagination', such as careful sifting of sources.[55] However, from local science journals to *Harper's New Monthly Magazine* and the daily newspapers, and even in *Science*, most critics, especially those from the United States, lauded Warren's earnest attitude, logical methodology, literary poise, and patient accumulation of data. They also refused to accede to his argument. In the representative words of the British *Spectator*, '[w]e do not suppose, indeed, that the world will be convinced by Dr. Warren's arguments; but every one must respect his learning, and the ingenuity with which he uses it'.[56] Even the otherwise sceptical *Atlantic Monthly* ended with a gentle recognition that 'the felicity' of the book's 'style' and its 'charming spirit of open-minded, frank enquiry' would appeal to 'the general reader'.[57]

The fact that so many readers praised Warren's literary technologies, all while refusing to be recruited as virtual witnesses and agree that 'LOST EDEN IS FOUND', hints at the cultural function of these syntheses: they symbolized, even if imperfectly, the survival of a respectable intellectual space in which palaeoscientific expertise represented just one route to knowledge of Earth's history. They acted

---

[52] John Tebbel, *A History of Book Publishing in the United States: Volume II: The Expansion of an Industry, 1865–1919* (New York: R. R. Bowker, 1972), 253 (Riverside); John Tebbel, *Between Covers: The Rise and Transformation of Book Publishing in America* (New York: Oxford University Press, 1987), 121 (Houghton).
[53] 'Where Is Eden', *New York Times*, 5 April 1885, 5.
[54] Winchell, 'Ignatius Donnelly's Comet', 106.
[55] 'Paradise Found', *Atlantic Monthly*, 56 (1885), 126–32 (131).
[56] Untitled review of *Paradise Found*, by William Fairfield Warren, *Spectator*, 58 (1885), 886.
[57] 'Paradise Found', *Atlantic Monthly*, 132.

as counterweights to the likes of *Nature*, which dismissed *Paradise Found* in a short review, and which was redefining what it meant to be a scientific researcher by downgrading, in Melinda Baldwin's words, 'the man of letters' from 'general intellectual' to 'literary specialist' and sowing doubt on scientific contributions by 'politicians' due to their vocation's reliance on 'rhetoric'.[58] Warren would accept no narrowing of the scope of learning. By ignoring 'historical and humanistic studies', and the learning of past centuries, modern naturalists were rejecting 'any kind of truth outside the limited range of their own specialized field' (316). In contrast, *Paradise Found* was addressed to 'truth-seeking spirits,—not less to the patient investigators of nature than to the students of history, of literature, and of religion' (xii). These readers were encouraged to respect the knowledge contained in those enduring works of literature that contrasted so dramatically with the fast-paced and transient world of contemporary scientific publishing. His monograph embraced insights from earlier eras, including, of course, those contained in the Bible and the classics, but also those of Dante Alighieri, seventeenth-century geotheorist Thomas Burnet, and eighteenth-century mythographer Jacob Bryant. Warren surely sympathized with Gladstone, who, as we heard in chapter 1, doubted that Georges Cuvier's Napoleonic-era views on geohistory had lost their authoritativeness by the 1880s.

Marshalling the results of wide secondary reading into a compelling, forensic, and unprovable argument, especially one that implied scientific specialists still had something to learn, was a skill too culturally valuable to disappear. Review columns of newspapers and generalist or otherwise literary periodicals remained a space where these arguments, at least when put forward by men of privileged position and reputation, and in forms legible to educated readerships, could be legitimately discussed. After all, allowing these figures a voice meant that generalist journalists and their readerships were numbered among the 'truth-seeking spirits' deemed able to judge the merits of works on emotive subjects from Earth's deep history like the submergence of Atlantis and Eden. Even if Warren's hypothesis was unpersuasive, rejecting his book wholesale meant conceding this final corner of the public sphere.

Enough of H. H. H

Meanwhile, in Britain, one prolific suppositional synthesizer enjoyed an even more dramatic brand of double-edged success. Like Donnelly, Henry Hoyle Howorth was a barrister turned politician; the son of a wealthy Lancashire family admitted to the Bar in 1867, he sat as a Unionist Conservative MP for South

---

[58] Melinda Baldwin, *Making* Nature: *The History of a Scientific Journal* (Chicago: University of Chicago Press, 2015), 77, 83.

Salford, Greater Manchester, between 1886 and 1900 (Figure 4.2). Howorth was politically close to Lord Salisbury, the Prime Minister; indeed, his opinionated letters to *The Times*, appearing under the alias 'A Manchester Conservative', led one Liberal periodical to dub him Salisbury's 'Manchester henchman'.[59] Even before entering politics, however, Howorth had been more interested in archaeology and ethnology, with a focus on Mongolia, than in his legal career. Ensconcing himself in scholarly societies, he built a reputation as an opinionated author on scientific and antiquarian topics before shifting focus, around 1880, to a project on flood-catastrophes. Writing in the technical *Geological Magazine* and in increasingly large books, Howorth attempted to prove that the ice age, a commonplace of scientific thought since the mid-nineteenth century, had been no ice age at all, but rather an immense deluge that had decimated the planet within the palaeolithic age. This was, he insisted, not a biblical argument, but rather a naturalistic interpretation, drawn from much of the same evidence used by Donnelly and Warren.

Mammoths lay at the project's centre, the existence of their remains in the northern hemisphere signifying not their adaptability to the cold, Howorth argued, but their extinction following the sudden obliteration of a warm climate by the titular flood. Howorth presented his work as an attack on Charles Lyell's uniformitarianism, the methodological assumption of the same rate of geological causes acting in the past as in the present. After his death, Lyell's humble precept had mutated into a cultural strawman, cited by disbelieving opponents as prohibiting the postulation of any catastrophic events in geohistory.[60] In the habit of a lifetime, the catastrophist Howorth rebelled against this (somewhat manufactured) orthodoxy. As recounted by William Boyd Dawkins, his old classmate at the fee-paying Rossall School in Lancashire (and the savant to whom the *American Naturalist* had directed credulous readers of Donnelly's *Atlantis*), Howorth's 'thirst for knowledge' had early on led him to reject 'the beaten track'.[61] The iconoclastic Howorth claimed to have absorbed 'the habit of always adopting a combative attitude towards individuals ... who are "cock shure" [sic]' from the scientific naturalist T. H. Huxley, motivating his opposition to the 'dominant school' of 'orthodox' geology.[62] He accrued a contentious Fellowship of the Royal Society in 1893, cementing his position as one of the most prominent scientific controversialists of the *fin de siècle*.

Howorth seems like precisely the species of literary-political figure the coming generation of scientific practitioners desired to exclude. Instead, he and his

[59] 'Gladstone or Salisbury', *Congregationalist*, 13 (1884), 715–26 (715).
[60] For a nuanced interpretation of uniformitarianism and catastrophism, see Adelene Buckland, *Novel Science: Fiction and the Invention of Nineteenth-Century Geology* (Chicago: University of Chicago Press, 2013), chapter 3, esp. 106–13.
[61] William Boyd Dawkins, 'Sir Henry Hoyle Howorth, K.C.I.E., D.C.L., F.R.S.', *Man*, 23 (1923), 138–39 (138).
[62] Henry Hoyle Howorth to T. H. Huxley, 2 February 1893, Papers of Thomas Henry Huxley, vol. 18, f. 308, Imperial College London, as filmed by the Australian Joint Copying Project (M876-M916) <http://nla.gov.au/ nla.obj-878,210,676>.

Figure 4.2 Undated photograph of a characteristically resolute Henry Hoyle Howorth, wearing ceremonial sword. Letters and Autographs of Zoologists with Biographies and Portraits, vol. 4, Ellen S. Woodward Collection, Blacker Wood Collection, McGill University Archives, Montreal.

diluvial articles became a perpetual presence in the *Geological Magazine*. The circumstance is partially explained by the needs of the editor, Henry Woodward, to keep this struggling commercial journal afloat. Woodward, as Gowan Dawson demonstrates, could not always afford to refuse eye-catching borderline research by a larger-than-life public figure.[63] Howorth appears to have recognized Woodward's pragmatism, encouraging the publication of one paper by pointing out that, even 'if you cannot quite agree with its conclusions', its 'scientific' argument would 'interest a good many people'.[64] General interest, Howorth knew, lubricated the acceptability of maverick content. His saturation of the precarious periodical did not go unchallenged. 'Life is too short to be squandered on a barren and interminable controversy', Howorth's exasperated former schoolfellow Dawkins complained in 1883, warning Woodward that '[y]ou will have enough of H. H. H. before you have done with him, if you don't let him destroy the G.M.'[65] Despite these ominous predictions, the 'G.M.' survived and Woodward continued to publish Howorth's writings into the twentieth century. For all his contrarianism, Howorth avoided the fate of the Duke of Argyll, a politician-geologist who shared some of the Manchester Conservative's marginal scientific views but who found himself shut out of scientific periodicals towards the end of his life. Whereas the Duke railed against this boundary work, Howorth insisted that he had 'no sympathy' with Argyll's 'diatribes about scientific conspiracies'.[66] As the Cambridge geologist Thomas McKenny Hughes noted, Howorth was never 'discourteous' or 'unfair'.[67]

Although the publication of numerous articles in the *Geological Magazine* indicates his surprisingly extensive access to late Victorian science, Howorth's massive books were more widely known. The first instalment in his deluge series, *The Mammoth and the Flood: An Attempt to Confront the Theory of Uniformity with the Facts of Geology* (1887), dealt solely with palaeontology and archaeology, allowing Howorth to 'remit all questions' of a geological nature to the 'subsequent volume'.[68] Within this qualified scope, the MP built his argument from the bottom up, expansively quoting in chronological order not simply all known global deluge lore, but also all textual evidence of knowledge about mammoths dating from antiquity to the present. Data included Harun al-Rashid's gift of a unicorn

---

[63] Gowan Dawson, '"An Independent Publication for Geologists": The Geological Society, Commercial Journals, and the Remaking of Nineteenth-Century Geology', in *Science Periodicals in Nineteenth-Century Britain: Constructing Scientific Communities*, ed. by Gowan Dawson, Bernard Lightman, Sally Shuttleworth, and Jonathan R. Topham (Chicago: University of Chicago Press, 2020), 137–71.

[64] Henry Hoyle Howorth to Henry Woodward, 3 May 1891, Letters and Autographs of Zoologists with Biographies and Portraits, vol. 4, Ellen S. Woodward Collection, Blacker Wood Collection, McGill University Archives, Montreal.

[65] William Boyd Dawkins to Henry Woodward, 11 August 1883, Letters and Autographs of Zoologists, vol. 2.

[66] Howorth to Huxley, 2 February 1893, Papers of Thomas Henry Huxley, vol. 18, f. 308. For Argyll, see Baldwin, *Making Nature*, 81–83.

[67] Thomas McKenny Hughes, 'The Causes of Glacial Phenomena', *Nature*, 48 (1893), 242–44 (244).

[68] Henry Hoyle Howorth, *The Mammoth and the Flood: An Attempt to Confront the Theory of Uniformity with the Facts of Recent Geology* (London: Sampson Low, Marston, Searle, & Rivington, 1887), xix. Subsequent references provided in text.

horn and griffon claws to the court of Charlemagne (5), Chinese descriptions of a burrowing ratlike creature called a 'Fyn Shu' (77), and the flood narrative in Ovid's first century CE poem *Metamorphoses* (435–37). In rallying this fleet of discrete facts, leaving no stone unturned, Howorth showed himself to be yet another discipline of Bacon. With his data, he painstakingly built up an 'induction' (293) showing the historical reality, refusing to 'burden my case with hypotheses' as to this flood's cause 'until all my facts are produced' (324). Given that readers would have to wait for his geological information, the *Zoologist* quipped that the book 'stops short . . . exactly as the first volume of a novel leaves the hero at the most critical period'.[69]

The two-volume *The Glacial Nightmare: A Second Appeal to Common Sense from the Extravagance of Some Recent Geology* (1892–93) followed, but Howorth's continued refusal prematurely to speculate, or to put all his cards on the table, took suppositional caution to extremes. The consistency of this strategy of quasi-serialized deferral meant that, with the three-volume *Ice or Water: Another Appeal to Induction from the Scholastic Methods of Modern Geology* (1905) left incomplete, the third volume being unpublished, Howorth never actually finalized the geophysical conclusions from his induction.[70] The main point was that the flood that obliterated the mammoths, mistaken by geologists for an ice age, had apparently been prompted by the rapid, catastrophic eruption of mountain ranges. In a sense, however, incompletion was the point of this project: absent-minded profusion contributed to Howorth's self-fashioning. One journalist, presumably writing with the MP's blessing, reported that 'he boasts that no one has ever read' his colossal *History of the Mongols* (1876–88) 'except himself and the printer's devil'.[71]

Although the unremitting density of its sequels presumably taxed most readers beyond endurance, the compilatory form of his best-known volume, *The Mammoth and the Flood*, gave the appearance of allowing a general reader to accrue all the information necessary to judge this part of the subject. Howorth's strategy here was recognized by critics like that of the *Quarterly Review*.[72] The professedly 'romantic' (1) appeal of his antiquarian content also attested to the book's readerly value. Howorth admitted to the argumentatively superfluous nature of much of his content: after a discussion of the historic association of mammoth bones with those of giants, he appended unrelated 'notices of giants which are not uninteresting in themselves' (14). In so doing, he rejected prevailing trends towards abbreviation in technical scientific writing. Alex Csiszar has shown that,

---

[69] Untitled review of *The Mammoth and the Flood*, by Henry Hoyle Howorth, *Zoologist*, 11 (1887), 438–40 (439).

[70] Albert R. Corns and Archibald Sparke, *A Bibliography of Unfinished Books in the English Language* (London: Bernard Quaritch, 1915), 115. For inductive presentation in geological writing, see Mott T. Greene, *Geology in the Nineteenth Century: Changing Views of a Changing World* (Ithaca, NY: Cornell University Press, 1982), 174.

[71] 'Salford—South', newspaper clipping, c.1895, Letters and Autographs of Zoologists, vol. 4.

[72] 'The Mammoth and the Flood', *Quarterly Review*, 166 (1888), 112–29.

following the rise of 'prolix' scientific memoirs in the eighteenth century, 'durable contributions to knowledge' in which 'literary merit' was 'valued', the influence of commercial journals on nineteenth-century scientific publishing encouraged novelty and 'condensation' over compilation, reducing 'contextual matter' like 'historical framing' and thus 'limiting the reading audience' of new contributions.[73] Howorth, in contrast, lauded the noble art of the 'compiler', inductively '*bringing together remote facts*' in the tranquillity of the 'closet' (xxiii). This angle helpfully downplayed the stigma attached to second-hand popularization by framing it as inductive methodology. Comprehensive, curious, and enduring, Howorth's books belonged to a backward-looking vision of scientific literature as fit for a gentleman's library.[74]

Adding a counter-current of urgency to this otherwise dilatory presentation was Howorth's presentation of his views as a Renaissance or Reformation sweeping away Gothic relics. Uniformitarian geologists, with their belief in gradual ice ages rather than a sudden inundation, were encumbered by the same 'metaphysical methods of reasoning' that 'caused the scientific sterility of the Middle Ages' (vii), in this case by slavishly repeating the 'dogmatic creed' (xi) of the 'Archpriest' Lyell (xiii). Their resulting 'glacial nightmare' (xx) left geologists pursuing 'mediæval ghosts in the shape of scholastic formulæ'.[75] Anti-Catholic dismissals of medieval scholasticism as anti-scientific clutter had been routine among Protestant thinkers, including Bacon himself, since the Reformation, and in Howorth's day liberal Protestant reformers were distancing Christianity from dogma like never before.[76] Howorth reinforced his enlightened credentials through derogatory references to the '*naïveté* of the Jews' and their 'infantile' Genesis cosmogony (viii), assuring Huxley that he lived in 'constant dread' of 'Noah'.[77] By drawing on stock attitudes shared by both scientific naturalists and liberal Protestants, and turning them against mainstream geologists, Howorth camouflaged his diluvial work's disturbing similarity to biblical literalism. After all, at times he sounded uncannily like young-earth geologists of earlier decades, who, in attributing the fossil record to the Deluge, had wielded against geohistorical speculations not just the Bible but also Baconian induction and common sense.[78]

---

[73] Alex Csiszar, *The Scientific Journal: Authorship and the Politics of Knowledge in the Nineteenth Century* (Chicago: University of Chicago Press, 2018), 31, 37, 211.

[74] James A. Secord, 'Science, Technology and Mathematics', in *The Cambridge History of the Book in Britain: Vol. VI 1830–1914*, ed. by David McKitterick (Cambridge: Cambridge University Press, 2009), pp. 443–74 (452).

[75] Henry Hoyle Howorth, *The Glacial Nightmare and the Flood: A Second Appeal to Common Sense from the Extravagance of Some Recent Geology*, 2 vols (London: Sampson Low, Marston & Company, 1893), I, vi ('ghosts').

[76] James C. Ungureanu, *Science, Religion, and the Protestant Tradition: Retracing the Origins of Conflict* (Pittsburgh, PA: University of Pittsburgh Press, 2019), 107–8, 113.

[77] Howorth to Huxley, 2 February 1893, Papers of Thomas Henry Huxley, vol. 18, f. 308.

[78] Ralph O'Connor, 'Young-Earth Creationists in Early Nineteenth-Century Britain: Towards a Reassessment of "Scriptural Geology"', *History of Science*, 45 (2007), 357–403 (368, 377).

For the out-of-touch Gladstone, diluvial theorists like Howorth and the Duke of Argyll were plainly 'serious scientific inquirers'.[79] Most general readers, however, could hardly miss that Howorth's challenge to uniformitarianism was professedly 'unorthodox'.[80] Much that applies to the response to Donnelly and Warren (both of whom were published in Britain by Howorth's publisher, Sampson Low)—criticism of forensic methods alongside respect for independent-minded diligence—applies to the extensive contemporary reception of Howorth's tomes. This time, however, the sheer wealth of his evidence meant that even geologists, like Hughes in *Nature*, felt obliged to 'welcome' the 'protest against the extravagant views of the extreme glacialists' contained in Howorth's 'valuable encyclopædia'.[81] From the sporting *Field* to *Nature* itself, critics attested, in the *Spectator*'s words, to each book's value as a 'storehouse of information ... for the purposes of reference'.[82] By devoting so much space to compilation, and so little to interpretation, Howorth made his books extremely repurposable: their amenability to literary replication meant that his works could be approvingly quoted by professional geologists and young-earth creationists alike. This susceptibility was the fate of many such tomes, recalling, for example, a similar quality of Frazer's contemporaneous *Golden Bough*, the compendiousness of which, in Steven Connor's words, enabled readers to 'refigure its evidential material ... without reference to its governing theory'.[83] Indeed, while journalists fondly described Howorth as 'nothing if not heterodox', and his ice age denialism enjoyed qualified endorsement at best, it was, in fact, his ability to toe the lines of scientific print culture that distinguishes him from most of the other thinkers I discuss in this book.[84]

Howorth's magisterial diluvialism was defused by generically recategorizing his monograph as a reference work. So doing enabled him safely to be eulogized as the noble relic of an age of versatile polymaths. Peter Burke finds in nineteenth-century Britain an intellectually specializing climate increasingly 'hostile to polymaths', on the one hand, and, on the other, an idealization of the 'all-rounder', a gentlemanly amateurism forged on the playing fields of Rugby School.[85] Such figures remained usefully elegiac symbols in the twentieth century. During the Great War, for instance, the liberal intellectual elite, from the British Association for the Advancement of Science to *The Times*, contrasted its own

---

[79] 'Mr. Gladstone on Recent Corroborations of Scripture', *Manchester Guardian*, 23 September 1890, 6.
[80] Untitled review, *Zoologist*, 439.
[81] Hughes, 'The Causes of Glacial Phenomena', 244.
[82] 'The Mammoth and the Flood', *Spectator*, 61 (1888), 275–76 (276).
[83] Steven Connor, 'The Birth of Humility: Frazer and Victorian Mythography', in *Sir James Frazer and the Literary Imagination: Essays in Affinity and Influence* (Basingstoke: Macmillan, 1990), 61–80 (77).
[84] 'Salford—South', newspaper clipping, c.1895, Letters and Autographs of Zoologists, vol. 4.
[85] Peter Burke, 'The Polymath: A Cultural and Social History of an Intellectual Species', in *Explorations in Cultural History: Essays for Peter Gabriel McCaffery*, ed. by David F. Smith and Hushang Philsooph (Aberdeen: Centre for Cultural History, 2010), 67–79 (74, 76).

humanistic individualism with the specialist technocracy of German society.[86] When Howorth died in 1923, his obituarists framed him as the romantic representative of an unspecialized past. As *The Times*, old outlet for the 'Manchester Conservative', fondly remarked, Howorth possessed a mind becoming 'increasingly rare with the growth of specialism'.[87] These obituarists usually had little to say about Howorth's actual contributions to geology. Dawkins framed his value as residing less in 'his scientific knowledge' than 'his rare and attractive personality'.[88] Significantly, observed the *Manchester Guardian*, Howorth 'made no mark' in parliamentary debates 'but was a centre of attraction in the lobbies and tearooms'.[89] As in the case of his constantly deferred argument about deluges, the journey was more important than the destination. *The Mammoth and the Flood* attested that Howorth was at least as interested in showing that rigorous science could be practised in the 'closet' (xxiii) as he was in proving any particular point about flood-catastrophes.

The Lure of the Subjective

These flood-catastrophist synthesists attempted to legitimize their borderline claims through a speculative restraint that, in the case of Howorth, verged on obsession. This restraint rendered the composition of imaginative depictions of catastrophe-engulphed prehistoric worlds, whether in prose or image, unacceptable. The non-pictorial approach lent sublimity to Donnelly's near-unseen Atlantis, Warren's Eden, and Howorth's deluge, but it compromised attempts to recruit virtual witnesses. Often, suggestive parts of these works were absorbed into more bombastic texts I have previously mentioned. What H. P. Blavatsky reframed as the 'rare intuition' of *Atlantis*, for instance, became valuable evidence for the more vivid discussion of that super-civilization in *The Secret Doctrine*, while *Paradise Found* was cannibalized into the mock-scholarly apparatus of Willis George Emerson's *The Smoky God*, in which the protagonists discover the Eden that Warren believed gone without a trace.[90] Authors of suppositional syntheses were, however, effective at attracting readerships hungry for their brand of science writing: scholarly enough to merit discussion in specialist periodicals, but presented to the judgement of general readers—enchanted, without surrendering

---

[86] Anna-K. Mayer, 'Reluctant Technocrats: Science Promotion in the Neglect-of-Science Debate of 1916–1918', *History of Science*, 43 (2005), 139–59.
[87] 'Sir Henry Howorth: A Life of Wide Interests: Politics, Science, and Art', *The Times*, 17 July 1923, 14.
[88] Dawkins, 'Sir Henry Hoyle Howorth', 138.
[89] 'Sir Henry Howorth', *Manchester Guardian*, 17 July 1923, 16.
[90] Helena Petrovna Blavatsky, *The Secret Doctrine: The Synthesis of Science, Religion, and Philosophy*, 2 vols (London: Theosophical Publishing Company, 1888), II, 782; Willis George Emerson, *The Smoky God; or, A Voyage to the Inner World* (Chicago: Forbes, 1908), 26, 41, 100, 106.

sobriety. As the twentieth century progressed, this balance came under increasing strain.

The strain was not due to any reduction in the supply of relevant scientific evidence about flooded continents, including Atlantis. As part of his long-standing efforts to provide historical backing for Genesis, John William Dawson speculated, in 1894, on the possibility of bringing 'the tradition of that perished continent into harmony with geology' by conceiving it as an antediluvian land frequented by the early humans known as Cro-Magnons.[91] In 1912, the same year that Alfred Wegener introduced continental drift to German geologists, Pierre Termier, Director of Service of the Geologic Chart of France, threw his weight behind Atlanticism in a dramatic lecture given to the Paris Institut Océanographique (and eagerly discussed by anglophone geologists after its 1915 translation).[92] Further afield, British radiochemist Frederick Soddy lyrically speculated that deluge myths had stemmed from a forgotten civilization's self-destructive misuse of atomic power, explicitly associating this civilization with Atlantis in a 1917 lecture.[93] Even more promisingly, from the perspective of suppositional synthesizers—despite modernizing trends towards 'ethnographicization' and 'dehistoricization'—hyper-diffusionism enjoyed a resurgence in academic anglophone anthropology during the 1910s and 1920s.[94] University of London anthropologist William James Perry, for instance, traced all world culture to Egypt in *The Children of the Sun* (1923). As dedicated Atlanticists knew, this was just one step away from endorsing the lost continent itself.

Amid this resurgence of interest in sunken continents and hyper-diffusionism, Lewis Spence, a Scottish journalist, scholar, and litterateur living in Edinburgh, rose to prominence. Spence came to the sunken continent via his extensive research into Mexican antiquities. Like the previous scholars I have discussed, this was chiefly a textual encounter: Spence claimed to have spent 'sixteen years in reading and digesting every line, ancient and modern, which had been written or printed' on Aztec religion, filling a 'fair-sized room' with books, keeping up a 'diligent periodical search among quarterlies and the journals of such learned societies' as were relevant, and employing 'press-cutting agencies' to keep track of newspaper items.[95] Like Donnelly, Spence found the connections between Indigenous American and Egyptian culture too similar to be coincidental. Consequently, his articles in the *Occult Review*, the high-end esoteric periodical published in Britain

---

[91] John William Dawson, *The Meeting-Place of Geology and History* (Chicago: Fleming H. Revell, 1894),157.
[92] De Camp, 164.
[93] Mark S. Morrison, *Modern Alchemy: Occultism and the Emergence of Atomic Theory* (Oxford: Oxford University Press, 2007), 160–65.
[94] Stocking, *After Tylor*, chapter 5.
[95] Lewis Spence, 'Investigating a Subject', *Library Review*, 26 (1933), 45–50 (47).

and the United States by William Rider and Son (a firm directed by the aristocratic occultist Ralph Shirley), turned Atlantis-wards.[96]

The initial culmination of this work, *The Problem of Atlantis* (1924), an elegant volume also published by Rider, was intended to place 'the whole problem on a more accurate basis' than that of 'unscientific' predecessors.[97] A moderate on paranormal matters and painfully aware of the problematic association of Atlantis with unsubstantiated clairvoyant pronouncements, Spence had expressed 'sympathy with the spirit of the esoteric societies' in his *Encyclopædia of Occultism* (1920) but also 'disbelief in the occult knowledge of the generality of their members', comparing the lost-continent research of Theosophists to that of 'pseudo-scientists' writing 'imaginative fiction'.[98] Rather than legitimize this fictive Atlantis, Spence sought a credible alternative not reliant on clairvoyant communications, albeit one still opposed to the debunking conclusions of 'official archæologists' (75). Unshakably confident that world myths always contained kernels of truth, he insisted that Plato's story was conspicuously free of 'fictional atmosphere', including any trace of 'allegorical character' (14). As I demonstrated in chapter 1, literary close reading skills had similarly been deemed integral to any sensitive interpretation of Genesis.

In *The Problem of Atlantis*, Spence argued that a huge Atlantic continent had begun to disintegrate at the end of the Miocene period, splitting into Atlantis proper, near the Mediterranean, and an island called Antilia in the West Indies. He concluded that, as Dawson had hinted, Plato's story was an exaggerated description of the palaeolithic Cro-Magnon or Aurignacian people, who had migrated into Europe from Atlantis before its final sinking around 9600 BC. Spence's explanation was indebted to certain palaeoanthropologists' lionization of Cro-Magnon people as having enjoyed an unsurpassed natural-cultural balance of artistic sensitivity and virile adaptability to harsh conditions. The best-known promulgator of this view was Henry Fairfield Osborn, whose book, *Men of the Old Stone Age* (1915), was one of Spence's main sources.[99] In Spence's view, the pioneering Cro-Magnons had spread flood myths across the Atlantic, along with practices like pyramid-building. As in most suppositional reconstructions of racial migrations, and following Osborn's example, Spence attributed civilization to Aryans, whose culture had diffused 'from West to East' (230). This was a pointed rejection of Theosophical Orientalism, with its lauding of ancient Egyptian and Indian cultures.

---

[96] Morrison, *Modern Alchemy*, 33.

[97] Lewis Spence, *The Problem of Atlantis* (London: William Rider & Son, 1924), v, 124. Subsequent references provided in text. For Spence's theory, see De Camp, *Lost Continents*, 91–98.

[98] Lewis Spence, *An Encyclopædia of Occultism: A Compendium of Information on the Occult Sciences, Occult Personalities, Psychic Science, Magic, Demonology, Spiritism and Mysticism* (London: George Routledge, 1920), xii, 50.

[99] Ronald Rainger, *An Agenda for Antiquity: Henry Fairfield Osborn & Vertebrate Paleontology at the American Museum of Natural History, 1890–1935* (Tuscaloosa: University of Alabama Press, 1991), 146, 174.

White supremacism was nothing new, but Spence's agenda was unusually precise. *The Problem of Atlantis* argued 'that Scotland's admitted superiority in the mental and spiritual spheres springs almost entirely from the preponderant degree of Crô-Magnon blood which runs in the veins of her people' (230). Spence's Atlantis thereby wrote Scottish nationalism into prehistory at a time when, he claimed, 'fear of racial extinction' via absorption by 'Englishmen' was inspiring 'Scotsmen ... to urge the necessity of self-government'.[100] A fervent advocate of Scottish culture, he founded his own Scottish National Movement, a breakaway from the larger Scots National League, a few years after *The Problem of Atlantis* was published. Spence was not always a valued ally in the nationalist cause. In the words of one later historian, he was a politically 'naïve' cultural nationalist whose 'bizarre ideas' undermined the credibility of independence.[101]

However 'bizarre' he may have appeared to fellow nationalists, Spence's *The Problem of Atlantis* displayed all the elegant conventions of the suppositional synthesis: summative, enumerative, forensic, and neatly structured into disciplinary chapters like 'The Evidence from Geology' and 'The Evidence from Central American Archæology'. Indeed, Spence emulated Donnelly's literary form and praised his predecessor's use of chapter titles as 'signposts' of 'the most important departments of Atlantean study'.[102] But, while Donnelly's readers had been invited to take the imaginative view of Captain Nemo, gazing at the toppled columns of Atlantis, Price found another way to transport his readers back to the lost city without undue speculation. Although no remains of Atlantis proper remained, vestiges of its probable former colonies did. Thus, when Spence described cyclopean Peruvian relics 'at some length', he did so because 'they preserve the atmosphere of what I believe Atlantis to have resembled' (190). Use of photographic reproduction technologies allowed for a more direct visual experience. A relatively generous supply of halftones of archaeological sites like these allowed readers, following Spence's lead, to absorb the 'atmosphere' of Atlantis through its analogical architectural descendants (Figure 4.3). Evidentiary prudence and tantalizing, romantic suggestion walked hand in hand.

Although Spence's work on lost continents always discussed geological evidence, he followed suppositional predecessors in valuing cultural tradition and archaeological analogies just as—or even more—highly. Unlike predecessors, he was able tactically to exploit the extreme polarization of interwar geological thought over the divisive issue of continental drift. In *The Problem of Lemuria* (1932), Spence remarked that disputes over the mobility of continents rendered geology 'chaotic', meaning that stably textual evidence could 'be

---

[100] Lewis Spence, 'The National Party of Scotland', *Edinburgh Review*, 248 (1928), 70–87 (75).

[101] Richard J. Finlay, *Independent and Free: Scottish Politics and the Origins of the Scottish National Party, 1918–1945* (Edinburgh: John Donald, 1994), 51–53.

[102] Lewis Spence, 'Ignatius Donnelly', in Ignatius Donnelly, *Atlantis: The Antediluvian World*, rev. edn, ed. by Egerton Sykes (London: Sidgwick and Jackson, 1950), xvii–xix (xvii).

To face p. 198. *PLATE XII.*

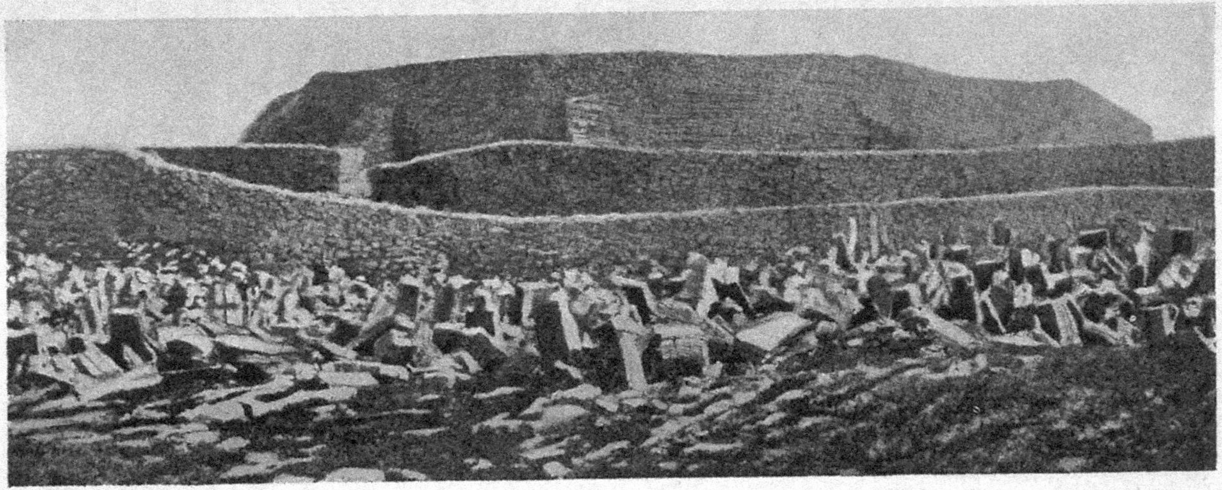

The Citadel of Dun Ængusa, Aran Islands.

(Photo by Dr G. Fogerty, R.N.)

Figure 4.3 Lewis Spence, *The Problem of Atlantis* (London: William Rider & Son, 1924), facing 198. This halftone of the circular 'Citadel of Dun Ængusa, Aran Islands' appeared opposite the assertion that this ancient Irish structure was 'undoubtedly built on the Atlantean plan' (198). Copy in author's collection.

appealed to, over and above the mere geological testimony'.[103] Similarly, evidence based on biogeography was 'much too contradictory to allow the layman to arrive at any conclusion'.[104] Scientific lack of consensus empowered generalists like Spence, who saw these disputes as undermining specialist authority. The ability of the layman to comprehend, moreover, was evidence of a sound theory. Spence's Lemuria, like his Atlantis, was considerably different from the versions seen via the Theosophical 'Akashic Records', 'wretched inventions which fall immeasurably beneath the avowed fictions of a Swift, an H. G. Wells, or an M. P. Shiel'.[105] In a 'hypothesis' shown to have 'gradually unfolded itself quite naturally' in his mind and based on findings 'much more susceptible of rational deduction than those of geology', Lemuria became an Oceanian component of the wider 'Atlantis Culture-complex'.[106] Happily for those more desirous of geological evidence, striking proof of Lemuria's former existence was encountered by the Sir John Murray Oceanographic Expedition to the Indian Ocean just a year after the publication of Spence's book on the subject.[107]

Spence was an eminent journalist with credentials in the world of folklore, and, as such, *The Problem of Atlantis* and its sequels received a hearing across major British and American scholarly and literary journals. Impressed but unpersuaded critics, like botanist Robert Neal Rudmose-Brown in the *Geographical Journal*, praised Spence for 'marshalling the evidence' for a case 'worth attempting', while anthropologist Thomas Athol Joyce concluded in the *Times Literary Supplement* that the book was 'the best yet put forward', if 'by no means conclusive'.[108] *Nature*, notably still reviewing works on Atlantis, admitted that Spence 'conveniently collects' evidence but derided his use of 'newspaper statements' as 'scientific evidence'.[109] In the unstratified world of Atlantis scholarship, home to such armchair detectives as Arthur Conan Doyle, 'newspaper statements' were a valid source of data. Atlantis was particularly liable to attract them: in Spence's wake, a letter to the *Observer* pointed to the bizarre migration patterns of eels as reflecting 'organic memory' of Atlantis, while one 'dilettante' contributor to the *Times of India* insisted that no one could possibly 'dismiss' Spence's hypothesis.[110] One reader, apparently in the mid-1920s, pinned newspaper clippings about the movements of the seabed to the flyleaf of their copy of *The Problem of Atlantis*, and

---

[103] Lewis Spence, *The Problem of Lemuria: The Sunken Continent of the Pacific* (London: Rider, [1932]), 7, 25.
[104] Spence, *Problem of Lemuria*, 180.
[105] Spence, *The Problem of* Lemuria, 91.
[106] Spence, *Problem of Lemuria*, 224, 235, 240.
[107] 'Story of Lost Lemuria', *Chronicle* (Adelaide), 28 December 1933, 3.
[108] Robert Neal Rudmose-Brown, 'The Problem of Atlantis', *Geographical Journal*, 64 (1924), 181–82; Thomas Athol Joyce, 'The Problem of Atlantis', *Times Literary Supplement*, 24 July 1924, 456.
[109] 'The Problem of Atlantis', *Nature*, 114 (1924), 409–10 (410).
[110] Robert Peart, 'The Problem of Atlantis', *Observer*, 10 August 1924, 14; 'Hy-Brasil', *Times of India*, 17 April 1925, 5.

diligently glossed the book in pencil (Figure 4.4). Spence had, after all, addressed readers of 'disciplined imagination' (229), of whom his anonymous annotator might be considered an ideal example.

By his own admission, however, the author's imagination was liable to misbehave. As early as his grounded *Encyclopædia of Occultism*, Spence had admitted that the 'romantic character' of occult phenomena thwarted attempts to assess them using 'purely scientific considerations'.[111] *Problem of Atlantis* ended with a spectacular confession: Spence admitted to possessing an 'intuition' about the continent's nature far stronger than mere 'apparatus of scholarship' (231–32) and far closer to the otherwise denigrated clairvoyance of Theosophists. 'Imagination, vision, if rightly interpreted and utilised', he contended, 'is one of the most powerful aids to historical and archæological understanding; and the ability to cast an eagle glance down the ages is, it seems to me, but one of the first steps in psychic progress' (232). Evidently, the naturalistic and academic presentation Spence thought would place Atlantean studies on 'a more

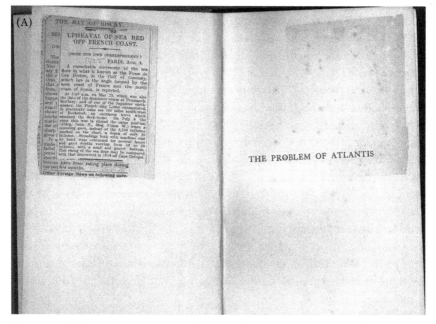

Figure 4.4 (A) Pertinent newspaper clippings, dated 1925 and 1926, attached to a copy of Spence, *The Problem of Atlantis*. (B) At the end of a chapter on 'The Evidence from Geology', the book's anonymous annotator complains that most of the scientific references are 'very old', pointing to the need for 'absolutely the latest scientific opinion' on this contested topic. Copy in author's collection.

---

[111] Spence, *Encyclopædia of Occultism*, ix.

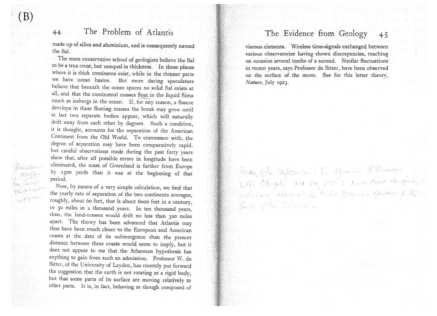

Figure 4.4 Continued

accurate basis' (v) had been adopted with some reluctance. Rather, he envied those occultists who had one-upped palaeoscientists' own time-transcending 'eagle glance'. In *The Problem of Lemuria* he pressed further, despite his protestations against recourse to the quasi-science-fictional *akashic* records, arguing that the 'day is passing when mere weight of evidence alone, unsupported by considerations which result from inspiration and insight, can be accepted'.[112]

Tensions between scholarly approach and deeper intuition resurfaced when Spence contributed to a modernized edition of Donnelly's classic *Atlantis*, published in 1950. The elderly Scottish nationalist now described himself as 'fatally open to the lure of the subjective' and thus painfully aware of the necessity of fighting this 'lure' and sticking, as he believed Donnelly's research had done, to 'the iron discipline of scientific detachment'.[113] The latter referred, in effect, to the suppositional method and genre I have been describing. Spence concluded sadly that 'there are two Atlantises—the Atlantis of fantasy and imagination and that of reality'.[114] A primitive Atlantis, home of Cro-Magnon humans, might be easier to air before the *Geographical Journal* and the *Times Literary Supplement*, but it

---

[112] Spence, *Problem of Lemuria*, 182.
[113] Spence, 'Donnelly', xviii.
[114] Spence, 'Donnelly', xviii.

did not satisfy him in the way Plato's advanced super-civilization did. Plausible suppositions did not scratch that itch.

Positive Facts

The work of one contemporary, James Churchward, demonstrates the kind of confident, occult text Spence may have *wanted* to write, one in which the self-imposed restrictions of the suppositional synthesis genre were shattered from within. Born in Devon, Churchward made a career as a tea planter in Ceylon (Sri Lanka), before moving to the United States around the 1890s. There, having apparently become rich on mechanical patents, he began a wealthy retirement in Mount Vernon, near Manhattan. In the early 1920s, Churchward began to advance the claim that an ancient continent called Mu, former home of a white, monotheistic super-race and the fount of civilization, lay submerged in the Pacific.[115]

Mu was a syncretic variant of Atlantis and Lemuria. The basic tenets of Churchward's theory fused the suppositional deluge tradition with Theosophy: ancient humans (which were pointedly not evolved from apes) had reached a high level of civilization as a white, monotheistic super-society, but this was destroyed by the same catastrophe that raised the mountain ranges—an event that geologists mistakenly interpreted as an ice age. Churchward's Mu became a surprisingly fashionable location in the interwar United States. Visitors to the California Pacific International Exposition, held in San Diego in 1935, could view a 'hydrographic relief map' of the Pacific Ocean in the Palace of Natural History; the Exposition guidebook hinted that this map revealed traces of Mu, which 'presumably existed thirteen thousand years ago . . . according to many scientific authorities'.[116] In the same year, singing cowboy Gene Autry encountered the remnants of Mu's high-tech culture in the twelve-part film serial *The Phantom Empire*.

Churchward's work showed how drastically the desire for textual—in addition to organic—evidence, and for fact over theory, could be wrested free of their scholarly suppositional moorings. Although there are various less impressive precedents for its discovery, Churchward claimed in the 1920s to have learnt about Mu from the so-called 'Naacal tablets', encountered while he was in India during the 1870s.[117] As explained in his first monograph, *The Lost Continent of Mu* (1926),

---

[115] For Churchward's life, see Jack E. Churchward, 'Resources', my-mu.com <my-mu.com/resources.html> [accessed 29 January 2022]. See also De Camp, *Lost Continents*, 47–50; and Card, *Spooky Archaeology*, chapter 6.

[116] *Descriptive Guide Book of the California-Pacific International Exposition at San Diego California 1935* (Chicago: American Autochrome, 1935), 27.

[117] James Churchward, *The Lost Continent of Mu: The Motherland of Man* (New York: William Edwin Rudge, 1926), 234. Subsequent references in text.

these 50,000-year-old tablets were translated with the help of an Indian 'high priest' (2). There is a clear parallel here with the Theosophists' access to ancient wisdom through the secretive Mahatma Letters, but Churchward's assured and consciously artless exposition gave little indication that readers might doubt inaccessible sources. Why would readers need to see the Naacal glyphs themselves, when these were reproduced and translated throughout his book? The ancient tablets were eloquent testimonials to deep history. Uninterested in constructing provisional hypotheses, Churchward rejected scientists' transient 'deductions' in favour of the 'positive facts' (40) that the tablets revealed. A pamphlet promoting his book noted that, while scientists had searched for 'the original home of man' in 'fossil remains' and 'geological formations', Churchward relied on 'more authentic sources'.[118] Even when Churchward acknowledged academic publishing conventions, he treated them with something like contempt, using footnotes to cite vague and unverifiable sources like 'Greek Record' and 'Various Records' (23), and providing his own highly irregular interpretations of the visual genre of the stratigraphic section (121, 253).

Another indication that Churchward's needs transcended suppositional constraints was his transgression of the convention that one's lost world must only be vicariously indicated. Instead, sacrificing caution, he invited virtual witnesses to experience textual and pictorial simulations of ancient Mu (Figure 4.5). Thus an 'Easter Island Tablet' informed readers that the Muvian soundscape resounded with the 'sweet lays' of 'feathered songsters', while 'Indian and Maya Records' (apparently directly quoted) told of 'herds of "mighty mastodons and elephants" flapping their big ears to drive off annoying insects' (24). Even when not quoting inaccessible sources, Churchward appealed to emotion, attempting 'to re-create, if we can' the Muvians' 'feelings of horror and despair' as their continent sank around 11,500 years ago (32). A competent artist working in a naïve style, Churchward filled his works with paintings of the ancient life of Mu and scenes from deep time more generally, some of the latter dated as early as the 1900s.[119]

The author's wealth meant that his idiosyncratic ideas enjoyed a polished public presentation. Churchward's first book on Mu was privately but luxuriously printed by William Edwin Rudge, a prestigious printer-publisher located in Churchward's own city, Mount Vernon, whose work made *The Lost Continent of Mu* into what one sceptical critic admitted was 'a fine specimen of modern typography'.[120] It also overflowed with quality halftones of Indigenous peoples, cryptic symbols, and

---

[118] Jack E. Churchward, ed., *Lost Gems of the Lost Continent of Mu: Reviews and Correspondence from James Churchward's Scrapbooks Volume 1* (Clearwater, FL: Churchward, 2006), 37.

[119] For illustrations of prehistoric animals see James Churchward, *Cosmic Forces of Mu* (New York: Ives Washburn, 1934).

[120] John Tebbel, *A History of Book Publishing in the United States: Volume III: The Golden Age between Two World Wars 1920–1940* (New York: R. R. Bowker, 1978), 343–44; *Lost Gems*, 21.

THE LOST CONTINENT OF MU

THE LAST MAGNETIC CATACLYSM. THE BIBLICAL "FLOOD"
AND THE GEOLOGICAL MYTH, THE GLACIAL PERIOD

42

Figure 4.5 James Churchward, *The Lost Continent of Mu: The Motherland of Man* (New York: William Edwin Rudge, 1926), 42. Churchward illustrates the moment of Mu's destruction, when the Muvians and their mammoth contemporaries flee a 'magnetic cataclysm'. In Churchward's interpretation, subsequent misconceptions about this cataclysm gave rise to notions of the biblical Flood and the scientific ice age.

ancient slabs that helped make the maverick author's prehistoric worldbuilding alluring. Rider, Spence's publisher, dealt with the British editions of Churchward's books.

Spence reviewed these editions in Rider's *Occult Review*, and this meant walking a careful line. After all, in his own books he worked hard to differentiate his lost-continent work from that of authors reliant, like Churchward, on occult evidence. Churchward's approach to Pacific sunken continents, he noted, 'adopts an altogether different method of proof' to that of John Macmillan Brown's *The Riddle of the Pacific* (1924)—a generically conventional suppositional synthesis on Lemurian subject matter more akin to the style of Spence's own *The Problem of Atlantis*—and, damningly, 'does not provide photographs of the original Naacal tablets'.[121] Although Spence was eager to demarcate these unscholarly practices from sounder work, his imaginative sympathy shone through, escalating in conviction even in mid-sentence. If not a rigorous work, and 'heterodox' even on the intrinsically heterodox 'Atlantean question', *The Lost Continent of Mu* was, for those who wished to 'peruse' it, 'an exciting story of what may probably have occurred in the Pacific area'.[122]

Those who wished to 'peruse' this suppositional story did so in significant numbers. By 1933, Churchward's two first books, sold for a not inconsiderable three dollars in the United States and fifteen shillings in the United Kingdom, had purportedly sold around twelve thousand copies.[123] Book sales were also just a small portion of the reach of Churchward's good word, which was spread widely by media technologies old and new. The expatriate Churchward was a regular speaker on New York's WYNC radio station, and copies of his books were sent to European monarchs, while his local newspaper, the Mount Vernon *Daily Argus*, was a reliable expositor of his views.[124] Given that a substantial portion of book reviewers in American newspapers during this period saw their job as simply to paraphrase content for the aid of consumers, many blandly repeated Churchward's theories without criticism, while interviews with the colourful Englishman provided excellent copy.[125] This journalistic dissemination of Muvian theory was part of a long tradition of mutually profitable arrangements with explorers in lost worlds. In the previous decade, British geographer Percy Harrison Fawcett had funded his doomed quest for Atlantean colonies by selling newspaper rights to the

---

[121] Lewis Spence, 'The Lost Continent of Mu: A Critical Appreciation', *Occult Review*, 55 (1932), 102–4 (102).
[122] Spence, 'The Lost Continent of Mu', 103.
[123] 'Finds All Religion Springs from Mu', *New York Times*, 3 April 1933, 8.
[124] *Lost Gems*, 3, 5, 7, 26; Andy Lanset, 'WNYC and the Land of Mu', *WNYC* <https://www.wnyc.org/ story/179,746-wnyc-and-land-mu/> [accessed 29 January 2022].
[125] For uncritical treatment of Churchward, see Lisetta Neukom Higgins, 'When Mu Ruled the World', *Washington Post*, 27 November 1932, n.p. For American newspaper reviewers, see Joan Shelley Rubin, *The Making of Middlebrow Culture* (Chapel Hill: University of North Carolina Press, 1992), chapter 2.

North American Newspaper Alliance.[126] Interpretive chasms on what constituted convincing science yawned between those newspaper critics who expressed judgement on Mu. For the critic in William Randolph Hearst's *Syracuse Herald*, *The Lost Continent of Mu* was 'dripping with erudition', yet, for the New York *Herald-Tribune*, its incoherence outclassed even that of Samuel Taylor Coleridge's famous opium-fuelled poem 'Kubla Khan' (1816).[127]

Those few scholarly critics who reviewed his books in specialist journals were unconvinced. One scholar of Latin American history deemed Churchward's books as belonging 'on the border line between fact and fiction'.[128] Nonetheless, academic discredit was of little importance, given that the Mu series was widely understood as a branch of science speaking directly to the public sphere, or rather a less elevated, more democratic, and intellectually demarcated 'popular' sphere. The *Daily Argus* characterized Churchward's 'cold logic' as appealing not to the scientific but rather to 'the popular mind', providing 'a popular solution which may meet with popular approval'.[129] In the *New York Times*, novelist and critic Edward A. Uffington Valentine explained that Churchward's 'easy logic' was addressed 'not to the scientific world but the common reader', a judgement intended as exoneration that was also voiced in a friendly *Amazing Stories* review by editor Conrad A. Brandt.[130] Evidence for this popular approval can be found in gushing letters to the author. Novelist Aida Rodman de Milt told Churchward that his book 'reads like a romance' and displayed 'a clarity of deduction' and 'logical sequence' that 'put me in touch with the Infinite', while New York electrical engineer Paul Francsek wrote to Churchward asking for further information, his curiosity counter-intuitively piqued by dogmatic criticism he had encountered in one newspaper review.[131]

Writing to announce Churchward's death in 1936, Brandt noted that, although 'I rarely agreed with him, I always admired the strength of his convictions'.[132] Back in the 1880s, critics writing in highbrow periodicals like the *Atlantic Monthly* had carefully explained why suppositional syntheses lacked sound scientific method, predicting that their educated readerships could not necessarily evaluate the reputability of a book like Donnelly's *Atlantis* on textual evidence alone. Over the following decades, opportunities to air suppositional syntheses in intellectual circles had shrunk, but their popularity had grown. By the interwar era, from the United States' most respected newspapers to its pulp science fiction magazines,

---

[126] Brian Fawcett, 'Epilogue', in Percy Harrison Fawcett, *Exploration Fawcett* (UK: Arrow, [1963]), 273–304 (278).

[127] *Lost Gems*, 19, 21.

[128] Curtis Wilgus, untitled review of *The Sacred Symbols of Mu*, by James Churchward, *Hispanic American Historical Review*, 14 (1934), 85–86 (86).

[129] *Lost Gems*, 51.

[130] Edward A. Uffington Valentine, 'Mu, the Lost Atlantis of the Pacific Ocean', *New York Times*, 7 June 1931, 8; Conrad A. Brandt, 'Overthrowing Old Traditions', *Amazing Stories*, 6 (1931), 762.

[131] *Lost Gems*, 6, 20, 31, 43.

[132] Conrad A. Brandt, 'In Memoriam', *Amazing Stories*, 10 (1936), 134.

critics might even puff the wildest suppositional syntheses as charming works of pseudoscience. After all, texts like Churchward's were no serious threat to the authority of palaeoscientists, because, given their razor-thin pretences of academic presentation, they could no longer be confused with scientific works. It was implied that only the uneducated 'popular' mob—unattuned to the ways scientific authority was indicated by ascribed expertise, professional training, institutional affiliation, primary investigation, and access to accredited publishing venues—would succumb to the superficially 'cold logic' of these works.

Unauthorized Notions

A further indication of the special place these suppositional scientific works held in interwar culture is the fact that their reviewers were often neither geologists nor ethnologists but rather poets and authors of fiction. Usually, these reviewers highlighted not the texts' factual accuracy but rather their power of verisimilitude or suggestion. For instance, Brandt in *Amazing Stories* 'recommended' Churchward's 'plausible' work as inspiration for authors of 'scientific fiction yarns'.[133] Some novelists went further, exhibiting the aforementioned 'popular' approach to scientific argumentation. Writing in the *Seattle Post-Intelligencer*, Kenneth Gilbert, veteran romancer of pulp magazines like *Western Story*, enthused that Churchward's 'well written' book communicated a 'convincing' theory that 'seems to the reviewer, the most satisfying one ever offered, of the origin of mankind'.[134] Adopting Gilbert's casually democratic attitude to scientific contributions, readers might deem themselves competent to evaluate complex palaeo-archaeological matters based on the internal consistency and attractiveness of an accessible book. His enthusiasm also evidences the continued strength of popular scepticism about the deflated place humanity had played in the geohistorical story since the palaeoscientific researches of the nineteenth century. Churchward's books on Mu projected white human civilization far back in time, telling many of his readers exactly what they wanted to hear.

To determine how these suppositional books were read beyond the world of professional journalism, we find a useful case study in one set of 'common readers': the circle of impecunious, self-educated American pulp writers clustered around H. P. Lovecraft, whose own interest in the related field of Theosophy I discussed in chapter 2. Lovecraft was pointed towards Spence in April 1928, and by June he had read a library copy of one of Spence's books on Atlantis.[135] In November 1930,

---

[133] Brandt, 'Overthrowing Old Traditions', 762.
[134] *Lost Gems*, 11.
[135] *Dawnward Spire, Lonely, Hill: The Letters of H. P. Lovecraft and Clark Ashton Smith: 1922–1937*, ed. by David E. Schultz and S. T. Joshi, 2 vols continuously paginated (New York: Hippocampus Press, 2017), I, 159, 161.

when his Californian correspondent and fellow contributor to fiction magazines like *Weird Tales*, Clark Ashton Smith, made an offhand reference to Churchward, Lovecraft expressed his curiosity; both had already read Donnelly's *Atlantis*, and, in 1933, Smith ordered a copy of Spence's *History of Atlantis* (1927) direct from Glasgow.[136] Over the following years Lovecraft sought with little success to get his hands on 'Churchwardiana', although difficulty attaining Churchward's '"nut" classic' *The Children of Mu* (1931) did not prevent him from bringing Muvian content into fantasy stories like 'Bothon' (co-written in 1932).[137] In the meantime he, Smith, and others were excitedly trading 'Theveninic' [sic] clippings: excerpts of lurid tabloid articles on lost continents, written by the French savant René Thévenin.[138] The publication of these articles in Hearst's syndicated Sunday *American Weekly* were overseen by the assistant editor, Abraham Merritt, yet another author of weird lost-world romances, including *The Moon Pool* (1919) and *The Face in the Abyss* (1931). Suppositional texts were grist to the mills of pulp-penning autodidacts obsessed with cataclysms, ancient tomes, and lost prehistoric civilizations.

Significant diversity of interpretation regarding the scientific persuasiveness of these texts existed even among so sympathetically minded a group as the Lovecraft circle. Evaluating the books' merits, this circle continued, in private, the kinds of lively discussions seen in science fiction magazines' correspondence sections: fan fora in which questions of literary verisimilitude were debated.[139] A well-read enthusiast of establishment science, Lovecraft prided himself on the ability to distinguish orthodox ideas from the formally convincing imitations produced by suppositional synthesists. He told Smith that Spence's painstaking books on Atlantis were 'by no means unscholarly' but produced 'no scientific convictions when viewed analytically', presumably referring to the dubiously forensic argumentation so often alluded to by expert reviewers of suppositional syntheses.[140] In his own eyes, the scientific consensus on Atlantean civilization, at least, was unequivocal. After attending an 'excellent' lecture on Mayan ruins in 1936, Lovecraft reiterated the speaker's prosaic assertion that 'nothing behind A.D. 68 can be traced', nullifying any claims that America had enjoyed a civilized prehistory as an Atlantean colony.[141] Despite the mass of conflicting scientific opinions over subjects like continental stability, and even despite his habitual fondness for placing white Europeans and their ancestors at the vanguard of culture, an authoritative orthodoxy against Atlantis was discernible to Lovecraft.

---

[136] *Dawnward Spire*, I, 267, 269; II, 470.
[137] *Dawnward Spire*, I, 361, 366; II, 512.
[138] For example, see *Dawnward Spire*, II, 510, 550.
[139] Michael Saler, *As If: Modern Enchantment and the Literary Prehistory of Virtual Reality* (Oxford: Oxford University Press, 2012), 97.
[140] *Dawnward Spire*, I, 161.
[141] *Dawnward Spire*, II, 638.

The ability to consume suppositional texts on subjects like Atlantis without becoming unduly captivated by their claims was essential to Lovecraft's rational self-image. He was thus surprised to learn that the otherwise 'sound and sensible' Merritt, rather than airing sensational content in the *American Weekly* out of an editor's commercial cynicism, was actually 'a trifle inclined to believe in pseudo-scientific extravagances like the Thevenin stuff'.[142] While Merritt's notions of scientific plausibility were, apparently, surprisingly close to the content of his own pulp romances about ancient civilizations, Lovecraft was all too conscious that delusions could be imbibed through credulous reading of fantastic fiction and suppositional non-fiction. After all, he regularly received letters from fans asking for information about his fictional tome, the Necronomicon, and corresponded with a 'Supreme High Pontiff of the Mayan religion' who 'distributes 72-page mimeographed catechisms in which he declares Central America to be the cradle of mankind—all other branches having dispersed through Atlantis & Lemuria'.[143] Churchward's similar disassociation from reality, Lovecraft admitted, likely made for a 'highly blissful' existence, but credulity was not always benign.[144] One conspiracy theorist, the Iowan G. P. Olsen, was, in Lovecraft's words a 'well-known pest of weird & scientifiction writers' to whom Smith attributed 'megalomania, dementia, mystic delirium'.[145]

The scientific and textual borderlines policed by Lovecraft, however, were less clear to his correspondents. Robert E. Howard, author of the 'Conan the Barbarian' stories, professed conviction in Spence's moderate, Cro-Magnon interpretation of Atlantis.[146] Smith, resident of the isolated town of Auburn, California, was also amenable to fringe science. Wishing to avoid 'the gullibility of Sir Arthur Conan Doyle' (a reference to the novelist's Spiritualism, rather than his more respectable Atlanticism), Smith admitted that 'the arguments of material science are pretty cogent', but he also called science 'a guessing-game' biased by the 'preconceived theories' of 'professional scientists'.[147] When speaking with his Catholic correspondent August Derleth, as opposed to the atheistic Lovecraft, Smith went further. Concerned that science would be 'the principal Mumbo Jumbo of the near future', he kept his mind 'open' to the 'weird, unknown and preternatural', including 'the *possible* development in man of those higher faculties of perception which mystics and adepts *claim* to develop'.[148] As such, he was less liable than Lovecraft to

---

[142] *Dawnward Spire*, II, 512.
[143] *Dawnward Spire*, II, 425 ('Necronomicon'); *Dawnward Spire*, I, 176 ('Mayan').
[144] *Dawnward Spire*, II, 512.
[145] *Dawnward Spire*, I, 372; II, 469.
[146] Rob Roehm and Rusty Burke, eds., *The Collected Letters of Robert E. Howard*, 3 vols (Plano, TX: Robert E. Howard Foundation Press, 2007–2008), I, 237.
[147] *Dawnward Spire*, II, 469 ('cogent'); David E. Schultz and Scott Connors, eds., *Selected Letters of Clark Ashton Smith* (Sauk City, WI: Arkham House, 2003), 152 ('guessing-game').
[148] *Selected Letters of Clark Ashton Smith*, 214, 211.

dismiss suppositional syntheses, even those drawing on intuition, as mere imaginative fuel. Smith considered Donnelly's *Atlantis* 'quite solidly done' and found the 'mass of data' in *The Lost Continent of Mu* 'truly interesting', admitting that he had 'no means of knowing how much reliance' could be placed on Churchward's translations of the 'Naacal tablets'.[149] Despite its apparent caution, Smith's phrasing implicitly conceded the existence of the controversial tablets.

These submerged continents appeared regularly in Smith's literary output, typically associated with obsession, erotic delights, and numinous experiences. The poetic fragment 'Atlantis', collected in *The Star-Treader* (1912), for instance, sensuously described the sunken city's 'closéd lips' and 'altars of a goddess garlanded/with blossoms of some weird and hueless vine', while the ageless speaker of 'In Lemuria', collected in *Ebony and Crystal* (1922), recalls the '[c]arnelians, opals, agates, almandines,/I brought to thee some scarlet eve of yore'.[150] Influenced by French Aestheticism and *fin-de-siècle* Decadence, Smith's prose and poetry aimed, in Brian Stableford's words, to give readers 'an experiential wrench' and 'permit the relief of "seeing" worlds of the imagination'.[151] For Lovecraft, absorption in suppositional works was a kind of pleasant self-hoaxing, as when, in chapter 2, we heard him play with the avowedly delusive concept of 'pseudo-memories'. Tellingly, what Lovecraft deemed 'imaginative pseudo-science' had significant overlaps in content with the genre that was gradually becoming known as science fiction.[152] Smith's mental voyaging was less consistently ironized. Moving from poetry towards fiction in the later 1920s and 1930s, he produced numerous weird short stories and artworks concerning Mu, Lemuria, and the fragment of Atlantis known as Poseidonis. His reputation for knowledge in the field of sunken-continent research was such that Smith even became the target of a Stanford University student's technical queries about 'Mu and Lemuria' (which Smith treated as the same Pacific location).[153] In the 1950s, when the science fiction author L. Sprague de Camp was compiling a series of articles that would become one of the earliest attempts to historicize the works of Donnelly, Spence, and Churchward, he wrote to the reclusive Californian for advice.[154]

Smith, an anti-materialistic and independent-minded aesthete, was sympathetic to those who teetered on the edge of other worlds. His short story, 'An Offering to the Moon', completed in 1930 but only published in *Weird Tales* in 1953,

---

[149] *Dawnward Spire*, I, 272.

[150] Clark Ashton Smith, *The Star-Treader and Other Poems* (San Francisco, CA: A. M. Robertson, 1912), 56; Clark Ashton Smith, *Ebony and Crystal: Poems in Verse and Prose* (Auburn, CA: Auburn Journal, 1922), 24.

[151] Brian Stableford, *Outside the Human Aquarium: Masters of Science Fiction*, 2nd edn (San Bernandino, CA: Borgo Press, 1995), 85.

[152] *Dawnward Spire*, II, 503.

[153] *Dawnward Spire*, I, 255.

[154] *Selected Letters of Clark Ashton Smith*, 367.

dramatizes the slide from suppositional examination of lost continents to absorption in intuited theories. Smith's story follows a businessman-archaeologist named Morley as he investigates Muvian ruins, located in the French Polynesian Marquesas Islands. His 'unimaginative' colleague Thoraway is curious that Morley's 'unauthorized notions' about the existence of lunar temples on sunken Mu seem guided by 'subconscious instinct' rather than material evidence, while Morley, whose 'dreamy, beardless, olive features . . . seemed to repeat some aboriginal Aryan type', admits that his own Churchwardian speculations have 'the authority of an actual recollection, rather than a closely reasoned inference'.[155] Perhaps reflecting the manner in which Smith himself imbibed Muvian stimulation, Morley's submerged memories began to resurface after he read 'an illustrated article which described the ancient remains on Easter Island'.[156] These remains were regularly considered remnants of Lemuria or Mu (as demonstrated by Churchward's dubious citation of an 'Easter Island Tablet', discussed above).

Morley's investigative recourse to 'instinct' over 'inference' becomes deadly when his former Muvian incarnation, the source of this 'instinct', takes over his body. He loses his grip on the present entirely, and, as time winds mysteriously backward, leaves the expedition's boat one night for a nearby island. There, lost in ecstasy on an inexplicably prehistoric Mu, he is ritually sacrificed to the lunar goddess. As Spence recognized, it was difficult to explore lost continents with 'the iron discipline of scientific detachment'.[157] For believers, there was a dangerous pleasure in getting lost.

Conclusion

Suppositional synthesizers contended that a probing analysis of manifold published sources could—without relying on revelation or clairvoyance—enable evidence from across the world to be correlated into an all-encompassing narrative about an early human culture submerged by a primordial catastrophe. These lawyers, politicians, and litterateurs were knowledgeable about comparative religion, anthropology, archaeology, geology, palaeontology, and palaeoanthropology, sometimes even publishing in specialist periodicals; their works were built from forbidding reams of conscientiously cited scholarship; their register and structuring emanated logic, hypothesis, or induction; their claims never went beyond the limits of methodological naturalism, nor were they seen to let imagination take free rein in prose or in artistic representation of their subject matter; and

---

[155] Clark Ashton Smith, 'An Offering to the Moon', *Weird Tales*, 45.4 (1953), 54–65 (55, 56).
[156] Smith, 'An Offering to the Moon', 58.
[157] Spence, 'Donnelly', xviii.

they sought to publish with quality presses who ensured that their books radiated dignity. Most trained scientific researchers and educated readers reviewing these books in periodicals typically indicated they were not fully to be trusted, pointing less frequently to the authors' lack of technical expertise than to their unacceptably forensic approach. Despite the growing sense that authoritative scientific writing was indicated by extra-literary factors that suppositional syntheses and their authors lacked, there were many for whom these books were scientific.

Crucially, however, only the most scoffing readers, and the most concerned savants, saw nothing of value in the suppositional project. The idea that a non-disciplinary book, diligently composed by an original mind—whether US populist or Scottish nationalist—could meaningfully contribute to the palaeoscientific conversation was an attractive notion to large, diverse audiences not entirely comfortable with a culture deferring to specialist expertise. Moreover, because these authors aimed to compile all the information necessary to pronounce upon their subject, they involved these audiences directly in their intellectual quest. In this sense, they resembled detective novels solved in reverse, providing the readerly satisfaction of following a self-contained argument that might persuade at least until the pages were closed. Their composition, too, was readerly. From Warren's defence of 'humanistic studies' to Spence's denigration of 'mere geological testimony', these thinkers rejected the notion that secondary reading was no substitute for hands-on investigation of nature. To Spence, a 'daily horizon bounded by book-lined walls' was no less thrilling than the lost worlds of the 'explorer', while fictional detectives like 'Sherlock Holmes ... assuredly have "nothing on" the literary bloodhound who has been tossing all night on his couch awaiting the moment when the doors of his library will open and he can resume the chase of that elusive reference'.[158] A literary method of scientific scholarship opened up bibliographical worlds of rational detection.

The suppositional synthesizers kept alive ideals of high-level, generalist participation in science that had been besieged since the early nineteenth century. However, as we have seen, authors employing this judiciously half-committal genre often buckled under methodological, metaphysical, and stylistic restraints that had been adopted less from conviction than from pragmatism. Suppositional synthesizers could not doubt that the continents and catastrophes they described were real, but nor could they—without considerable sacrifice of intellectual credibility in the manner of Churchward—locate any smoking gun of discursive proof. The inconclusiveness attached to their authors' literary labours made the contents of these books particularly easy to detach and repurpose for very different ends. As chapter 5 shows, the literary technologies of synthesizers like Howorth were easily incorporated into the rhetorical armoury of young-earth creationism.

---

[158] Spence, 'Investigating a Subject', 45.

5
Geohistory, Unimagined

Towards the end of his *magnum opus* textbook, *The New Geology* (1923), George McCready Price reflected on the imaginative obstacles facing anyone attempting to explain the most stupendous geological phenomena:

> As one stands on the brink of the Grand Cañon, or looks down at the seething torrent at the base of a Niagara, or gazes at the lone spires of splintered rock rising through the thin upper air on any mountain top, there are very many phenomena which seem beyond the reach of any explanations we may offer. I am not insensible to such an appeal.[1]

Price's performative humility before near-unfathomable geological forces recalled the experience of an author he knew only too well. Charles Lyell had similarly stood in pious awe of sublime geological power, especially that displayed by Niagara Falls. 'The geologist may muse and speculate', Lyell declared in 1845, 'until, filled with awe and admiration, he forgets the presence of the mighty cataract itself'.[2] While Lyell channelled his 'awe' into the conclusion that the famous waterfalls had been carved out of the rock over a colossal timescale, Price resisted this kind of response to geological spectacle. A Canadian-born member of the Seventh-day Adventist Church who spent most of his life in the United States, Price believed that most rocks had rapidly been stratified during the biblical Flood. This event, which took place several thousand years after the universe was created in six 24-hour days, was also responsible for hewing out features like Niagara Gorge. Any notion of evolution in this short timespan was, of course, out of the question. Eloquent geologists like the late Lyell were dangerous opponents to those wishing to spread the tenets of 'Flood Geology'. Working to rehabilitate a literalist interpretation of Genesis, Price and fellow young-earth creationists were pitting themselves against a culturally, intellectually, and imaginatively compelling palaeoscientific tradition.

The diverse thinkers I have hitherto discussed were united by a confidence about the human mind's ability to unearth scientific knowledge from chasms of

---

[1] George McCready Price, *The New Geology: A Textbook for Colleges, Normal Schools, and Training Schools; and for the General Reader* (Mountain View, CA: Pacific Press Publishing Association, 1923), 692. Subsequent citations of Price's books exclude his first names.

[2] Charles Lyell, *Travels in North America, in the Years 1841–2; with Geological Observations on the United States, Canada, and Nova Scotia*, 2 vols (New York: Wiley and Putnam, 1845), I, 53.

time and space. Early twentieth-century young-earthers, chief among whom was Price, believed that this confidence, and even the more moderate form expressed by professional geologists, was unscientific. These fiercely empirical creationists consequently opposed the potent literary technologies by which palaeoscientific authors and their fringe counterparts alike had peddled geohistorical content to virtual witnesses. To discredit these technologies, young-earthers enforced traditionalist notions of scientific method that characterized the writings of respected naturalists as no more reliable than the most aberrant Theosophical rhapsodies. Historical sciences like geology were, in their eyes, vulnerable to this attack, as they fitted uneasily into conservative philosophies of science modelled on experimental or mathematical testability.[3] If the provisionality of claims about Earth's deep history was what made palaeoscience conveniently malleable for many borderline thinkers, it discredited the endeavour in the eyes of young-earth creationists. Their approach both to biblical interpretation and to geological writing was thus very different to that of the day-age creationists with which I began, let alone that of occultists and hollow-earthers. Nonetheless, we shall see that visionary power remained central in various ways, while young-earthers' rhetoric was indebted both to earlier geological writings and to aspects of the suppositional synthesis tradition, discussed in chapter 4.

Price himself has been the subject of sustained attention by scholars of Christianity and science, most especially Ronald L. Numbers, as one of the leading architects of modern young-earth creationism and anti-evolutionism.[4] Important aspects of his output, however, remain overlooked. Price was not a regularly active field or laboratory researcher, but rather a prolific reader and writer who was considered a 'highbrow' litterateur—even 'a poet by nature'—among the Adventist community (Figure 5.1).[5] With little reverence for truth claims made by authors like Lyell, he treated classic scientific texts as *texts*, arguing that their literary technologies harboured unwarranted, illogical geohistorical presuppositions. This probing attitude not just to content but also to form may be seen as a logical extension of the bibliocentrism of Christian fundamentalism, which Kathleen C. Boone calls a '*verbal* system ... which takes a text [the Bible] as its starting point'.[6]

---

[3] For the undermining of historical sciences, see Michael D. Gordin, *On the Fringe: Where Science Meets Pseudoscience* (Oxford: Oxford University Press, 2021), 7.

[4] For major examples, see Ronald L. Numbers, *The Creationists: From Scientific Creationism to Intelligent Design*, rev. edn (Cambridge, MA: Harvard University Press, 2006); Carl R. Weinberg, '"Ye Shall Know Them by Their Fruits": Evolution, Eschatology, and the Anticommunist Politics of George McCready Price', *Church History*, 83 (2014), 684–722; Kurt P. Wise, 'Contributions to Creationism by George McCready Price', *Proceedings of the International Conference on Creationism*, 8 (2018), 683–94; and Abraham C. Flipse, 'The Origins of Creationism in the Netherlands: The Evolution Debate among Twentieth-Century Dutch Neo-Calvinists', *Church History*, 81 (2012), 104–47.

[5] Harold W. Clark, *Crusader for Creation: The Life and Writings of George McCready Price* (Mountain View, CA: Pacific Press Publishing Association, 1966), 21, 68.

[6] Kathleen C. Boone, *The Bible Tells Them So: The Discourse of Protestant Fundamentalism* (Albany: State University of New York Press, 1989), 12.

Figure 5.1 College of Medical Evangelists Library, c.1912. Loma Linda University Photo Archive, LLU00691, Department of Archives and Special Collections, Loma Linda University, Loma Linda, California. Surrounded by books, as was his wont, George McCready Price is barely visible to the right. This photo was taken when Price was working at the Adventist Institution now known as Loma Linda University.

Despite the importance of textuality in Price's work, scholars have not paid close attention to the literary facets of his vast output, nor has there been concerted examination of his assault on the visual language that, as Martin J. S. Rudwick influentially argued in 1976, was so integral to geology's intellectual development.[7] Price's textual strategies, moreover, have not been related to those of his intellectual predecessors or his sympathizers. These omissions reflect an absence of attention to young-earth creationism as aesthetic literature more generally, with this field of enquiry chiefly represented by Ralph O'Connor's article on early nineteenth-century British young-earthers.[8]

In this chapter, therefore, I tie early twentieth-century young-earth creationism into the larger literary tradition of borderline palaeoscience. I argue that those

[7] Martin J. S. Rudwick, 'The Emergence of a Visual Language for Geological Science 1760–1840', *History of Science*, 14 (1976), 149–95.
[8] Ralph O'Connor, 'Young-Earth Creationists in Early Nineteenth-Century Britain: Towards a Reassessment of "Scriptural Geology"', *History of Science*, 45 (2007), 357–403.

anti-evolutionists who insisted on twenty-four-hour creation days, with Price at their head, refused to participate not only in the epistemological assumptions of palaeontology and geology, but also in their literary and visual genres. They undermined the imaginative techniques and the diagrammatic tropes used to communicate geohistorical ideas, which Price saw as offensive to the traditions of his own Seventh-day Adventist denomination and to those of Protestantism more generally. He attempted to defuse the romantic attraction of deep timescales and return readers' admiration to the Genesis narrative, adopting a satirical tone and drawing upon registers of Baconianism, common sense, law, detective fiction, and populism. These literary techniques were bolstered by a photographic aesthetic, co-developed by Price and other pioneering creationists like Byron C. Nelson and Harry Rimmer, that presented strict biblical literalism as the only sensible response to casting one's eyes upon the natural world. In the process, these authors agitated long-standing cultural anxieties about the ultimately unseeable happenings of the prehistoric past and the hubris of assembling a secular geohistorical narrative. Their goal in so doing was to place faithful readers of the Bible on higher scientific ground than any credentialed geologist.

The Holy Bible and Common Sense

Price's career was closely linked to the rise of Christian fundamentalism, although, for reasons I will establish, it must be stressed that fundamentalism is not coterminous with young-earthism. The movement was named for *The Fundamentals* (1910–15), a widely publicized series of evangelical essays espousing Protestant orthodoxy and denouncing foes like theological liberalism and evolutionary biology. In the United States it grew in strength dramatically towards the end of the Great War, peaking in influence in the mid-1920s, when the so-called Scopes Monkey Trial of 1925 brought populist politician William Jennings Bryan to Dayton, Tennessee, to denounce the teaching of evolution in high schools.[9] As George M. Marsden (and many others) observe, the philosophy of science professed by twentieth-century fundamentalists, including Price, was an extreme continuation of what had, a century prior, been the dominant ideology in American Protestantism and anglophone science more generally: a fusion of Scottish common-sense realism and Baconian inductive reasoning. After the arrival of common-sense realist ideas in American in the eighteenth century, the 'democratic or anti-elitist' notion that 'truth was a single unified order and ... all persons

---

[9] George M. Marsden, *Fundamentalism and American Culture*, 2nd edn (Oxford: Oxford University Press, 2006). For the Scopes Trial, see Edward J. Larson, *Summer for the Gods: The Scopes Trial and America's Continuing Debate Over Science and Religion*, rev. edn (New York: Basic Books, 2006).

of common sense were capable of knowing that truth' speedily became 'unquestionably *the* American philosophy'.[10] Helpfully, Protestant faith in the individual's ability to understand the text of the Bible, due to its 'perspicuity', was here given a parallel in the 'perspicuity of nature', rendering the interpretation of natural phenomena a commonsensical affair that could never conflict with scripture.[11] Nature, after all, was conventionally seen as God's second 'book'.[12]

Common-sense ideas dovetailed with those derived from contemporary interpretations of the ideas of Francis Bacon, whose work was routinely invoked as mandating the patient and empirical collection of facts from God's second book, rather than the premature construction of hypotheses about it. We have already heard, in chapter 4, about Baconian induction's controversial invocation by Ignatius Donnelly and Henry Hoyle Howorth in support of non-creationist interpretations of the biblical Flood. In less radical form, Baconianism had been a salient ingredient in what Adelene Buckland describes as the anti-narrative sensibility found in much early and mid-nineteenth-century geological writing, epitomized by the work of the Geological Society of London in its early decades. This sensibility discouraged practitioners from overspeculation (or, worse, imaginativeness) in the face of insufficient scientific data about geohistory, as well as—in an age when old-earth creationism was the norm—screening one's work from being interpreted in an evolutionary manner. Unspeculative early geologists, as Buckland observes, were 'wary of too much plot'.[13] This resistance to the notion that human knowledge could reconstruct any reliable, global geohistorical narrative characterized much of Lyell's career, and it survived in later essays like Herbert Spencer's 'Illogical Geology' (1859).[14]

In the second half of the nineteenth century, however, Bacon's star and that of Scottish common-sense realism were already falling in both British and US science. With them declined the sway of those stern cautions levelled against anyone claiming to know too much about the shape of Earth's deep history (a shape which could defy common-sense interpretations). The establishment of a widely agreed-upon geological column—evincing progress from the simplest fossil life forms to complex being like reptiles, birds, and mammals—gradually appeased even doubters like Lyell, who were likewise compelled also to concede to evolutionary theory, generally accepted by the front ranks of the scientific community in the 1860s and 1870s. Within conservative theological circles, however, common-sense realism and Baconianism remained valued tools. In Britain, the Victoria Institute, the gentlemanly evangelical science organization, became a forum in which both

---

[10] Marsden, 14.
[11] Marsden, 16.
[12] Ronald L. Numbers, *Science and Christianity in Pulpit and Pew* (Oxford: Oxford University Press, 2007), chapter 3.
[13] Buckland, *Novel Science*, 20.
[14] Buckland, *Novel Science*, 227–28.

methods were appealed to in exposing the fallaciousness of liberal ideas like evolution and the German higher criticism. John William Dawson was only the most scientifically reputable evangelical geologist to appeal to 'our old Baconian mode of viewing nature' and 'the domain of common sense and sound induction'.[15] These words, in similar formations, can be found in the majority of Price's young-earth publications across the first half of the twentieth century.[16]

And yet, for all his Baconianism, Dawson was, of course, a day-ager who accepted that Earth was immensely ancient (as was his son, the engineer W. Bell Dawson). So too was the fundamentalist William Jennings Bryan, whose denunciation of evolution at the Scopes Trial has become the stuff of legend.[17] Bryan's belief that Earth was immensely ancient was shared by many fundamentalists who pronounced upon the subject. Even those unpersuaded by day-age theory could opt instead for gap theory: this alternative appeared in many convoluted permutations, but, by positioning the long geohistorical narrative *before* the creation week, it allowed the former to be, if desired, conveniently ignored.[18] Rather than geological time, it was evolutionary theory, upon which modern evils such as immorality and German militarism were blamed, that lay beyond the pale for most fundamentalists.[19] Sheer commonsense Baconianism alone, then, did not necessarily lead Protestants to reject, as Price did, the geological column and deep time. In fact, his doing so made him stand out even among otherwise like-minded conservatives. Day-agers and gap theorists were serious obstacles to the acceptance of Flood Geology. Price's extreme scepticism, which entailed a rejection of even these commonplace Genesis-geology concordance schemes, hearkened back to—and rejuvenated— the anti-narrative geological science that had been current in the early nineteenth century, before any detailed consensus upon evolution or even geohistory had been established.

The Geological Society's Baconian heyday had also roughly coincided with the emergence of Price's intellectual predecessors: the first-wave of young-earth creationist geologists. Young-earthers rose up to make their case in the 1820s and 1830s, when writerly geologists like Gideon Mantell were introducing general audiences to old-earth geohistory in books like *The Wonders of Geology* (1838).[20] The biblical Flood, rapidly losing its scientific status elsewhere, was still usually seen by these scriptural literalists as the all-purpose explanation for the

---

[15] Quoted in Stuart Mathieson, *Evangelicals and the Philosophy of Science: The Victoria Institute, 1865–1939* (London: Routledge, 2021), 8.

[16] See, for instance, Price, *God's Two Books; Or Plain Facts about Evolution, Geology, and the Bible* (South Bend, IN: Review and Herald Publishing, 1911), 88.

[17] Numbers, *Creationists*, 58.

[18] Tom McIver, 'Formless and Void: Gap Theory Creationism', *Creation/Evolution*, 8 (1988), 1–24.

[19] Numbers, *Creationists*, 55–56.

[20] For an overview by a young-earth creationist, see Terry Mortenson, *The Great Turning Point: The Church's Catastrophic Mistake on Geology—Before Darwin* (Green Forest, AR: Master Books, 2004).

phenomena of extinction, fossilization, and stratification. Their response to the development of geology *per se* was not necessarily antagonistic: O'Connor has shown that individuals in this heterogeneous group were often as enthusiastic about this new science as old-earth competitors like Mantell, inventing literary conventions like the 'panorama' metaphor to enchant readers with their monster-filled shallow time.[21] While many, admittedly, fashioned themselves as drily Baconian induction machines, the more passionate early literalists left themselves open to an attack by Adam Sedgwick, President of the Geological Society, for their undisciplined 'inventive power' and ignorance of 'the method of our inductions'.[22] By the mid-1830s, the most influential organs of the periodical press in Britain had joined elite scientific organizations in siding decisively with old-earth geology.[23] As the repute of young-earth creationism sank, its authors increasingly took recourse to a sarcastic Baconianism, rather than sugaring their biblicist pill with romance.

A similar process was taking place in the United States, exemplified by the sternly inductive writings of the Calvinist brothers, Eleazar and David Nevins Lord, of New York—the latter's *Geognosy; or, The Facts and Principles of Geology Against Theories* (1855) being a representative title.[24] Writing in the influential *Southern Presbyterian Review*, a periodical based in Columbia, South Carolina, pastor Edwin Cater thundered that old-earth geologists suffered from a 'hallucination' that has 'given them new eyes' through which to wildly misinterpret geological evidence.[25] Indeed, in 1841, the Massachusetts geologist Edward Hitchcock had characterized his conversion to belief in the glacial origins of phenomena formerly attributed to the Flood as 'a *new geological sense*' that allowed him to look upon 'our smoothed and striated rocks, our accumulations of gravel ... with new eyes'.[26] Geologists' 'diseased vision', recalled, in Cater's essay, the bathetic implications of the '*dioramic* hypothesis' in Hugh Miller's *The Testimony of the Rocks* (1857), whereby the vagueness of the Genesis text was explained through the Mosaic seer's blundering inability to see clearly through the distorting 'atmospheric phenomena' of Earth's early day-ages.[27]

---

[21] Ralph O'Connor, *The Earth on Show: Fossils and the Poetics of Popular Science, 1802–1856* (Chicago: University of Chicago Press, 2007), 207; O'Connor, 'Young-Earth Creationists', 376.

[22] Adam Sedgwick, 'Proceedings of the Geological Society', *Philosophical Magazine*, 7 (1830), 309–10 (289–315).

[23] O'Connor, 'Young-Earth Creationists', 389.

[24] Richard Perry Tison II, 'Lords of Creation: American Scriptural Geology and the Lord Brothers' Assault on "Intellectual Atheism"', unpublished PhD Thesis, University of Oklahoma (2008).

[25] Edwin Cater, 'Geological Speculation and the Mosaic Account of Creation', *Southern Presbyterian Review*, 10 (1858), 534–73 (535).

[26] Edward Hitchcock, 'First Anniversary Address before the Association of American Geologists', *American Journal of Science*, 41 (1841), 232–75 (253).

[27] Cater, 544, 547, 566.

By the time this 1858 article was published, however, young-earth geology was relatively marginalized on both sides of the Atlantic. In Britain, naturalist Edwin Lankester claimed, in a more favourable review of Miller's book, that the young-earth school of geology 'has at present few public advocates'.[28] If we accept Lankester's judgement, however, his qualifying term 'public' is critical. Vast demographics may well have sustained a quiet faith in the traditional young-earth timescale. 'No doubt many Christians, perhaps most, remained unpersuaded' by old-earth geology, suggests Numbers, speaking of the nineteenth-century United States, 'but these people rarely expressed their views in books and journals'.[29]

Those few young-earthers of the later nineteenth century who made their voices heard continued to scoff at geologists' self-proclaimed powers. One pungent example is *An Explanation of the Author's Opinions on Geology* (1867), a booklet written by former Halesowen gun-barrel-maker William Rose. Rose insisted that fossils represented not former living things but mere sports of nature, a stance that had been scientifically heterodox since the eighteenth century. Invoking 'the holy Bible and common sense' against the Babelesque hubris of geohistorical claims, he flipped the tables on what he called 'believers in geology'.[30] In addition to placing geology on the same footing as faith—a matter of belief—Rose pounced upon practitioners' visionary and Orientalist self-fashioning. The theory of the molten origin of the planet was 'one of the most flimsy and elastic imaginations ever conceived by any man', outvying 'talk of Sinbad the Sailor, or Aladdin and his Wonderful Lamp', while palaeontologists' claims to describe animals never seen by human eyes were as absurd as the predictions of 'astrologers and soothsayers'.[31] Rose spent much time with Mantell's by then outdated *The Wonders of Geology*. The book's speculative engraving of Earth as seen from the moon could only, Rose joked, have been produced if 'this gentleman' visited the moon 'in a trance', while Mantell's 'fairy-like' climax, which imagined the experiences of an extraterrestrial visitor observing Earth at various geological periods—a notion inspired, according to Mantell himself, by 'the metaphor of an Arabian writer'—was the product of a man who 'has just woke up from an enchanting dream'.[32] The reach of the gunsmith's polemic, likely limited to what is now the West Midlands of England, did not reignite the young-earth cause, but this kind of rhetoric was to grow in importance in the century to come.

---

[28] [Edwin Lankester], untitled review of *The Testimony of the Rocks*, by Hugh Miller, *Athenæum*, 1536 (1857), 429–31 (430).

[29] Numbers, *Creationists*, 30.

[30] William Rose, An *Explanation of the Author's Opinions on Geology: Showing the Fallacy of the Extreme Believer in That So-Called Science, and How Contradictory Geology Is When Compared with the Bible* (Birmingham: Martin Billing, Son, and Co., 1867), 2–3, 5.

[31] Rose, 51–52, 63.

[32] Rose, 118, 126; Gideon Mantell, *The Wonders of Geology; or, A Familiar Exposition of Geological Phenomena*, 2 vols (London: Relfe and Fletcher, 1838), I, 373.

Just Like Every Other Week

Unlike Britain, the United States acquired an enduring stronghold of belief in an Earth created in six literal days, around six thousand years ago: the Seventh-day Adventist Church. An isolated Protestant denomination, Seventh-day Adventists were an offshoot from the Millerites, an apocalyptic group awaiting Christ's imminent return (and whose beliefs had been mocked by Lyell on his visit to the United States in the 1840s).[33] Baptist preacher William Miller's precise dating of the Second Coming to 22 October 1844 was based on a firmly Protestant and American application of common-sense realism and Baconian methods to the interpretation of biblical prophecy, but the 'Great Disappointment' of Christ's failure to arrive proved his calculations inaccurate. As Malcolm Bull and Keith Lockhart demonstrate, Miller's mid-century optimism regarding human access to truth underlay the visionary direction in which, influenced by Ellen G. White, the offshoot that became Seventh-day Adventism subsequently moved.[34]

White (née Harmon), a former Methodist, rose to prophetic prominence thanks to her spectacular visions. The most important of these, purportedly experienced in Ohio in 1858, confirmed the literal accuracy of Genesis. 'I was then carried back to the creation', White recalled in the third volume of her *Spiritual Gifts* (1864), 'and was shown that the first week, in which GOD performed the work of creation in six days and rested on the seventh day, was just like every other week'.[35] This was an attractive notion to Seventh-day Adventists, who placed exceptional value on the institution of the Sabbath (celebrated on Saturday, rather than Sunday), as descended from God's rest on the seventh creative day. Less interested in the logistics of vision than predecessors like Miller, White also wrote graphically about the destruction of the 'very large, powerful animals' of the antediluvian world by the Flood, which also wrecked the world's climate.[36] Her words on Genesis rapidly became second in authority only to Scripture among Adventists, and they travelled across the globe thanks to the Church's zealous publishing and missionary industries.[37]

---

[33] Charles Lyell, *A Second Visit to the United States of North America*, 2 vols (London: John Murray, 1849), I, 86–92.

[34] Malcolm Bull and Keith Lockhart, *Seeking a Sanctuary: Seventh-day Adventism and the American Dream*, 2nd edn (Bloomington: Indiana University Press, 2007), 26–27. See also Ann Taves, 'Visions', in *Ellen Harmon White: American Prophet*, ed. by Terrie Dopp Aamodt, Gary Land, and Ronald L. Numbers (Oxford: Oxford University Press, 2014), 30–48.

[35] Ellen G. White, *Spiritual Gifts III: Important Facts of Faith, in Connection with the History of Holy Men of Old* (Battle Creek, MI: Steam Press of the Seventh-day Adventist Publishing Association, 1864), 90. For context, see Cornelis Siebe Bootsman, 'The Nineteenth Century Engagement Between Geological and Adventist Thought and Its Bearing on the Twentieth Century Flood Geology Movement', unpublished PhD thesis, Avondale College of Higher Education (2016), 88–89.

[36] White, *Spiritual Gifts III*, 92.

[37] Arthur Patrick, 'Author', in *Ellen Harmon White: American Prophet*, ed. by Terrie Dopp Aamodt, Gary Land, and Ronald L. Numbers (Oxford: Oxford University Press, 2014), 91–107.

Although Numbers and Rennie B. Schoepflin note that White's catalysation of young-earth creationism in the twentieth century was 'largely accidental', late nineteenth-century Adventists were already thinking about the implications of her statements and favourably citing older works by American young-earth literalists.[38] Indeed, the movement's unusually comprehensive rejection of deep time can be seen as built into its attitude to time itself. Bull and Lockhart argue that 'Adventist theology is primarily concerned with time—with the end of time, the correct timing of the Sabbath, the prophetic interpretation of time', and thus 'Adventists have been discouraged from keeping up with other forms of popular culture that might offer a rival understanding of the structure and significance of time'.[39] This was also one of the reasons White spoke out against almost all forms of fiction. Even quality novels, she argued 'contain statements and highly wrought pen-pictures that excite the imagination and give rise to a train of thought which is full of danger, especially to the youth', by encouraging 'superficial reading, merely for the story'.[40] Competing chronologies, even those of novels, but especially that of old-earth geology, distracted Christians from the pressing question of the looming end of time.

White may have occasionally dismissed wholesale the geohistorical narrative, but the subject was not a pressing concern for her. In Price's works, the anti-narrative tradition of geology dovetailed with the literalist and anti-fiction components of Adventist eschatology to generate a much more sustained assault on the notion of reconstructing Earth's deep history. Price, whose early life in Canada and the United States was one of painful financial insecurity, turned to White's writings as an immense comfort.[41] Her visions were among the few instances of time travel, real or figurative, that Price would credit: his fond recollections of White's 'revealing word pictures of the Edenic beginning of the world' point to the prophetess's immunity from the accusations of delusion he directed to authors of imaginative or secular narrative.[42] Price's own limited formal scientific education was imbibed as part of a teacher training course at the Provincial Normal School of New Brunswick in 1896, and his initial career was literary (and precarious), ranging from amateur poet to Adventist colporteur (book-peddler) to a failed attempt at New York hackwork. By the 1910s, he gained regular employment as a teacher of miscellaneous subjects, including English literature, Greek, and

---

[38] Ronald L. Numbers and Rennie B. Schoepflin, 'Science and Medicine', in *Ellen Harmon White: American Prophet*, ed. by Terrie Dopp Aamodt, Gary Land, and Ronald L. Numbers (Oxford: Oxford University Press, 2014), 196–23 (217); Bootsman, *passim*.

[39] Bull and Lockhart, 230, 230.

[40] Ellen G. White, *Counsels to Teachers, Parents and Students regarding Christian Education* (Mountain View, CA; Pacific Press Publishing Association, 1913), 383. For an overview, see Gary Land, *Historical Dictionary of Seventh-day Adventists* (Lanham, MD: Scarecrow Press, 2005), 171–75.

[41] Unless otherwise stated, discussion of Price's biography is indebted to Numbers, *The Creationists*.

[42] Price, *Genesis Vindicated* (Washington, D.C.: Review and Herald Publishing Association, 1941), 300.

chemistry, moving between Adventist institutions across the Western United States, especially California, and the Midwest (alongside an appointment in Watford, England, between 1924 and 1928).

Price was denouncing old-earth geology long before he gained secure employment. Between the publication of what is probably his earliest geological article in the Adventist *Review and Herald* in 1901 and his death in 1963, Price penned many hundreds of anti-geohistorical and anti-evolutionary items and dozens of books, alongside assorted theologically and socially conservative essays. These were part of a huge Adventist evangelizing machine: a Japanese edition of his *Back to the Bible* (1916) was published in Tokyo in 1927, for example, and Price's works were promoted in Seoul, Shanghai, and Lucknow (although their translation and reception still requires scholarly attention).[43] Writing on 'The Significance of Fundamentalism' in a 1927 issue of the *Review and Herald*, Price neatly summed up his core belief that Adventists' ability to retain 'a true view of geology', combined with the 'consistent and harmonious system of truth' revealed by Ellen White, had been instrumental in protecting the denomination from succumbing to fundamentalism's great foe: the 'modernism' that encouraged symbolic, historicized, liberal interpretations of the Bible.[44]

Flood Geology and anti-evolutionary sentiment in early twentieth-century fundamentalism have been discussed at length elsewhere and so their scientific arguments will be passed over fairly briefly here: intellectually, Price's broad positions were not drastically different from those formerly adopted by literalists within and without the Adventist Church.[45] He rejected all interpretations of Genesis that incorporated deep time and was frustrated by the fact that, even among fundamentalists, few held this hard line (to which Adventists felt compelled by White's plain vision).[46] As he saw it, geologists dated rocks based only on their comparative fossil content, and palaeontologists dated fossils based solely on their position in the rocks, a case of 'reasoning in a circle' that licensed him to entirely discard the geohistorical narrative they had onerously constructed.[47] His favourite evidence for this was that of 'overthrust', the unintuitive phenomenon whereby distorted, upturned strata appeared in what geologists saw as non-chronological order, and which Price saw as discrediting the idea of any such order. He argued that the global geological column of fossils represented not superposed time periods but

---

[43] Unpaginated advertisement for *God's Two Books*, by George McCready Price, *Oriental Watchman*, 15 (1912); W. L. Burgan, 'Newspaper Work in Korea', *Advent Review and Sabbath Herald*, 101 (1924) 21–22; and W. P. Henderson, 'Working Among the Troops in Shanghai', *Advent Review and Sabbath Herald*, 105 (1928), 19–20.

[44] Price, 'The Significance of Fundamentalism', *Advent Review and Sabbath Herald*, 104 (1927), 13–14 (13).

[45] For an overview, see Wise, 'Contributions to Creationism'.

[46] Numbers, *Creationists*, 60.

[47] Price, *The Geological-Ages Hoax: A Plea for Logic in Theoretical Geology* (New York: Fleming H. Revell, 1931), 10.

rather antediluvian biogeography. The Flood itself, misunderstood as an ice age by geologists, had exterminated animals such as the dinosaurs and stratified their remains in sediment, while more physically familiar extinct animals, like mammoths, were simply climatic variations of modern species. To stress the recency of the antediluvian world, Price encouraged the proto-cryptozoological belief that it was 'entirely possible' that gigantic reptiles like *Mosasaurus* 'may have survived to our time'.[48] Price's bête noire was Lyell, whose uniformitarianism, for Price and many conservative contemporaries, represented the philosophical backbone of evolutionism and anti-miraculous scientific naturalism. By neutralizing uniformitarian geology, evolution, with its unsavoury connotations of brute struggle and animal descent, would also be neutralized. The Flood would wash them all away.

Price's view of Lyell as an architect of the deep geohistorical narrative was based on the lionized caricature of contemporary history of science, rather than the man who had, in fact, made considerable contributions to anti-narrative geology. Ironically, Price signalled his allegiance to this tradition in his early work *Illogical Geology* (1906), named for Spencer's essay of the same name—an essay which Price cited with enthusiasm throughout his career for its scepticism about the construction of the geological column.[49] He contrasted the hypothetical, a priori reasoning of geologists with mature sciences like astronomy, which had been made mathematically exact by 'the magic call of common sense and true Baconian methods'.[50] Although, as we have seen, this language was an established part of conservative Protestant science, Price's writings suggest he imbibed its most florid phraseology from Howorth, whose anti-uniformitarian *The Glacial Nightmare and the Flood* provided *Illogical Geology* with a scathing epigraph denouncing premature theory-mongering. His reliance on the Manchester Conservative went far deeper than epigraphs: Howorth's encyclopaedic sources about the Flood and belligerently inductive rhetoric, discussed in chapter 4, were shorn of their naturalism and trotted out across Price's *oeuvre*.[51] Price went to heights even Howorth had never envisaged. To purge geology, he envisioned an ambitious citizen science project: 'a large army of patient observers' working to forge 'a geology stripped of all speculation' and 'securely laid on facts alone'.[52] To a limited extent, this can be seen as a serious statement of a research programme, given that Price sometimes worked in nascent anti-evolutionary organizations like the Religion and Science Association (founded 1935).[53] On the whole, however, his approach was more that of the denialist seeking to muddy the waters with problems scientists could not

---

[48] Price, *New Geology*, 517.
[49] Price, *Illogical Geology: The Weakest Point in the Evolution Theory* (Los Angeles, CA: Modern Heretic Company, 1906), 11–12.
[50] Price, *God's Two Books*, 88.
[51] For example, see *Illogical Geology*, 12, 36, 39, 56, 64, 67–69, 83.
[52] Price, *Geological-Ages Hoax*, 118.
[53] Numbers, 239–85.

answer, rather than of the founder of a positive young-earth geology.[54] This was especially the case in the later decades of his life, when Price accepted that evolution represented an end-times apostasy—indeed, allegiance to it was potentially related to the Mark of the Beast.[55]

In *The Geological-Ages Hoax* (1931), Price complained that 'professional geologists and palæontologists always resent any criticism', implying that they had shut their doors upon him.[56] In reality, after some civil conversations with scientists in the 1900s, Price appears to have mostly bypassed their opinions, preferring to address evangelical and fundamentalist audiences through the Adventist press and publishers like Fleming H. Revell.[57] That said, Price's indifference to persuading different demographics should not be overstated. During his little-studied years in Britain in the mid-1920s, teaching at the Adventist-run Stanborough School in Watford, Price experimented with a more engaged attitude towards the establishment. This was the height of the fundamentalist furore in the United States, and Price's criticisms of geology in recent works like *The Phantom of Organic Evolution* (1924) were significant enough to prompt Arthur Smith Woodward, former Keeper of Geology at the Natural History Museum, to refute them in a three-page *Nature* article in January 1926. Although Woodward did not name Price, tell-tale allusions to anti-evolutionists accusing geologists of 'artificial' reasoning and 'phantom' pursuits, and his own admission that the history of geology contained some instances of 'reasoning in a circle' left no doubt as to his target.[58] Price's own attention to *Nature* is suggested by the fact that he immediately penned a letter to Woodward, noting it was 'quite obvious' to whom the article alluded and thanking the geologist for his 'courteous' discussion of the subject. He repeated his typical arguments against old-earth geology, and overthrust in particular, including the claim that the assumptions of geologists implied 'supernatural' knowledge that contradicted 'our eyesight and common sense'.[59] He sent a copy of this letter to *Nature*, although it was not published, nor did he expect it to be.

In May of 1926, Price wrote to Woodward's successor at the museum, Francis Arthur Bather. Requesting recent literature on 'over-thrust' and asking about his own eligibility for the upcoming International Geological Congress in Madrid, Price insisted that he was not intending to 'rope' Bather into his own 'peculiar views'.[60] Bather politely directed Price to an expert on overthrust and told him

---

[54] For denialism, see Gordin, *On the Fringe*, 93.
[55] Clark, *Crusader for Creation*, 93; Price, *Modern Discoveries Which Help Us to Believe* (New York: Fleming H. Revell, 1934), 204–7.
[56] Price, *Geological-Ages Hoax*, 55.
[57] Numbers, *Creationists*, 110.
[58] Arthur Smith Woodward, 'The Relative Age of Rocks containing Fossils', *Nature*, 117 (1926), 21–23. For Price's allusions to 'artificial' reasoning, see *The New Geology*, 17, 19, 223, 282.
[59] Price to Arthur Smith Woodward, 1 January 1926, Natural History Museum Archives DF PAL/100/180/4, Palaeontology Department Correspondence. Subsequently NHM. The letters cited below are included in the same envelope.
[60] Price to Francis Arthur Bather, 9 May 1926, NHM.

that it was unlikely he would face 'the smallest difficulty in becoming a member' of the Congress, especially as Price was, as his letterhead ambitiously claimed, 'LATE PROFESSOR OF GEOLOGY' at Union College in Nebraska.[61] It was from Watford that Price, the following year, wrote to *Science* to defend himself against charges of copyright violation in his reuse of images from geological textbooks in *The New Geology*.[62] These exchanges show Price's renewed attempts, in mid-career, to discuss his ideas with leading palaeoscientists, as well as their occasionally diplomatic approach to dealing with him. Price's visibility in the 1920s was such that one of the decade's bestselling popular science publications, *The Science of Life*—serialized in 1929–30 and co-written by Julian Huxley, H. G. Wells, and Wells's son, G. P.—was obliged to refute the claim made by unnamed 'opponents of Evolution' that, when palaeontologists used fossils to determine the age of strata, they were 'arguing in a vicious circle'.[63] This was one of Price's catchphrases.

The Fingerprints Found after a Homicide

In generic terms, Price's publications were typically, but not always, polemical essays, most of which assumed little technical knowledge from the reader. They dissected mainstream scientific concepts piece by piece, quoting heavily from elite naturalists (many of them, like Lyell, long dead) and from borderline figures such as Howorth, often resembling suppositional syntheses with their strategy of arranging logical theses into numbered lists, unceasingly signposting the direction of the argument, and refusing to satisfy the imagination in any easy sense with speculative word-painting. Price regularly recycled quotations, metaphors, and arguments, making his books distinctly repetitive, which is why this chapter does not strictly follow the development of his thought in chronological order.

A legalistic language of evidence, witnesses, impartiality, corruption, and justice prevailed throughout all his works. Indeed, it was drawn both from the ethos of Adventism, which extolled fair debate and evidenced argumentation, and from the geological and evolutionary literature Price mined for content.[64] This register—sometimes, as in the case of the Scopes Trial, applied in actual courtrooms—was a defining characteristic of fundamentalism, and, as we heard in the periodical debates discussed in chapter 1, creationism in general.[65] However, despite using legalistic and evidentiary language themselves—Lyell had been a trained lawyer

---

[61] Francis Arthur Bather to Price, 11 May 1926, NHM.
[62] For this controversy, see Numbers, *Creationists*, 108–10.
[63] H. G. Wells, Julian Huxley, and G. P. Wells, *The Science of Life: A Summary of Contemporary Knowledge about Life and its Possibilities*, 3 vols (London: Amalgamated Press, [1930]), I, 212.
[64] Bull and Lockhart, 106–7. For the legalistic language of evolutionists, see Thomas H. Huxley, *American Addresses, with a Lecture on the Study of Biology* (New York: D. Appleton, 1877), 12–14.
[65] Michael Ruse, *The Evolution-Creation Struggle* (Cambridge, MA: Harvard University Press, 2005), 250.

and was sometimes criticized for his forensic argumentation—most geologists and biologists were keen to stress the expertise that made a scientific authority's opinion weigh more heavily than that of a layperson.[66] For the most part, Price was unwilling to recognize distinctions between scientific and legal methods. The jury, after all, was the ultimate judicial enshrinement of non-specialist common sense in the public sphere. Despite their foregone conclusions, the books in which Price aimed for his largest audiences were regularly presented as fair trials presented before commonsensical juries. In this vein, *Illogical Geology*, self-published under the ambiguous label of the 'Modern Heretic Company' and distributed to a more diverse scientific readership than his later works, kept its theistic stance relatively quiet until the end.[67] A blind reading may have led to the surprisingly gentle review the 'earnest' Price's book received in *Nature*.[68] Across his far more extensive tome *The New Geology*, the inevitable attribution of almost the entire fossil record to the Flood was built up in slow Howorthian manner, as if via neutral induction.

Juries, however commonsensical, can be vulnerable to rhetorical eloquence, and many of the nineteenth-century geologists so regularly cited by Price were masters at conjuring up literary vistas of the prehistoric world as a sublime, romantic, and enchanted place.[69] Knowing that he had to strip prehistory of its enticements, Price conflated these geologists' imaginative literary form with their scientific content and dismissed both from the court of creation. Geologists could 'grow eloquent on the long millenniums required for' the generation of coal, but this provided nothing like hard evidence.[70] He deromanticized the 'phantom world' or 'fairy world' described so vividly by old-earth geologists, declaring that it was 'simply an older state of our present world', and modestly professed his inability to 'undertake' any 'fairylike excursions into the unknown and unknowable', instead contenting himself 'with prosaic statements regarding the present'.[71] As for evolutionary thinking, the nonsensical logic of '*Alice in Wonderland*, the *Wizard of Oz*, or any other form of "Jabberwocky" would be about as scientific'.[72] Old-earth geology had gained cultural prestige in the nineteenth century through association with myth and fairy tales, but young-earth geologists in the twentieth could balance the scales by turning these associations against it.

Jettisoning romance in favour of decidedly quotidian figurative language, and stressing not scientific expertise but universally accessible logic, Price's essays are

---

[66] Sherrie Lynne Lyons, *Species, Serpents, Spirits, and Skulls: Science at the Margins in the Victorian Age* (New York: State University of New York Press, 2009), 30–31.

[67] Price, *Back to the Bible*, 4; Price, *Illogical Geology*, 89.

[68] Grenville Arthur James Cole, untitled review of *Illogical Geology: The Weakest Point in the Evolution Theory*, by George McCready Price (1906), *Nature*, 74 (1906), 513.

[69] O'Connor, *Earth on Show*, 126–27, 244; Melanie Keene, *Science in Wonderland: The Scientific Fairy Tales of Victorian Britain* (Oxford: Oxford University Press, 2015), 21–53.

[70] Price, *Geological-Ages Hoax*, 116–17.

[71] Price, *Geological-Ages Hoax*, 76; Price, *New Geology*, 338.

[72] Price, *A History of Some Scientific Blunders* (Chicago: Fleming H. Revell, 1930), 123.

ideal examples of Adventist aesthetic principles: plainness, familiarity, humour.[73] His language appealed to the imagined jury with minimal ornamentation, skewering pretentious intellectuals with homely witticisms. Thrusts at the expense of the eighteenth-century mineralogist Abraham Gottlob Werner are typical. Werner's purported belief that the geological formations around Saxony extended all around the world (another oversimplification taken from the era's Anglocentric histories of geology), resembled those 'three little green peas in the little green pod' who supposed that 'the universe' was also 'green'.[74] Even Price's preferred replacement term for the prehistoric world in *The New Geology*, the 'olden time', was self-consciously quaint.[75] Japanese Adventist Shohei Miyake, who met Price in 1906, while the latter was a handyman at Loma Linda College of Evangelists in California, recalled having been 'impressed that a man like Professor Price, who was well educated, and the author of two books at the time, would engage in such a menial task as constructing a building'.[76] This image of Price as a published intellectual unpretentious enough to work humbly with his hands was precisely the authorial figure his books implied.

In connection with his courtroom register and rejection of romance in geological writing, Price's most characteristic analogy was homicide. Rather than a wondrous narrative of former life, the fossil record was a crime scene. Geologists were, or ought to be, 'world-coroners'.[77] Uniformitarians like Lyell, however, were 'poor coroners' who attributed all deaths to natural causes even if there existed 'ample evidence that the poor fellow under consideration had been shot'.[78] Price's *un*natural cause, the bullet that killed the dinosaurs, was an unrepeatable and therefore non-uniformitarian event: the miraculous biblical Flood. Unlike the enchanting, necromantic rhetoric employed by palaeontologists, Price's postmortem register made geology into an unrelentingly morbid affair. A typical instance is his description of

> fossil bivalves, whose valves are quite often found applied, as if buried alive, before the shells had time to open; or the hosts of brachiopods, usually found with the valves closed, and the interior often hollow, showing that no mud had time to work through the hole near the hinge-joint which always exists after the animal dies. These telltale conditions should be to us what a bullet hole found in the skull, or arsenic in the stomach, of a subject of a *post mortem* ought to be to those holding an inquest.[79]

[73] Bull and Lockhart, 222, 227.
[74] Price, *God's Two Books*, 69.
[75] Price, *New Geology*, 52.
[76] Clark, *Crusader for Creation*, 66–67. These are Clark's words, rather than Miyake's.
[77] Price, *Modern Flood Theory*, 19.
[78] Price, *God's Two Books*, 96–97; Price, *New Geology*, 645.
[79] Price, *God's Two Books*, 158.

Price, probably working on commonsensical assumptions about the merely rhetorical nature of his analogy, did not think it meaningful that this made God akin to a murderer.

This postdiluvian '*post mortem*' recalls the language not only of a coroner but of a detective, as did a comparison between the evidence represented by fossils and 'the fingerprints found after a homicide'.[80] Price, whose career peaked during the interwar Golden Age of Detective Fiction, the era of Agatha Christie and Dorothy L. Sayers, even sounds like Sherlock Holmes when he derides geologists' Watson-like 'clumsy reasoning'.[81] Arthur Conan Doyle himself, fascinated by palaeontologists' detective-like powers, had memorably compared Holmes's ability to draw grand deductions from minimal evidence to Georges Cuvier's purported, and subsequently discredited, ability to reconstruct an animal from a single bone.[82] Just as he had weaponized geology's association with fairy tales, Price refabricated Doyle's flattering comparison to ridicule Cuvier's geohistorical presuppositions: 'Sherlock Holmes might attempt to diagnose a disease by a mere glance at his patient's boots, but even this gave him more data and was a more logical proceeding than the facts and methods of Cuvier supplied for constructing a scheme of organic creation'.[83] Price—who, following White's anti-fiction precepts, may well have never actually read a Holmes story—blamed Cuvier for the notion that extinct animals had existed during different geological periods, rather than in different but contemporaneous ecological zones across the short-lived antediluvian world.

Price encouraged readers of the rocks to apply the detective's eye without succumbing to yarn-spinning Cuvierian excesses. As Srdjan Smajić has shown, the prevalent analogy for *seeing* in the late nineteenth- and early twentieth-century detective novel was *reading*: accurate seeing meant correctly reading what one saw.[84] Price was an expert at this: for a Baconian common-sense realist, reading God's Book of Nature was unproblematic. Nature was legible to Price in much that same way that the classic detective novel, as Elaine Freedgood has argued, depicts a 'utopian' world in which 'the meaning of materiality' attains 'generically enclosed plenitude'.[85] Fossil bivalves were as indisputable as signifiers as the scraps of evidence infallibly interpreted by Holmes. The caveat, of course, was that these signifiers could only be correctly read thanks to the revelations contained in God's

---

[80] Price, *Modern Flood Theory*, 12.
[81] Price, *Modern Flood Theory*, 51.
[82] Gowan Dawson, *Show Me the Bone: Reconstructing Prehistoric Monsters in Nineteenth-Century Britain and America* (Chicago: University of Chicago Press, 2016), 359; Price, *Illogical Geology*, 18.
[83] Price, *Illogical Geology*, 18.
[84] Srdjan Smajić, *Ghost-Seers, Detectives, and Spiritualists: Theories of Vision in Victorian Literature and Science* (Cambridge: Cambridge University Press, 2010), 71.
[85] Elaine Freedgood, *The Ideas in Things: Fugitive Meaning in the Victorian Novel* (Chicago: University of Chicago Press, 2006), 152.

'guidebook', the Bible.[86] This caveat explains why Price could favourably allude, in *Modern Discoveries Which Help Us to Believe* (1934), to the controversial '*closed circuit*' creationist theory of naturalist P. H. Gosse's *Omphalos* (1857).[87] Gosse's argument that, since God necessarily created the world with signs of age, like the misleading rings of newly created trees, it was impossible to disprove young-earth creationism, had been much derided. That this argument nonetheless appealed to the detective mind is indicated by the grudging respect given to it by Dorothy Sayers herself, who, in a 1942 lecture, admitted that, since it is 'deducible from the evidence', Gosse's argument was 'quite real, whether or not it ever was actual'.[88] In Price's interpretation, Gosse showed that empiricism was effective only when it took both of God's two books into account.

Clearly, for all his Holmesian dedication to logical reasoning based on infallible seeing, Price was by no means in complete sympathy with Doyle's agnostic detective. After all, Holmes's techniques were indebted not just to Cuvier's methods of reconstructing fossil animals, but also to aspects of palaeoscientific practice that Price resented. It is significant, given Price's emphasis on common-sense interpretations, that both nineteenth-century detective fiction and old-earth geology, including uniformitarian methods, have been seen as espousing an alternative common sense. Lawrence Frank argues that naturalists like T. H. Huxley and fictional detectives alike opposed their counter-intuitive reconstructions of past events—whether happenings in prehistoric ecosystems or crime scenes—to 'naïve empiricism'.[89] These savants claimed that scientific knowledge allowed them to plumb depths inaccessible to the uneducated eye. Price, however, wished to reinvigorate a levelling, literalist version of common sense seeing that enabled one to detect everywhere traces of the recent Flood. Rocks and fossils, 'tombstone inscriptions', could be 'put together into an intelligible and truthful narrative' that was far more intuitive than geologists' contingent story of deep time, upheaval, and erosion.[90] Much as the most reasonably literal interpretation of Scripture was best, Price asked readers to take geological evidence at 'face value' rather than assume that it was '*written in code*'.[91] Notably, a hieroglyphic code, translatable only after prolonged study, was exactly what many geologists compared the geological record to.[92] For Price, no Huxley and no Holmes need be called upon to accurately interpret this record, even if most Christians had allowed their common sense to rust in the previous century.

---

[86] Price, *Back to the Bible; or, The New Protestantism*, 3rd rev. edn (Taokoma Park, Washington, D.C.; Review and Herald Publishing Association, 1932), 19.
[87] Price, *Modern Discoveries*, 134.
[88] Dorothy L. Sayers, *Unpopular Opinions* (London: Victor Gollancz, 1946), 55.
[89] Lawrence Frank, *Victorian Detective Fiction and the Nature of Evidence: The Scientific Investigations of Poe, Dickens, and Doyle* (Baskingstoke: Palgrave Macmillan, 2003), 168.
[90] Price, *New Geology*, 15.
[91] Price, *New Geology*, 637.
[92] Frank, *Victorian Detective Fiction*, 162.

The Camera Does Not Lie

The importance of nature's legibility for fundamentalists brings me to the visual element of their work. Here, Price was not the only innovator, although he did set a new standard for anti-geohistorical and anti-evolutionary imagery. In the instances when Price had the capacity to reproduce illustrations, as in his expensive *The New Geology* (which sold for a hefty $3.50 even at a discount, in contrast with his other works, ranging between $1.25 and 25 cents), we see him at his most ambitious. Throughout this textbook, Price took advantage of the public domain status of photographs taken by the US Geological Survey to encourage readers to exercise commonsensical seeing. For example, below a halftone depicting the Medina sandstone of Lockport, New York, Price explained that the rocks revealed 'ripple marks' potentially caused by 'a gigantic tidal wave' (Figure 5.2). These images were sometimes accompanied by patches of speculation about landscapes in former times, a technique Price used very sparingly. Here, a few pages after a halftone of Colorado's Garden of the Gods, he envisions its sedimentary shapes being carved out by the Flood:

> It is an impressive sight to see such a mesa or a butte rising for many feet above a great plain with its strata in a horizontal position, and to realize that all of the country, perhaps as far as the eye can reach, must at one time have been filled up with rock material at least as high as the top of this solitary monument, and that all this surrounding material has since vanished and been carried away by the waters.[93]

Images and texts worked in unison to encourage readers to see the records of recent catastrophe.

The New Geology heralded a general enthusiasm for the evidential virtue of photographic reproduction among early twentieth-century creationists. In this period, the reduced cost of portable cameras and of halftone reproduction meant that creationist authors could bedeck their work with marks of photographic objectivity like never before.[94] Their desire to present the meanings of photographs as self-evident can be understood through the history of objectivity as a concept. Lorraine Daston and Peter Galison have memorably argued that, from the mid-nineteenth century, savants increasingly aimed at ideals of unmediated objectivity when producing images of nature, rhetorically rejecting 'the temptations of aesthetics, the lure of seductive theories, the desire to schematize'.[95] This was changing by the

---

[93] Price, *New Geology*, 128.
[94] Miles Orvell, *American Photography* (Oxford: Oxford University Press, 2003), 35–37.
[95] Lorraine Daston and Peter Galison, *Objectivity* (New York: Zone Books, 2007; repr. 2010), 120.

398   *THE NEW GEOLOGY*

Fig. 247. Remnant of a giant ripple mark, in Medina sandstone (L. Silurian), New York Central Railway, Niagara Gorge, New York. This is nearly 20 miles west of the locality shown in Fig. 248. Probably these phenomena prevailed in the Medina sandstone over all this region. (Gilbert, U. S. G. S.)

Life

Very many of the Silurian rocks are packed full of fossils; but as a rule, they do not show any great variety of types, except a profusion of marine invertebrates, chiefly *corals, crinoids,*

Fig. 248. Giant ripple marks in Medina sandstone (L. Silurian), Lockport, New York. These are manifestly quite abnormal. They could have been caused at the bottom of very deep water — perhaps a mile or more deep — if a gigantic tidal wave was moving through the ocean. It is difficult to see how else they could have been made. (Gilbert, U. S. G. S.)

Figure 5.2 George McCready Price, *The New Geology: A Textbook for Colleges, Normal Schools, and Training Schools; and for the General Reader* (Mountain View, CA: Pacific Press Publishing Association, 1923), 398. Price's halftone helps readers to notice signs of the Flood in New York sandstone. His caption's conclusion that '[i]t is difficult to see how else' these marks 'could have been made' discourages contradiction.

beginning of the next century, Daston and Galison contend, when 'trained judgment, subjectivity, artisanal practice, and unconscious intuition' were being seen by a more confidently established scientific community as superior ways of visually representing nature.[96] While this generalization remains controversial, it makes sense that twentieth-century young-earthers, who kept other nineteenth-century conceptions of scientific method current, would also preach besieged ideals of objectivity that encouraged humble attentiveness to the nature's self-evident facts.

For Harry Rimmer, a Presbyterian minister from California, photographic reproduction brought evolutionary debates into a genuinely democratic public sphere. During the interwar decades he, rather than Price, was, Numbers notes, the 'antievolutionist' with the larger audience 'among American evangelicals', while mid-century reprints of his early books, published by William B. Eerdmans, sold over 100,000 copies.[97] Rimmer was, admittedly, a gap theorist rather than a strict young-earth creationist, but this distinction was more important to Price than to creationist colleagues. As gap theory, unlike day-age theory, allowed for a literal seven-day creation week, the existence of deep geohistory before the gap must have seemed of negligible significance to Rimmer. As a result, although it denounced all concessions to old-earth viewpoints, including gap theory, *The New Geology* was nonetheless lauded by Rimmer as 'a masterpiece of REAL Science'.[98] His own work exudes Pricean stylings. The product of fragmented educational history, Rimmer was another ultra-Baconian who tirelessly espoused a DIY approach to science, first constructing a laboratory and photographic studio in his garden and then using membership fees from his Research Science Bureau, founded in Los Angeles in 1921, to fund research and the publication of cheap, polemical pamphlets.[99]

One of these pamphlets, *Monkeyshines* (1926), which was addressed to 'the average lay reader who can use his eyes and brain', foregrounded both the uses and abuses of scientific photography.[100] Quipping that '[t]he camera does not lie; but liars use the camera', Rimmer complained that photographs reproduced in evolutionary textbooks betrayed evidence of retouching that blurred the differences between human and ape skeletons.[101] With typographical flamboyance, he attested in quasi-legal manner to the virtuous objectivity of his own photographs: '**these pictures were all taken by the author in his own laboratory, from specimens in his own private museum, and the author takes oath they are not retouched**

---

[96] Daston and Galison, 342.
[97] Numbers, *Creationists*, 76. For sales, see Larson, 232.
[98] Harry Rimmer, *Noah's Ark, Modern Science and the Deluge* (Los Angeles, CA: Research Science Bureau, 1925), 28.
[99] For biographical information, see Roger Daniel Schultz, 'All Things Made New: The Evolving Fundamentalism of Harry Rimmer, 1890-1952', unpublished PhD thesis, University of Arkansas (1989); and Numbers, *Creationists*, 76-87.
[100] Harry Rimmer, *Monkeyshines: Fakes, Fables, Facts Concerning Evolution* (Los Angeles, CA: Research Science Bureau, 1926), 2.
[101] Rimmer, *Monkeyshines*, 5.

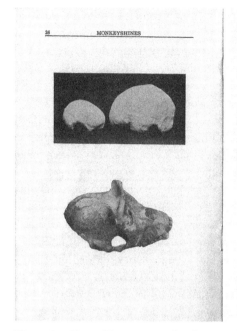

Figure 5.3 Harry Rimmer, *Monkeyshines: Fakes, Fables, Facts Concerning Evolution* (Los Angeles, CA: Research Science Bureau, 1926), 26–27. Reproducing photographs taken in his laboratory, Rimmer advises readers to flip back and forth in his pages to learn the detectible differences between human and ape skulls. Copy in author's collection.

or doctored in any way. **EXAMINE THEM FOR THOSE STRUCTURAL EVIDENCES THAT PROVE RELATIONSHIP!**' (bolding and full capitalization in original).[102] Aside from the fact that readers were required to trust his word about the untouched nature of his images, Rimmer's booklet, as the injunction implied, encouraged the same kind of active readership we have seen facilitated across different fields of borderline palaeoscience. He pestered readers of *Monkeyshines* to distinguish human from ape skeletons both fossil and recent, rather than simply accepting his word (Figure 5.3). For most of this pamphlet, Rimmer implied that the book's expository voice was his own, but an unexpected interlude introduced a jarring shift to the third person and past tense. 'At this point in the discussion', we are told, 'a student arose', enquired about the 'missing links' and was answered by 'the Speaker'.[103] By reframing the foregoing exposition as a 'discussion', Rimmer's unconventional segue further stressed that this was a dialogic text.

[102] Rimmer, *Monkeyshines*, 5.
[103] Rimmer, *Monkeyshines*, 31.

In the final section, Rimmer contrasted the photographic evidence of his Los Angeles laboratory with artistic reconstructions of missing links housed at the American Museum of Natural History. In so doing, he attacked one of that prestigious institution's potential weak spots. Lukas Rieppel, drawing on Daston and Galison's scholarship on objectivity, has shown how the institution's leading palaeontologists vigorously stressed that the artistic recreations of extinct animals on display were near-watertight extrapolations made from undisputed fossil material.[104] While this argument was easy enough to defend with regard to the dinosaurs Rieppel discusses, it was more contested in the field of palaeoanthropology. Rimmer recognized that important specimens like those of Piltdown Man and Java Man, considered by scientists to be prehistoric humans, were based on extremely fragmentary remains. The Museum's lifelike busts of these poorly understood proto-humans, Rimmer told readers, were based on speculative assemblies of shattered bones found in entirely separate locations. The necessary train of indexical evidence linking the excavated bones with the realistic models on display in the American Museum's galleries was entirely absent: these missing links were, in his words, 'made up of plaster of Paris and imagination!'[105] In contrast with Rimmer's research, based on specimens personally photographed for the scrutiny of fundamentalist readers, the United States' premier natural history museum camouflaged the scanty justifications for its speculative exhibits.

A similarly interactive spirit, and visual flair, was displayed by Byron C. Nelson, whose anti-evolutionary monograph, 'After Its Kind' (1927), was intended 'to induce the reader to become his own authority'.[106] Nelson, a Lutheran minister from Wisconsin with experience of university-level science training, was one of Price's earliest allies outside the Adventist Church.[107] He was converted to young-earth Flood Geology by reading *The New Geology*, and that work's aesthetic and intellectual influence is visible in Nelson's subsequent publications, especially *The Deluge Story in Stone* (1931). This was a historical and technical monograph, heavily indebted not just to Price's arguments in favour of Flood Geology, but, in Nelson's words, to Henry Hoyle Howorth's scientifically 'outstanding' (but, he admitted, 'tedious') books.[108] Nelson brought together centuries of scholarship to argue that the association of stratification with the Flood had been a respectable theory until the early nineteenth century, when it was unjustly

---

[104] Lukas Rieppel, *Assembling the Dinosaur: Fossil Hunters, Tycoons, and the Making of a Spectacle* (Cambridge, MA: Harvard University Press, 2019), 184–85, 214–15.

[105] Rimmer, *Monkeyshines*, 45,

[106] Byron C. Nelson, *'After Its Kind': The First and Last Word on Evolution* (Minneapolis, MN: Augsburg Publishing House, 1927), 10.

[107] Biographical information on Nelson is indebted to Paul A. Nelson, 'Introduction', in *The Creationist Writings of Byron C. Nelson*, vol. 5 of Ronald L. Numbers, ed., *Creationism in Twentieth-Century America: A Ten-Volume Anthology of Documents, 1903–1961* (New York: Garland, 1995), ix–xxi; and Numbers, *Creationists*, 125–30.

[108] Byron C. Nelson, *The Deluge Story in Stone: A History of the Flood Theory of Geology* (Minneapolis, MN: Augsburg Publishing House, [1931]), 118, 128.

discarded. The book, called 'exceedingly valuable' by Price, was published by Augsburg, a Lutheran firm which, John Tebbel notes, was 'one of the better denominational publishers', known for 'handsomely printed volumes'.[109]

In this polished book, Nelson followed Price in reproducing photographs available from the US Geological Survey and retraining readers' eyes to see in them rocky evidence of the Flood. *The Deluge Story in Stone* begins with an anecdote about Nelson's own quasi-epiphanic introduction to Flood Geology, which happened as he was travelling through the Rocky Mountains in Alberta. Approaching a river 'to slake his thirst', he looked up to see a 'horizontal stratum full of shells' which 'gave the impression of having been formed by the settling of vast amounts of sediment in a moving, disturbed condition of great waters'.[110] His book's many photographic images prompted readers to feel the same intuitive recognition of geology's diluvial origins. The caption for a halftone of the Petrified Forest of Arizona, for example, asserted that '[e]very stratum speaks of water'.[111] Needless to say, the term 'water' did not refer to the water of uniformitarian sedimentation. The empirical Flood Geologist, who listens to how each rock plainly 'speaks', was contrasted with the typical evolutionary geologist, who 'merely needs to be told in his study' what fossils exist in a stratum before immediately assuming its age with reference to an a priori geological column.[112] Even visualization of the catastrophic Flood itself could be facilitated using entirely 'objective' imagery. Perhaps recalling Price's occasional calls (in prose) for readers to imagine Flood-lashed canyons and mesas, Nelson reproduced a photograph of the Grand Canyon in Arizona eerily filled with fog, providing the illusion that water raged between its walls.[113] The cleverly in-camera nature of this unretouched special effect mitigated against accusations of artistry.

The ostensible honesty of photographs like these contrasted with the theoretical assumptions built into schematic diagrams. *The New Geology* reproduced many diagrams by mainstream geologists, not, as one might expect of a diagram, for the purposes of evidencing, but rather to satirize them and discredit their theoretical content. Price thundered that these diagrams—depicting physically unseen depths, or, more abstractly, implied events taking place across deep time—were 'unworthy of serious attention from truth-loving people who are not committed to an *a priori* theory of how the strata *ought* to be found'.[114] In the name of combating such a pernicious visual language, Price was, on occasion, not above misleadingly

---

[109] Price, untitled review of *The Deluge Story in Stone*, by Byron C. Nelson, Ministry, 5 (1932), 25; John Tebbel, *A History of Book Publishing in the United States: Volume III: The Golden Age between Two World Wars 1920–1940* (New York: R. R. Bowker, 1978), 246.
[110] Nelson, *Deluge Story in Stone*, xiv.
[111] Nelson, *Deluge Story in Stone*, 8.
[112] Nelson, *Deluge Story in Stone*, 141.
[113] Nelson, *Deluge Story in Stone*, 65.
[114] Price, *New Geology*, 635.

altering these images to exacerbate their fallacies.[115] Reproducing diagrams from sources like the US Geological Survey, he added scathing captions to denounce as 'imaginary' the phenomena they depicted (Figure 5.4). Nelson, too aimed to dispel the 'hypnotic influence' of palaeoscience's visual languages.[116]

The dismissal of theoretical diagrams as delusive had a long history in literalist polemic. Back in 1858, Cater had similarly mocked the unfortunate terminology employed in William Buckland's *Geology and Mineralogy, Considered with Reference to Natural Theology* (1836). Buckland's book, which endorsed gap theory, featured an illustrated 'IDEAL SECTION', or 'imaginary section', of Earth's strata, which Cater derided as too aptly named, voicing his own opposition to the '*phantoms*' of '*ideal*' and '*imaginary*' forms of geology.[117] However, Cater, unlike the later Nelson and Price, had been unable to reproduce these images in the name of their demolition.

The care Price took in satirizing this visual tradition further suggests his awareness of the threats geohistorical thought posed to Adventist conceptions of time. The circulation of charts schematizing the system of Bible prophecies that linked patriarchal times to end times had been central to nineteenth-century Adventist practice.[118] Price himself was a dedicated scholar of prophecy throughout his life, as demonstrated in his late work, *The Greatest of the Prophets* (1955). Like fiction and its distracting secular chronologies, diagrams of the geological column or of geohistorial content, with their easily communicated portrayal of extra-biblical ages, disseminated ways of thinking at odds with Adventist eschatology. Despite its age, Price was likely familiar with Buckland's aforementioned 'IDEAL SECTION' of 1836, one of the earliest geological diagrams to schematize the entirety of deep time, given that he repeatedly cited the work in which it was found (in line with his habit of drawing selective quotations from classic nineteenth-century geological books).[119] Buckland's geological ages and their respective life forms progressed rightwards not, as in a prophetic chart, towards the apocalypse, but simply to the present era (Figure 5.5). Even worse, Price would have seen evolutionary variations on this genre in eminent scientific manuals like Louis V. Pirrson and Charles Schuchert's *A Text-Book of Geology* (1915), from which he had taken many of his own figures. Even the Christian framing of a chart like that of J. W. Grover (Figure 1.1), would have been made equally unappetizing to Price by the artist's depiction of deep time.

---

[115] For redrawing, see Bootsman, 223–25.
[116] Nelson, '*After Its Kind*', 40.
[117] Cater, 570–71.
[118] David Morgan, *Protestants & Pictures: Religion, Visual Culture, and the Age of American Mass Production* (Oxford: Oxford University Press, 1999), chapter 5.
[119] For example, see Price, *God's Two Books*, 153. For the origins and context of Buckland's image, see Jonathan R. Topham, *Reading the Book of Nature: How Eight Best Sellers Reconnected Christianity and the Sciences on the Eve of the Victorian Age* (Chicago: University of Chicago Press, 2022), 160–65, 205–6.

260   *THE NEW GEOLOGY*

tinction can be made between these so-called block mountains and what are called *folded* mountains, nor yet between either of these classes and ordinary *mountains of erosion*, such as we have already described. Too much theory as to the relative ages of the strata composing these mountains enters into the distinctions made between them.

The Jura Mountains, between France and Switzerland, are usually pointed out as typical examples of mountains composed of a series of wavelike *folds*. They consist of a number of parallel ranges trending northeast and southwest, each range appearing to be bowed up like an arch, and the valleys being like the troughs of these wavelike folds. Much of the material once composing the peaks of the arches has been eroded away and now partly fills the intervening troughs. But the whole range is composed of strata once laid down horizontally in the sea, for an abundance of sea fossils are contained within them, though now they are many thousands of feet above sea level.

But by far the greater number of mountain ranges are so highly complex in their structure that they are usually classed together as *complex mountains*. Such are the Alps, the Hima-

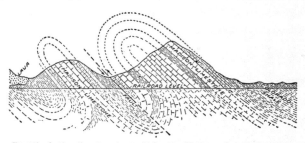

FIG. 158. Section (imaginary) near Livingston, Montana, where Madison (Mississippian) limestone is repeated three times, with strata of different age intervening. These huge folds are imagined to have taken place here, in order to explain this serial repetition of the same kind of beds. We must remember that all the drawing below the horizontal line is absolutely imaginary, as are also the folds pictured above. (U. S. G. S.)

layas, and many of the individual ranges of the Rockies. As already remarked, there is no sharp line of distinction to be drawn between the mountains of this type and mountains like the Juras, which are obviously made up of wavelike folds. And these complex mountains present so many diverse types of structure that it is difficult to do more than point out a few characteristics common to very many of them. Their majestic peaks constitute the most elevated portions of the earth, and

Figure 5.4 Price, *The New Geology*, 260. Reproducing a figure from the US Geological Survey depicting the processes of folding and erosion, Price labels its subject matter 'imaginary' three times.

Speaking of the maps, stratigraphic cross sections, and other pictorial genres that have been central to communicating geology since the early nineteenth century, Rudwick argues that this visual language 'embodies a complex set of tacit

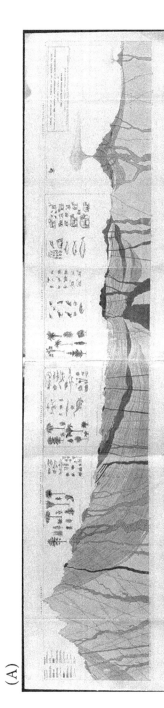

Figure 5.5 (A) William Buckland, *Geology and Mineralogy Considered with Reference to Natural Theology*, 2 vols (London: William Pickering, 1836), fold-out chart. © The British Library Board, W5/7294. (B) *Bible Readings for the Home Circle* (Battle Creek, MI: Review and Herald Publishing Company, 1889), 428. Buckland's influential chart mapped geohistory onto the stratified rocks, from the earliest formations (on the left) to those of the present era (on the right). Seventh-day Adventist charts instead schematized the chronological implications of various cryptic scriptural prophecies. While Adventist charts ventured back far less deeply in time than geohistorical ones, they went further into the future, when time itself would be no more.

212 CONTESTING EARTH'S HISTORY IN TRANSATLANTIC LITERARY CULTURE

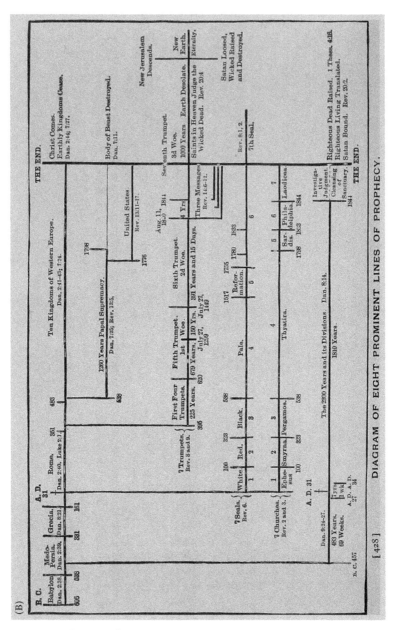

Figure 5.5 Continued

rules and conventions that have to be learned by practice' and implies 'the existence of a social community which tacitly accepts these rules'.[120] Flood Geologists

---

[120] Rudwick, 'Visual Language', 151.

refused to 'tacitly' accept the rules, instead reducing palaeoscience to little more than an assemblage of disingenuous literary technologies. Here it is pertinent to recall that Steven Shapin and Simon Schaffer, from whom I take this term, see not just literary 'technologies' but also social and material ones as necessary tools in the construction of cogent scientific truth claims.[121] For early twentieth-century creationists with limited access to scientific networks or specimens (through both choice and necessity), literary technologies took on an outsized importance because they could be critically engaged with. When reduced significance was ascribed to extra-textual factors involved in the production of scientific claims— Price's occasional correspondence with geologists, Rimmer's backyard museum of simian skulls, and Nelson's insistence on listening to the rocks notwithstanding— then the literary nature of these claims showed up in high relief. In *Contesting Earth's History* I have often shown literature to be the main battleground of borderline scientific practice, a place where figures on the fringes could fight on home territory, but it was in young-earth creationism that mainstream palaeoscience was itself placed on an almost purely literary plane.

A Moving Picture Show of Creation

This dismissal of palaeoscientific images and the concepts they expressed was tied to young-earthers' oft-repeated point that no scientists had ever *seen* past events first-hand, nor had evolution been observed in real time. In an early denunciation of overthrust, Price observed 'that none of the authorities who report this circumstance can testify as eye-witnesses of this marvellous event'.[122] Like Rose, the gun-barrel-maker, Price mockingly asserted that the certainty of geologists' explanations required a 'supernatural knowledge of the past', rendering their work 'no more logical or scientific than the methods of astrology or theosophy'.[123] This quip alluded to the Theosophical clairvoyants I examined in chapter 2. Price, who travelled extensively in California, working at the Loma Linda College of Evangelists between 1906 and 1912, likely knew of the large Theosophical community at nearby Point Loma. They certainly knew of him: in characteristically syncretic style, one member, writing in Point Loma's *Theosophical Path* in 1922, cited Price's demolition of the geological column as support for occult claims about the existence of humanity throughout deep history.[124] If he was unwilling to credit Theosophical evidence in return, Price by no means disbelieved in what he called 'supernatural knowledge of the past'. Access, however, was restricted to

---

[121] Steven Shapin and Simon Schaffer, *Leviathan and the Air-Pump: Hobbes, Boyle, and the Experimental Life*, new edn (Princeton, NJ: Princeton University Press, 2011), 25.
[122] Price, *Illogical Geology*, 27.
[123] Price, *God's Two Books*, 99; Price, *Geological-Ages Hoax*, 9.
[124] Charles James Ryan, 'Archaeological Notes', *Theosophical Path*, 23 (1922), 61–68 (64).

all but Ellen White, who died in 1915, and the author of Genesis, neither of whom could expand on their statements. Anyone else claiming to explain how strata had contorted over eons was simply writing a 'geological fable, invented solely for the purpose of evading the plain evidence of our eyesight and common sense'.[125] Geologists' 'fantasy of total visibility', to return to Ralph O'Connor's phrase, was turned into evidence against their scientific integrity.[126]

In Price's day, the geohistorico-evolutionary narrative, that detested 'fable', was culturally ubiquitous. The story of the succession of life on Earth was the quotidian subject matter of literature ranging from the countercultural cosmogonies of Theosophists to the prose epics of science journalists. The *Illustrated London News*, for instance, dubbed one 1912 survey of Earth's palaeontological history for general readers 'a piece of literary cinematography' constituting 'an almost awesomely realistic *revue* of the history of the world'.[127] The technologies that enlisted virtual witnesses in support of particular interpretations of prehistory were reaching new heights of complexity in literally cinematographic form too: American films like Willis O'Brien's *The Ghost of Slumber Mountain* (1918) even began to depict extinct animals in action using early stop-motion animation. Price, however, refused to be made into a virtual witness and scorned 'leaders in natural science' who 'openly boasted that science . . . could almost give us a moving picture show of Creation in the making'.[128] This was, he implied, the height of hubris. Conceding in *A History of Some Scientific Blunders* (1930) that 'if we could have . . . watched the real creation taking place, we should probably have called it an orderly and beautiful process', Price doubted that it was 'scientifically profitable to spin any more imaginary processes about the origin of things'.[129]

For Price, any extra-biblical narrative about Earth's antediluvian history was an artificially diachronic arrangement of synchronic fragments. He delighted in this conceit in *The New Geology*, repurposing Charles Darwin's metaphor of the geological record as a mutilated book into a metaphor of the geological column as an arbitrarily arranged library:

> A card catalogue or index is a very useful thing in a library. But it would strike us as highly absurd if some antiquary should come along and solemnly assure us that this card catalogue showed the real history of the order in which the books listed had been published. ... But the geological classification as currently taught, based on the grade of fossils contained in the various rocks, is just as purely artificial as the card catalogue of a library, and has no more time-value to its subdivisions.[130]

---

[125] Price, *Modern Flood Theory*, 59.
[126] O'Connor, *Earth on Show*, 432.
[127] 'The Comfortable Word "Evolution"', *Illustrated London News*, 20 April 1912, 582.
[128] Price, *Back to the Bible*, 76.
[129] Price, *Scientific Blunders*, 33.
[130] Price, *New Geology*, 294.

As in his accusations of their circular reasoning, Price here denied geologists any rationale for the dating and ordering of fossils and rocks. An 'ingenious antiquary might devise a brilliant theory about the history of the development of printing all over the world, based on this catalogue index', but their arbitrary theory would be no more logical than the methodology of the 'evolutionary geologists' who 'have given us a moving picture of the evolution of life, based on their geological series'.[131] His scornful reference to old-earth geology as an all-too-fluent 'moving picture' modernized the metaphoric dismissal, routine in the nineteenth century, of suspect science writing as novelistic.

Much of this creationist concern about the reconstruction of narratives centred around the American Museum of Natural History, upon which we have previously heard Rimmer heap scorn for its palaeoanthropological reconstructions. This institution has been described by scholars as a machine honed for courting virtual witnesses, the fossil skeletons and paintings in its galleries being intended optically to instil the views of its chauvinistic president, Henry Fairfield Osborn, into the minds of visitors.[132] Osborn saw the march of evolutionary progress as having culminated in the prehistoric Cro-Magnons—the aforementioned inhabitants of Lewis Spence's Atlantis—to whom he looked back with fondness for their ability to achieve high levels of pure-blooded culture in nature's fierce arena, especially compared with the racially diverse and (to his mind) enervated inhabitants of modern American cities. By the 1920s, Osborn had almost industrialized the translation of fossil bones unearthed in the field into lifelike images on the page and spectacular exhibits in the museum gallery. Marianne Sommer discusses how the media-savvy Osborn, himself a vocal opponent of fundamentalism, assisted by other palaeontologists, technicians, and artists, created a 'serial workflow' in which 'traces of lost worlds were collected and processed, from whence the resulting reconstructions of ancient creatures', whether Cro-Magnon or dinosaur, 'were immediately fed into a multi-media distribution network'.[133]

Price referred to Osborn's multimedia machine when he complained that any 'fragment ... of a 'million-dollar' pig's tooth from Nebraska' was 'sufficient excuse to set the telegraph wires humming or the presses turning out a lifelike picture of what happened ten or a hundred million years ago'.[134] Even with his prolific output in the evangelical and fundamentalist press, Price could never hope to attain such domination of the channels of scientific communication. He could, however, turn his sceptical eye to the processes by which fossil discovery became museum

---

[131] Price, *New Geology*, 615.
[132] Victoria E. M. Cain, '"The Direct Medium of the Vision": Visual Education, Virtual Witnessing and the Prehistoric Past at the American Museum of Natural History, 1890-1923', *Journal of Visual Culture*, 9 (2010), 284-304.
[133] Marianne Sommer, 'Seriality in the Making: The Osborn-Knight Restorations of Evolutionary History', *History of Science*, 48 (2010), 461-82 (461-62).
[134] Price, *Geological-Ages Hoax*, 17.

exhibit. As part of his opposition research for *The New Geology*, Price visited the American Museum in 1920 and 1922 and paid particular attention to its palaeoanthropology displays. He was far from the kind of receptive visitor Osborn would have wished for. In the finished book, a caption underneath a photograph of the museum's 'serial but artificial arrangement of some skulls of apes and men' compared the display's arbitrary implications of progress and descent to an absurdist 'serial' display of 'the evolution of the American house' (Figure 5.6). The Museum's seriality—the linking together of individual fossils across the globe into broader geohistorical narratives, and the linking of specimens into a genealogical, evolutionary tree of life—was severed by Price's destructive satire. His criticisms of serially arranged displays and diagrams were particularly potent against iconographic representations of evolution as a linear process, the likes of which would have been familiar to all his readers in the United States of the 1920s.[135]

If Price generally refused to proffer his own imaginative glimpses of past events, he could, on occasion, appropriate those penned by historical geologists. This option was enabled by his extensive reading—whether in the originals or through secondary quotations—of early and mid-nineteenth-century geologists,

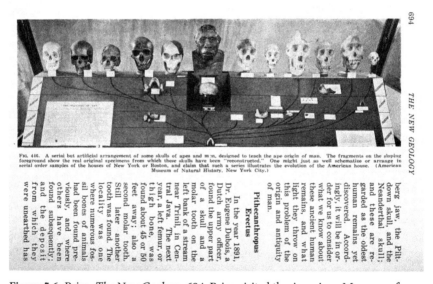

Figure 5.6 Price, *The New Geology*, 694. Price visited the American Museum of Natural History in New York when researching his textbook. These visits alerted him to the way that what he saw as arbitrary 'serial' attributions of geological age and evolutionary descent to fossil specimens were made seductively concrete in displays.

[135] Constance Areson Clark, '"You Are Here": Missing Links, Chains of Being, and the Language of Cartoons', *Isis*, 100 (2009), 571–89 (584).

from whose work Price permitted himself to borrow verve by recycling striking quotations. A key example of this tactical appropriation of imaginative authority is Price's quotation, in many of his books, of Miller's awesome description of the mass death of ancient fish in *The Old Red Sandstone* (1841).[136] It is easy to see why this passage attracted him. While Miller was by no means referring to the Flood, the catastrophic passage was easily interpreted as such; moreover, although Miller was best known for textually revivifying prehistoric worlds, his description of the fishes' entombment was a sustained exercise in the deathly, sepulchral imagery that was Price's preferred means of evoking antediluvian natural history. Price's quoting technique could lead to suggestive bricolages. In its appearance in *God's Two Books*, the Miller quotation was bisected by an engraving of *Pterodactylus crassirostris*, appended with the uncharacteristically hypothetical observation that this reptile, which 'probably lived on fruit and nuts', corresponded 'more nearly to the "serpent" in the garden of Eden than anything yet discovered'.[137] The quotation of Miller followed another from Buckland's *Geology and Mineralogy* and the pterodactyl's proximity to this passage was no coincidence (Figure 5.7).[138] Price's infernal non sequitur, comparing a pterodactyl to Satan, was drawn from a figurative flight of fancy in Buckland's book that had become one of the most oft-quoted passages of any nineteenth-century geological work—taken seriously, as I showed in chapter 2, by H. P. Blavatsky herself. Price, usually hostile to speculations about antediluvian life, here compromised his empirical approach, apparently succumbing to the power of Miller's prose and Buckland's satanic simile.

Nonetheless, these moments were few and far between. It is apt that Price, in a rare attempt to pen his own verbal journey backwards in time, pointedly refused to go back very far. At the conclusion of *Back to the Bible*, chiefly a Protestant polemic rather than a work solely focused on Flood Geology, he begged permission to 'go back in my memory almost to the extreme eastern limit of the continent' to depict his former idyllic 'Christian home', which was suffused with the Bible and the seventeenth- and eighteenth-century poetry of 'Addison, Watts, Cowper, Wesley, Milton, Young'.[139] From here he journeyed back 'into the very midnight of the Dark Ages'.[140] When the Protestant story was traced to its murky origins, Price refused to go further. His book's account of the Flood was consigned to a subsequent appendix, in which Price's empiricism performatively strained against even his own geological argumentation. Here, he speaks in the third person:

---

[136] Nelson extended the act of literary replication by repeating the lines in his own work, for which see *Deluge Story in Stone*, 42.
[137] Price, *God's Two Books*, 155.
[138] Price, *God's Two Books*, 153.
[139] Price, *Back to the Bible*, 202–3.
[140] Price, *Back to the Bible*, 204–7.

218 CONTESTING EARTH'S HISTORY IN TRANSATLANTIC LITERARY CULTURE

> 154   GOD'S TWO BOOKS
>
> of Fishes." Dr. David Page, after enumerating nearly a dozen genera, says: —
>
> These fishes seem to have thronged the waters of the period, and their remains are often found in masses, as if
>
> GANOID FISHES FROM THE DEVONIAN ROCKS
>
> they had been suddenly entombed in living shoals by the sediment which now contains them.
>
> Hugh Miller's classic description of these rocks as they occur in Scotland may be taken as representative of numerous other Devonian rocks in all parts of the world: —

Figure 5.7 Price, *God's Two Books; Or Plain Facts about Evolution, Geology, and the Bible* (South Bend, IN: Review and Herald Publishing, 1911), 154–55. Across these two pages, Price recycles and transforms material from the literary history of palaeoscience. An imagined description of a catastrophic flood, quoted from Hugh Miller, is bisected by a caption comparing a pterodactyl to the serpent of Eden.

GOD'S TWO BOOKS 155

We read in the stone . . . a wonderful record of violent death falling at once, not on a few individuals, but on whole tribes. . . .

At this period of our history, some terrible catastrophe involved in sudden destruction the fish of an area at least a hundred miles from boundary to boundary, perhaps much

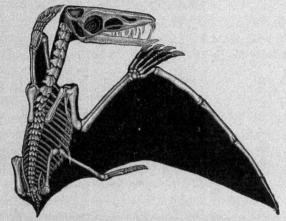

A reptile *(Pterodactylus crassirostris)* about a foot long, with about three feet spread of wing. It had a large brain, and as it probably lived on fruits and nuts, it is interesting from corresponding more nearly to the "serpent" in the garden of Eden than anything yet discovered. From the Jurassic of Europe. Other species of Pterodactyl were many times larger.

more. The same platform in Orkney as at Cromarty is strewed thick with remains, which exhibit unequivocally the marks of violent death. The figures are contorted, contracted, curved; the tail in many instances is bent round to the head; the spines stick out; the fins are spread to the full, as in fish that die in convulsions. . . .

In this attitude nine tenths of the Pterichthes of the lower old red sandstone are to be found.—" *Old Red Sandstone,*" *pages 48, 221, 222.*

Figure 5.7 Continued

The writer has so great a dislike for geological speculations that only the entreaty of friends has induced him to try his hand as this usually very unprofitable business. But as some seem still to have a difficulty in forming a mental picture of how a universal Deluge could possibly take place, the following facts are enumerated to show that such a thing accords perfectly with possibility, and even probability.[141]

He proceeded to speculate that the Flood may have been caused by a sudden alteration to Earth's axis that had had catastrophic consequences. For a precise description of those consequences, readers would have to turn to Ellen White's *Spiritual Gifts* or to Genesis itself. Price wished to upstage neither. His books were no story.

Real Literary Treatment

Given all that I have covered, it may come as a surprise that Price did, eventually, publish a work of imaginative literature. *Some Scientific Stories and Allegories* (1936) compiled a selection of short pieces, a few previously published in Adventist periodicals, spanning a variety of genres, including parable, science fiction, literary criticism, *conte philosophique*, dream vision, and even neo-Victorian drama. Price's own explanation for the volte-face was glibly conventional: his paternalistic suggestion that many 'prefer their science well disguised and somewhat sugar-coated', as the well-read author likely knew, dated back to *De Rerum Natura (On the Nature of Things)*, a didactic epic poem by the ancient Roman proto-evolutionist Lucretius.[142] Considering that—over the two millennia since Lucretius—evolutionists had 'made use of almost every imaginable literary device', while anti-evolutionists had 'been remiss in both the scientific accuracy and literary form of their opposition', Price offered 'a change from the argumentative form of most of my other works' (5). This new approach was aimed at those unable to follow 'formal scientific presentation' (6). Looking closely at this book's content explains its apparent betrayal of White, whose stern words against the dangers of even high-quality fiction had so often factored into Price's literary-scientific philosophy.

To begin, it must be reiterated that Price had begun his career as a poetry afficionado. Poetry was less problematic than fiction in White's view, especially the eminently safe early modern Protestant poets Price adored.[143] Indeed, some of his own poems appeared in Adventist periodicals, and towards the end of his life he collected them in a typescript, *Poems of My Long Ago* (1959), held by Andrews University in Berrien Springs, Michigan. At the age of 89, Price explained his decision to publish or republish these youthful poems as motivated by desire to return to his 'apprenticeship at verse' after an adult career

---

[141] Price, *Back to the Bible*, 225.
[142] Price, *Some Scientific Stories and Allegories* (Grand Rapids, MI: Zondervan, 1936), 5. Subsequent references included in text.
[143] Land, *Historical Dictionary*, 174.

of 'polemical prose in [sic] behalf of an unpopular cause'.[144] Although most of Price's 'strictly literary efforts' (3) were produced during the nineteenth-century *fin de siècle*, he was, unsurprisingly, no proponent of the then-current aesthetes' creed of art for art's sake. Instead, he used poetry primarily as a religious exercise, expressing spiritual struggles as well as addressing political and ecclesiastical topics in occasional poems. He called his poetry 'Victorian' (2): in both diction and content it appears indebted both to Robert Browning's dramatic monologues and to the rhythmic devotional poetry of Christina Rossetti. Biblical images of maternal love, life's crooked path, and Christ as the bridegroom recur, combined with Pricean scepticism about progress and apocalyptic fears of the global church federation that foreshadows the Antichrist's arrival.

Revealingly, Price's verse attempted to reconcile his desire to be known as a literary intellectual with the unworldly piety required of Adventists. In one characteristic passage in the dramatic monologue 'John in Machaerus' (*c*.1900), John the Baptist, imprisoned and awaiting death in King Herod's fortress, experiences a vision of a future. John sees that 'man' has abandoned God in favour of secular humanism, going to the grave

> as did his ancestors
> No nearer meeting the Creator's plan
> Than do the shepherds on Judea's hills.
> The latter may produce as good results
> In character, in mental, moral growth,
> As cultured science in those latter days. (56)

The poem 'Soul Culture' likewise lamented that Christianity teaches 'much that's deemed uncultured by the world' (38). The collection's recurring sense of cultural backwardness, worldly failure, and spiritual toil was likely fuelled by the difficulties Price faced when most of the poems were being written: his bleak, Thomas Gray–inspired 'Elegy Written in a Country School House' (1898) was penned while the ambitious Price languished in unrewarding teaching positions, including a post in what Numbers calls the 'dreary, remote village' of Tracadie in maritime New Brunswick, while allusions to religious 'Doubts' in the undated 'Afterward' may dramatize his challenging early encounter with evolution, or the suicidal thoughts that assailed him as he struggled to hold down a job in the early 1900s.[145] In their variety, Price's verses constitute strong evidence that strategy, rather than mere literary laziness, shaped his writing's later repetitiveness and antipathy to creativity.

---

[144] Price, *Poems of my Long Ago*, unpublished typescript, Center for Adventist Research, Andrews University (1959), 2–3. Subsequent references included in text.
[145] Numbers, *Creationists*, 91, 94.

In *Some Scientific Stories and Allegories*, Price evaluated the poetry of others. The essay 'Some Literary Aspects of the Evolution Theory' showed that Price's literary criticism was as realist as his geology: for him bad science created bad poetry. Citing the 'ineffectual literary garb' of poems like Langdon Smith's *Evolution* (1906), Price declared that only 'objective realities ... are capable of real literary treatment' (69). Here uninterested in questions of poetic form, despite the offhand references to 'ineffectual' presentation, Price saw a poem's philosophical insight, and, in a more unformulated manner, its literary greatness, as confirmed by its ability to endure through generations. Thus, while the 'mechanical construction' of one poem cited was 'fairly good', it was not 'a real poem' due to its evolutionary and thus immoral and transient subject matter (72). Real poems, such as the classics on which he had grown up, were such that a 'young mother' might recite to provide her children with 'moral sinew' against evil (75). Just as Price's new book was a form of 'sugar-coated' science (5), good poetry was sugared morality. Poetry, for Price, was only secondarily an aesthetic experience.

If poetry was, for all these qualifiers, a space where Adventists could experiment with aesthetic creativity, a reading of *Some Scientific Stories and Allegories* makes it clear that Price remained uneasy with plain fiction in modern genres. The book is best understood less as a collection of short stories than as a Menippean satire, the ancient genre described by Northrop Frye as dealing 'less with people as such than with mental attitudes', employing a 'loose-jointed narrative form' and often based around 'a dialogue or a colloquy'.[146] We have already heard, in the discussion of hollow-earthers in chapter 3, how this arch tradition of satire combines 'genres' and 'tones' in order 'to combat a false and threatening orthodoxy'.[147] Evolution and uniformitarian geology were Price's dominant orthodoxies to be dismantled. The Menippean satire's minimal requirements for fictional verisimilitude would also have endeared it to Price, who desired to engage only in the minimum worldbuilding and characterization necessary to sugar-coat the pill of his Flood Geology content.

In line with this disdain for readerly absorption in fiction, the book's individual items are, with the exception of his poetic criticism, highly pragmatic participants in their respective genres. Take, for example, 'The Man from Mars', originally published in *Signs of the Times* in 1930. To all but the most insular of Adventist readers, this title would have conjured up Mars's role as one of the premier settings of science fiction, as depicted in H. G. Wells's *The War of the Worlds* (1897) or Edgar Rice Burroughs' *A Princess of Mars* (1912). Price's story, cautiously described as '[a]n imaginary interview based on solid facts', quickly quashed any

---

[146] Northrop Frye, *Anatomy of Criticism: Four Essays* (Princeton, NJ: Princeton University Press, 1957), 309–310.

[147] Howard D. Weinbrot, *Menippean Satire Reconsidered: From Antiquity to the Eighteenth Century* (Baltimore, MD: Johns Hopkins University Press, 2005), xi.

such expectations.[148] 'The Man from Mars' was a twentieth-century revival of an older convention, one common to eighteenth- and nineteenth-century scientific didacticism, in which a knowledgeable sage, demon, or alien shares insights about the universe through dialogue with an inquisitive human. Moreover, rather than basing his alien ecology on Wells or the pulp magazines, Price worked within the traditions endorsed by White, who followed various evangelical thinkers in conceiving the inhabitants of other planets as unfallen humanoid beings. White had confirmed the accuracy of this tradition in a vision.[149]

The story's 'imaginary interview', framed in the third person, is further distanced from fictive illusionism by its status as a dream, dreamt by a liberal and geologically well-informed clergyman visiting the Grand Canyon. The dreaming clergyman, U. S. Brown, converses with the titular 'Man from Mars', who, speaking exactly like Price, challenges Brown's old-earth geological knowledge and calls Earth's scientists 'queer reasoners' (10). This Martian is a firm Christian who tells Brown that Genesis and its Flood narrative should be taken at 'face value', reducing the initially sceptical clergyman to 'mystified incredulity and wonder' (17). He does this not by drawing upon the superhuman knowledge of the past displayed, for instance, by the long-lived extraterrestrial depicted at the close of Gideon Mantell's *The Wonders of Geology*, a character from similar literary traditions, but through pious science alone. The Martian's confidence in young-earth geology is exclusively the result of belief in biblical inerrancy combined with Baconian inductive reasoning, related using all of Price's usual arguments about the artificial nature of the geological column. Refusing the Martian any knowledge of geohistory superior to that which might be accrued by a terrestrial fundamentalist, Price drained the story of all but the barest rudiments of imaginative technique. This was anti-fictive fiction.

Most of the other items in the book are similarly superficial repackagings of Price's tried-and-tested scientific and theological arguments. The book's most ambitious item, split across three chapters, is slightly different, but it returns us to familiar territory. 'An Old Discussion Modernized' is a three-scene play starring Huxley and Gladstone, whose debates over day-age theory we encountered in chapter 1. It reignites these debates amid the fundamentalist controversies of the 1920s, incorporating 'the increased knowledge of the intervening forty years' (35). Uninterested in explaining how this posthumous debate takes place, Price also rejects basic theatrical conventions of recognizable characterization: both Victorians speak, in their turn, like Price. Price's Gladstone even draws attention to his own use of homely Americanisms, begging forgiveness for using 'American slang' after invoking P. T. Barnum's maxim that 'there is a "sucker" born every minute', accusing Huxley of 'blowing hot and cold' (41), and later

---

[148] Price, 'The Man from Mars', *Signs of the Times*, 57 (1930), 2–3, 10, 15 (2).
[149] Bull and Lockhart, 73.

observing that 'Lyell's theory looks like two cents, to use the American slang phrase' (63). While Gladstone was a popular British prime minister in the United States during his lifetime, the twentieth century, in Price's view, would have Americanized the scholarly statesman yet further.[150] Price, it should be added, was capable of consistent characterization when he wanted to be, as demonstrated by his 1893 poem 'Lucifer's Soliloquy', a satanic blank verse monologue. In, 'An Old Discussion Modernized', he simply showed his disdain by refusing to employ it.

Price's Gladstone, as Huxley observes, follows 'the teachings of the Fundamentalists of America' (67), by which Price, in a sleight of hand, means Flood Geology rather than the day-age theory that Gladstone actually shared with a majority of his fundamentalist successors. The politician becomes even more improbably Price-like when he alludes, bizarrely, to a nephew in Alberta who has studied the same aspects of the Rocky Mountains that fascinated Price.[151] This nephew, Mr Cavendish, is Price's ideal naturalist: not a 'professional geologist' but rather a 'gentleman-farmer' of 'good education' who has inspected the rocks and, in the process, reignited his once-waning faith (47). Unlike authors of fiction and drama who submerge themselves in their characters to create illusions of autonomous worlds, Price refused to disguise the machinations of authorship. The debate ends as an intrigued Huxley asks for creationist reading recommendations, to which Gladstone responds with books authored by Price himself.

The closing chapter of *Some Scientific Stories and Allegories* sees Price take on the coveted role of visionary prophet. For all his humble protestations, Price was not averse to portraying himself as the Moses of young-earth creationism, prophesying (in the third person) that from 'Pisgah's top he may catch a glimpse of the promised land of undogmatic geology' that 'others' would reach.[152] In 'Two Pictures of the Future' his role was more like that of Isaiah, who predicted both destruction for the faithless and (Christians contended) the coming of Christ. First, Price described the dystopian future civilization of evolutionary morality that secularists and communists wished to create. Then, in one of the boldest passages of his career, Price begged to be allowed to 'draw a little on my imagination, and try my hand at what I believe to be a more truthful portrayal' of what the future would look like (100). Price's subsequent description, written in the past tense, explicitly evoked the accounts of prophetic biblical visionaries with which Adventists were so familiar. In the vision, Price travels among a vast crowd who are carrying 'a mass called CIVILIZATION' up a mountain (102). The 'mad enterprise' results in a scene of 'horror' when this 'final attempt' at creating paradise on Earth fails and the civilizers are sent plummeting to their doom (105–6). The

---

[150] Stephen J. Peterson, *Gladstone's Influence in America: Reactions in the Press to Modern Religion and Politics* (Cham: Palgrave Macmillan, 2018), 190.
[151] See, for example, Price, *The New Geology*, 258.
[152] Price, *Geological-Ages Hoax*, 118.

secular worldly meliorism, epitomized, for Price, by evolutionary theory, was the twentieth-century's deadliest threat to salvation. To combat it, and redirect attention from this world to the next, he was even willing to sign his name to a book of fiction—but not to lose his literary principles.

Conclusion

The aesthetic conventions and strategies of early twentieth-century young-earthers—including anti-Romanticism, photographic objectivity, and backhanded participation in fictive genres—are not obligatory characteristics of young-earth geology in the early twenty-first century. If figures like Price and Nelson turned budgetary constraints and a conservative epistemology of science into rhetorical virtues, their wealthier, more confident descendants today have re-established the subject's appeal to imagination and to wonder in sophisticated museum exhibits, illustrated popular media, and films.[153] The online bookshop for *Creation Today*, for instance, sells professionally illustrated children's works with titles like *Dragons: Legends & Lore of Dinosaurs*, the likes of which exploit the same nineteenth-century notions of science's overlap with myth that Price worked hard to deflate. Although the presentation of young-earth creationism, during its rise to political prominence in the 1960s and 1970s, was influenced by Price in his role as its Baconian John the Baptist (and I will touch on this in my epilogue), many of the movement's subsequent marketing strategies have been in opposite directions to those Price favoured. These strategies are attempts to compete in an evangelical marketplace in which young-earth creationism once again faces challenges from more liberal creationisms.[154]

Building on studies that have situated Price in his theological and social contexts, this chapter has revealed the complexity of his textual contexts and those of his few allies. We have seen that these men portrayed mainstream palaeoscientists as both thinking, writing, and arranging their museum displays in an unwarrantedly imaginative and visionary manner intended to bamboozle Christian publics. Price's own distinctive style invoked the courtroom, the mortuary, and the coroner, in order to disenchant mainstream views of geohistory and instead exalt the words of Genesis and Ellen White. This represented a belated revival of attitudes prevalent in the geological community a century prior, coupled with a profound reaction to the ways that wider publics had, since then, been taught to see through

---

[153] For example, see Ronald L. Numbers and T. Joe Willey, 'Baptizing Dinosaurs: How Once-Suspect Evidence of Evolution Came to Support the Biblical Narrative', *Spectrum*, 43 (2015), 57–68 (62–65); and Benjamin L. Huskinson, *American Creationism, Creation Science, and Intelligent Design in the Evangelical Market* (Basingstoke: Palgrave Macmillan, 2020).
[154] Huskinson, 17.

deep time. Price, Rimmer, and Nelson encouraged pious evangelicals and fundamentalists to become their own experts, training their eyes to see truths of nature—and, above all, humanity's central place in the universe—rather than be seduced by the textual tricks to which mainstream scientists resorted. To again quote Price's 'John in Machaerus',

> The rolling desert wastes, the silent stars,
> May teach more wisdom to the humbler soul
> Who has the key, an honesty of thought
> Between himself and God, to see himself
> A child that the Almighty would instruct,
> Than ages in a proud scholastic hall,
> The storehouse of the garnered thought of man. (57)

Epilogue
To See and To Make Others See

For those figures who challenged prevailing conceptions of Earth's deep history, literary technologies mattered. The figures I have discussed, including John William Dawson, Cuthbert Collingwood, the Denton family, H. P. Blavatsky, John Uri Lloyd, Ignatius Donnelly, Lewis Spence, and George McCready Price, took very different routes to their truth claims; all, however, engaged with the visual and textual techniques developed by geologists and palaeontologists to recruit virtual witnesses to their cogent interpretations of the planet's formation and inhabitation in the prehistoric past. This meant recontextualizing, appropriating, satirizing, rejecting: scenes from deep time were woven into poetry; figurative language was literalized; iconic diagrams were mischievously edited; readers were challenged with baroque framing devices; the writing of some of the world's leading scientists was deemed no more capable of truth-telling than imaginative literature. Along the way, compelling new depictions of primordial worlds very different from those found in most scientific textbooks captured the imaginations of novelists like Edgar Rice Burroughs, E. T. Bell, and H. P. Lovecraft, who kept these ideas alive in fiction.

In addition to attending to the methods by which scientific authority was written into being—and exploiting these methods to even the odds on a playing field increasingly dominated by well-equipped agnostics, specialists, professionals—the believers and visionaries among these authors were united by a cluster of attributes. These included a refusal to grant primacy to insights derived from organic rather than textual or artefactual sources, an insistence that serious scientific claims could be made outside serious scientific journals and without a basis in practical investigation, and an assertion of the outsized importance of humanity in the context of Earth's history. They wished to transcend time and to promote stories of the planet with more meaning, immediacy, and clarity than they found written into books and museum galleries by the world's most esteemed naturalists.

After all, the greatest savants of the age, for all their figurative claims to peer through time's abyss, merely saw all that *could* be seen through the probabilistic haze of a distorted rock, fossil, and archaeological record. Much could never be infallibly known, or seen, in a historical science, and what could be seen threatened to be less enchanting than one might wish. For occultists, hollow-earthers,

Contesting Earth's History in Transatlantic Literary Culture, 1860–1935. Richard Fallon, Oxford University Press.
© Richard Fallon (2025). DOI: 10.1093/9780198926191.003.0007

catastrophists, and creationists of all kinds, this was not enough. Did it truly reflect the extent of scientific knowledge? Writing in the Seventh-day Adventist periodical *Signs of the Times* in 1885, theologian Alonzo T. Jones skewered the hedgy language of Archibald Geikie's *Text-Book of Geology* (1882), one of the most widely used manuals of the late nineteenth century:

> When men laud this as 'advanced science', we have to say that it is simply a 'probability' linked with a 'likelihood' and sustained by a 'perhaps', and all supported by a 'must have operated', with not a fact to underlie any of it, because it is all concerning periods of which there is no 'visible record'.[1]

In contrast with such disenchanted guesswork, truths about geohistory had to be certain, accessible, rationally deducible, or unchanging, whether seen through one's own eyes, through the eyes of a trusted witness, through experimental verification, or, if these options were unavailable, recorded with accuracy by authorities long predating modern science. Borderline palaeoscience in its diverse forms expressed a conviction that humanity had not urgently required palaeontology or geology to be invented to understand the deep history of the planet.

* * *

'This work may be summed up as an attempt *to see* and *to make others see* what happens to man'.[2] So declared French Jesuit palaeoanthropologist Pierre Teilhard de Chardin in his posthumously published evolutionary epic *Le phénomène humain* (1955), written in the 1930s and translated into English as *The Phenomenon of Man* in 1959. Teilhard's *seeing* did not quite resemble the kinds I have discussed. After all, thanks to physicists' discovery of the observer effect, Teilhard reflected, no longer could scientists 'look down from a great height upon a world which their consciousness could penetrate without being submitted to it or changing it', nor did Teilhard perceive anything other than 'a cosmic contradiction in imagining a man as spectator of those phases which ran their course before the appearance of thought on earth'.[3] So much, then, for clairvoyance and day-age theory. Nonetheless, humans still could make time and space their own. 'I doubt', the Catholic naturalist stated, 'whether there is a more decisive moment for a thinking being than when the scales fall from his eyes and he discovers that he is not an isolated unit lost in the cosmic solitudes, and realises that a universal will to live converges and is hominised in him'.[4] True vision did not mean grossly optical access to prehistory, but rather meant identifying and accepting humanity's central place

---

[1] Alonzo T. Jones, '"Evolution" and Evolution (*Concluded*)', *Signs of the Times*, 11 (1885), 404.
[2] Pierre Teilhard de Chardin, *The Phenomenon of Man*, trans. by Bernard Wall (London: Harper Perennial, 1959), 31.
[3] Teilhard, 32, 35.
[4] Teilhard, 36.

in God's teleological evolutionary designs. The French edition of Teilhard's book sold 70,000 copies within just two years, while the English translation sold 17,000 copies in six months.[5] Clearly, there was immense hunger in the post-war decades for Teilhard's new way of seeing.

Despite this continued popular zest for forms of meaningful human agency in geohistorical terms, the middle decades of the twentieth century were also a period in which the themes I have clustered under the term 'borderline palaeoscience'—religious meaning, paranormal powers, non-specialist participation, and literary experimentation—were being demarcated out of elite science with more thoroughness than ever before. In the affluent United States, Cold War tensions with the Soviet Union substantially increased science's funding and prestige, helping to make the boundaries between science and 'pseudoscience' an extremely sensitive matter with geopolitical implications.[6] In Britain, increased post-war funding for scientific posts and the construction of new universities also worked to ensure that the somewhat fluid nature of geological practice in earlier decades, during which time amateurs and non-disciplinary figures of various kinds had continued to contribute to high-level research, was being replaced by a far more rigid professional and institutional structure.[7] These conditions were unfavourable for the reception of a new generation of Dawsons and Howorths, let alone Collingwoods, Dentons, Blavatskys, and Prices.

Alongside these shifts, some of the most fruitful mysteries formerly latched on to by borderline practitioners were being solved, to much fanfare, by elite scientists. In anglophone science, evolutionary theory was entering a new era. Biologist Julian Huxley, who wrote the English preface for Teilhard's *Phenomenon of Man*, was also a major architect of the evolutionary 'modern synthesis' of the mid-century.[8] This synthesis combined Darwin's notion of natural selection with Mendelian conceptions of heredity and the new field of population genetics, marginalizing the plethora of other competing mechanisms that had been used to explain evolution in the late nineteenth and early twentieth centuries. Evolutionary sceptics like Price had made much of this lack of consensus about mechanisms, tactically implying that evolution in general was defunct, while non-Darwinian models of evolution, typically theistic or directional ones, had been overwhelmingly preferred by Theosophists and other scientific mavericks. The intellectual

---

[5] Alfred Irving Hallowell, untitled review of *The Phenomenon of Man*, by Pierre Teilhard de Chardin, *Isis*, 52 (1961), 439–41; Susan Kassman Sack, *America's Teilhard: Christ and Hope in the 1960s* (Washington, D.C.: The Catholic University of America Press, 2019), 51.

[6] Michael D. Gordin, *The Pseudoscience Wars: Immanuel Velikovsky and the Birth of the Modern Fringe* (Chicago: University of Chicago Press, 2012), especially chapter 3.

[7] Jean G. O'Connor and A. J. Meadows, 'Specialization and Professionalization in British Geology', *Social Studies of Science*, 6 (1976), 77–89.

[8] Joe Cain, 'Synthesis Period in Evolutionary Studies', in *The Cambridge Encyclopedia of Darwin and Evolutionary Thought*, ed. by Michael Ruse (Cambridge: Cambridge University Press, 2013), 282–92.

coherence of the modern synthesis (or other developments like radiocarbon dating) did not, of course, halt this opposition—the enterprising Price, in a late essay, even quoted G. G. Simpson's synthesis-era classic *Tempo and Mode in Evolution* (1944) to support an anti-evolutionary stance—but it did force opponents to develop new arguments and strategies for a newly challenging climate.[9]

Meanwhile, a question dear to the hearts of other borderline thinkers was being resolved to the satisfaction of scientists. Sunken land bridges and even continents had been acceptable theoretical speculations for savants in an era of uncertainty over continental stability, a situation favourable to investigators of Atlantis and Lemuria like Spence, who cast doubt on geoscientists' authority due to their lack of clarity on this vital issue. Spence could even feel that he, a folklorist-mythographer-poet, might shed light on such a tenebrous scientific conundrum. The pluralism of the field was, however, coming to an end. James A. Secord has described how, from the 1930s, the tradition of cross-disciplinary geological theorizing—the mountain-top view enjoyed by savants like Alexander von Humboldt and Eduard Suess, in which Romantic interest in lost worlds was part of the quest to synthesize a global geology—was being phased out of the earth sciences in favour of a utilitarian focus on exact mathematical measurement, often in service of resource extraction or military demands. When Alfred Wegener's continental drift theory was finally vindicated by the development of plate tectonics in the 1960s, the idea that the field had once offered 'philosophical, theological and literary resonances' must have seemed quaint.[10] Besides, no sunken continent in the Atlantic, Pacific, or Indian Oceans could hide from rigorous sea-floor mapping.

Of course, as Spence's case suggests, much of the evidence cited by lost continent theorists had been not geological or oceanographic but was instead drawn from plausible archaeological and anthropological supposition. While this speculative approach could formerly survive on the borders of these fluid disciplines, it was becoming almost entirely detached from the methods of professionals. Just as geologists were eschewing the purportedly fanciful speculations of their predecessors in favour of cold, hard science, anthropologists were, in George W. Stocking's words, rejecting 'diachronic' and 'conjectural' approaches in favour of 'the empirical study of particular ethnographic entities existing in the present'.[11] Speaking to the American Anthropological Association on 'The Rising Quality of New World Archeology [*sic*] in 1944, retiring president Neil M. Judd reflected fondly on the passing of an age when 'speculation' had 'substituted for fact' in

---

[9] George McCready Price, *The Man from Mars* (Washington, D.C.: Review and Herald Publishing Association, 1950), 108. Despite its title, this is a collection of essays.

[10] James A. Secord, 'Global Geology and the Tectonics of Empire', in *Worlds of Natural History*, ed. by H. A. Curry, N. Jardine, J. A. Secord, and E. C. Spary (Cambridge: Cambridge University Press, 2018), 401–17 (416).

[11] George W. Stocking, Jr., *After Tylor: British Social Anthropology 1888–1951* (London: Athlone Press, 1996), 230–31.

American archaeology, a period characterized by pronouncements on 'Atlantis' and 'Mu'.[12] Elegizing the 'old-time naturalist' in an age of mundane 'specialization', Judd fondly contended that the former generation had erred only in its overzealous application of 'imagination', which, properly applied, was 'the spark-plug of archeology'.[13] Imagination itself was not to be abandoned, but the unbridled forms represented in the work of Donnelly, Spence, and Churchward, precarious enough figures in their own heyday, had, for Judd, no place in the future of these fields.

Even before the 1940s, many other trends that this book has been tracking were either on the wane or transforming to meet new circumstances. Walter Kafton-Minkel observes that Marshall B. Gardner's hollow-earth monographs of 1913 and 1920, with their rear-guard denial that the North and South Poles had truly been discovered, constituted the final major representatives of hollow-earth writing in its Symmesian form.[14] Theosophy, too, was changing. In 1928, the world membership of the Theosophical Society reached its peak at 45,000 and subsequently began to decline.[15] The embarrassing 1929 abdication of the so-called 'World Teacher', Jiddu Krishnamurti, whose epochal importance for human evolution had been championed by Annie Besant and Charles Webster Leadbeater, inflicted serious damage upon the centralizing coherence of the organization. Meanwhile, Christian fundamentalism in the United States was being reconfigured. In the 1930s, fundamentalism and anti-evolutionism were turning inwards, and, if by no means defeated during the sensational Scopes Trial of 1925, as folklore has implied, the movement was no longer seeking, or receiving, substantial media interest.[16] Price wrote on for decades more, but his brief period of public visibility was behind him. His young-earth crusade to rout rampant belief in the day-age and gap theories among fundamentalists had enjoyed only limited success.

Around the mid-twentieth century, then, borderline palaeoscience entered a new era. If this was an era of more clearly demarcated fringe status for enchanted and alternative prehistories, it can hardly be considered one in which mass audiences simply fell in step with the ideas of respected scientists. To begin with, new publishing developments meant that fantastic tales of Symmesian worlds and lost continents were peddled to wider and wider readerships. Not long after his death

---

[12] Neil M. Judd, 'The Rising Quality of New World Archeology', *Scientific Monthly*, 63 (1946), 391–94 (391).

[13] Judd, 'The Rising Quality of New World Archeology', 392.

[14] Walter Kafton-Minkel, *Subterranean Worlds: 100,000 Years of Dragons, Dwarfs, the Dead, Lost Races & UFOs from Inside the Earth* (Port Townsend, WA: Loompanics Unlimited, 1989), 133.

[15] Gregory Tillett, *The Elder Brother: A Biography of Charles Webster Leadbeater* (London: Routledge and Kegan Paul, 1982), 140.

[16] Edward J. Larson, *Summer for the Gods: The Scopes Trial and America's Continuing Debate over Science and Religion*, rev. edn (New York: Basic Books, 2006), 229–33. For creationist activities in the decades after the Scopes Trial, see Ronald L. Numbers, *The Creationists: From Scientific Creationism to Intelligent Design*, rev. edn (Cambridge, MA: Harvard University Press, 2006), chapters 6 and 7.

in 1950, Edgar Rice Burroughs' works were sought after by enterprising American publishers of affordable fantasy and science fiction paperbacks, including Ace Books and Ballantine Books, especially when it was discovered that around half of the author's titles had apparently fallen out of copyright. The result was what critic Richard A. Lupoff calls a 'flood' of Burroughs reprints, often in 'two, three, four competing editions'.[17] According to one source, '[b]etween 1962 and 1963 Burroughs books accounted for almost one thirtieth of all U. S. paperback sales'.[18] Beginning with *At the Earth's Core* in 1962 and ending with the obscure *Savage Pellucidar* in 1964, the Pellucidar series was fully reprinted by Ace. From then on, the saleability of Burroughs' ageing pulp fictions would routinely be enhanced by the art of Frank Frazetta, whose paintings usually stressed the sexuality of the stories' female characters as well as the ferocity of his dinosaurs. In 1969, Lin Carter—whose lost-continent novel, *The Wizard of Lemuria*, was published by Ace in 1965—began to edit Ballantine Adult Fantasy, an influential series that reprinted hard-to-find works.[19] Among these was *Poseidonis* (1973), a compilation of Clark Ashton Smith's Atlantean and Lemurian short stories and poetry. In most cases, these interwar adventures now became more accessible than they had ever been during their original printings, fuelling an ever-larger fan culture for the borderline palaeoscience of the early twentieth century.

It would be utterly inaccurate, however, to imply that borderline palaeoscience was simply relegated to popular fiction. The psychometric methods pioneered by Joseph Rodes Buchanan and the Denton family in the mid-nineteenth century, for instance, had become common currency in counter-establishment science by the mid-twentieth. While psychometry had diverse applications, explicated in works like Herbert Bland's *Psychometry: Its Theory and Practice* (1937), the possibility of reactivating the prehistoric past in Dentonian style remained especially tantalizing.

One strand of updated psychometry particularly suited to this task was the 'radionics' of George de la Warr, an English civil engineer and alternative healer. As the title of a book co-written with fellow enthusiast Langston Day, *New Worlds beyond the Atom* (1956), suggests, this was a psychometry for the nuclear age, positing the possibility of tuning in to the radiations left behind by long-extinct life forms. The book argued that 'the skeleton of a Megalosaurus or of a pterodactyl is in resonance with the creature which lived millions of years ago, and by analysing the radiations we should be able to discover more about it than can be gleaned

[17] Richard A. Lupoff, *Master of Adventure: The Worlds of Edgar Rice Burroughs*, new edn (Lincoln: University of Nebraska Press, 2005), xxxviii.
[18] Harry L. Rinker, ed., *Warman's Americana & Collectibles*, 5th edn (Radnor, PA: Wallace-Homestead Company, 1991), 471.
[19] Jamie Williamson, *The Evolution of Modern Fantasy: From Antiquarianism to the Ballantine Adult Fantasy Series* (Basingstoke: Palgrave Macmillan, 2015), chapter 1.

from the fossil bones'.[20] The parapsychologist T. C. Lethbridge, former employee of the Cambridge University Museum of Archaeology and Ethnology, evaluated claims like these in his book *Ghost and Ghoul* (1961). By no means a sceptic, Lethbridge was intrigued by the suggestion that 'by mechanical application we should be able to see a "Carboniferous" swamp by the agency of a lump of coal'.[21] His own experiments with psychometric sensitives, however, were unspectacular. While initially the psychometers' readings of ancient objects seemed persuasive, Lethbridge, perusing 'reports' of these narratives, soon found that 'I could identify things that I had seen in pictures, or read in books', even recalling images from a particular 'book of fairy tales'.[22] He thus experienced first-hand the disenchanting possibility that psychometers, just as sceptical readers of *The Soul of Things* had suggested a century prior, conjured up distant worlds from books they had read. Psychometry was, he concluded, merely subjective imagination fuelled by literary content.

Despite these setbacks, clairvoyant palaeontology elsewhere enjoyed striking success. At the dawn of the apartheid era, South African scientist John Talbot Robinson was introduced to Theosophy by his colleague Robert Broom, the highly regarded but also highly unorthodox palaeontologist of the Transvaal Museum.[23] Together, Broom and Robinson had excavated prehistoric hominin remains, but Robinson's growing occult interests encouraged him to look to stranger sources for a more intimate acquaintance with these beings. He sent fossils to the New Zealander clairvoyant Geoffrey Hodson and was so impressed by Hodson's insights into their past lives that, in 1960, he invited the man to test his abilities in person. As Hodson put fossil after fossil to his forehead in psychometric readings, Robinson found that Hodson 'never misidentified a specimen' and gave 'a strong impression of complete reliability', even providing previously unknown information on the extinct *Paranthropus* that later turned out to be true.[24] Robinson's article, 'An Investigation of Clairvoyance as Applied to the Study of a Chapter in Primate Evolution', co-written with the Theosophist Margaret Donnelly, was published in the Theosophical Society's *Science Group Journal* in the same year and later appeared, anonymized, in the edited collection *Psychism and the Unconscious Mind* (1968). This new victory for clairvoyance did not entail a corroboration of occultists' earlier attempts. Robinson and Donnelly admitted that the 'information provided by the Secret Doctrine [sic] and the Besant-Leadbeater investigations appears to be

---

[20] Langston Day and George De La Warr, *New Worlds beyond the Atom* (London: Vincent Stuart, 1956), 136.

[21] Thomas Charles Lethbridge, *Ghost and Ghoul* (London: Routledge and Kegan Paul, 1967 [1961]), 65.

[22] Lethbridge, 70-71.

[23] For Broom's unorthodoxy, see Jesse Richmond, 'Design and Dissent: Religion, Authority, and the Scientific Spirit of Robert Broom', *Isis*, 100 (2009), 485-504.

[24] Quoted in Christa Kuljian, *Darwin's Hunch: Science, Race and the Search for Human Origins* (Johannesburg: Jacana, 2016), 128.

highly incompatible with the views of modern scientific students of mammalian and human evolution'.[25]

Perhaps the most unexpected transformations of borderline palaeoscience took place in Asia. James Churchward's Mu, the ancient homeland of an advanced white race, for instance, became integral to the mythos of a post-war Japanese cult. The Mahikari or 'New Light' movement was founded in Tokyo in 1959 by Yoshikazu Okada (also called Kōtama Okada), yet another recipient of divine knowledge. According to Mahikari teachings, Japan itself is a vestige of Mu, and the Japanese people one of Mu's leading races. The Mahikari version of Mu led to the creation of what one scholar has called 'a Japanized version of the "root races" commonly found in the writings of the Theosophists', allowing for a prehistoric origin story that gave Okada's disciples 'the dominant place in the unfolding plot of the universe'.[26] During these febrile decades of rapid economic growth in Japan, Lemuria—Mu's sometime synonym—was surprisingly topical. It even featured in the work of the so-called 'father of manga', Osamu Tezuka. His series *Mitsume ga Tōru* (*Three-Eyed One*), serialized in the teenage *Weekly Shōnen Magazine* between 1974 and 1978, follows the eponymous protagonist, a Japanese schoolboy with mysterious connections to the three-eyed Lemurians.

Through the processes of literary replication, even the most shadowy theorists of sunken continents could enjoy a new lease of influence, as seen in the South Indian afterlife of William Scott-Elliot's *The Lost Lemuria*. Sumathi Ramaswamy has shown that, when Tamil historians refashioned scholarship on Lemuria into evidence for the existence of Kumarinātu, their ancestral homeland in the Indian Ocean, the obscure Theosophist was repeatedly cited as an authority. Thus the name of Scott-Elliot, who died in 1919 and left few records of his life behind, was invoked alongside that of the famous Ernst Haeckel in a 1959 school textbook by the celebrated Tamil scholar T. V. Kalyanasundaram.[27] A 1975 college textbook used Scott-Elliot's work to demonstrate that support for Lemuria's existence was '*not just aired by literary scholars*', the likes of whom were hardly to be relied upon, but also by '*the foremost geologists*', while one exam question even asked pupils '[w]hat are the true researches of Scott, Elliot, [*sic*] and others?'[28] As the latter quotation implies, Scott-Elliot's biography was no better known to Tamil educationalists than it is to modern scholars of esotericism. The problem with citing such an elusive figure had been pointed out within his own lifetime. In a footnote to a 1910 paper on the Tamil language, printed in the Shaivist Hindu journal

---

[25] John Talbot Robinson and Margaret Donnelly, 'An Investigation of Clairvoyance as Applied to the Study of a Chapter in Primate Evolution', *Science Group Journal*, 4 (1960), 9–16 (9).

[26] Winston Davis, *Dojo: Magic and Exorcism in Modern Japan* (Stanford, CA: Stanford University Press, 1980), 82.

[27] Sumathi Ramaswamy, *The Lost Land of Lemuria: Fabulous Geographies, Catastrophic Histories* (Berkeley: University of California Press 2004), 105.

[28] Quoted in Ramaswamy, *Lost Land of Lemuria*, 178–79.

The Light of Truth, the periodical's editors objected to the author's vague reference to 'Professor Elliot', 'an unknown figure in the geological world' whose dating of the catastrophes that ravaged Lemuria 'will need to be critically examined, before his observations can be accepted by Hindus'.[29] Ironically, Scott-Elliot's 'observations' had, in fact, merely polished up the clairvoyant research of Leadbeater, who penned an essay on astral powers in the very same volume of *The Light of Truth*.

The Lemuria of nineteenth-century naturalists became even less recognizable in twentieth-century science fiction, where it was syncretized with hollow-earth geology and other, newer conspiracy theories. The most sensational example, Richard S. Shaver's novella 'I Remember Lemuria!', was published in the March 1945 issue of *Amazing Stories*, accompanied by footnotes and rewritten by the magazine's publicity-hungry editor, Raymond Palmer. Palmer and the enigmatic Shaver's use of the terms Lemuria and Mu bore little connection to sunken continents, instead referring to vast underground caverns populated by 'detrimental robots' or 'Deros', the degenerate remnants of a once glorious super-civilization who now manipulate the surface world's affairs. This was, author and editor claimed, no mere fiction. A stream of conspiratorial content flowed from the paranoid and Theosophically influenced Shaver, boosting *Amazing Stories'* circulation by 50,000 and increasing the number of letters to the editor by a factor of fifty before his controversial content was phased out in 1948.[30] When Palmer compiled Shaver's material in a later quarterly, *The Hidden World* (1961–64), which no longer presented these notions in the guise of fiction, he even topped up one sparse instalment by republishing an obscure book I have discussed: De Witt C. Chipman's Christian hollow-earth romance, *Beyond the Verge*. Palmer admitted that this was a somewhat incongruous last-minute addition, inserted when the original plan for the quarterly's contents went 'astray', but nonetheless insisted that Chipman's Symmesian spin on the legend of the Lost Tribes of Israel 'agrees with so much of the evidence that has come forth from the Shaver Mystery', exhorting readers to 'watch for the links' that would connect 'the individual bits' of the expanded subterranean mythos.[31] Chipman's attempt to prove that the early books of the Bible were to be taken as literally as his own theory of the hollow constitution of the planet was repurposed as support for Richard Shaver's countercultural conspiracy theory.

One major figure among the consumers of Shaverian-Palmerian material was Raymond Bernard, aka Walter Siegmeister, an American health foods entrepreneur. Bernard was immersed in Theosophy and its German offshoot,

---

[29] Satappa Ramanatha Muthiah Ramaswami Chettiar, 'Tamil Language—A Phase of Its History and an Aspect of Its Modern Requirements', *The Light of Truth*, 11 (1910–11), 87–93, 134–44 (88–89).
[30] For an overview of the Shaver affair, see Kafton-Minkel, 137–47.
[31] Raymond Palmer, 'Editorial', *Hidden World*, A–6 (1962), 958–63 (958, 960).

Anthroposophy, which formed the subject of his New York University PhD thesis.[32] By the late 1950s he could be found in Brazil, investigating claims made by Theosophists that UFOs were emerging from a tunnel to an interior world inhabited by Lemurians and Atlanteans. Palmer, too, was investigating the UFO craze, which had exploded just as the Shaver frenzy started to wane. In a December 1959 issue of yet another of his magazines, *Flying Saucers*, Palmer announced that the hollow earth was, as Bernard suspected, the source of UFOs. This was an eventuality that Symmes had never foreseen. Palmer's evidence for an access route between interior and outer worlds was the 1947 account of US Admiral Richard E. Byrd, who—as told through Palmer's convoluted series of second-hand readings and literary misprisions—had flown into an opening at the North Pole and encountered lush vegetation and gigantic animals that were possibly mammoths.[33]

As Palmer continued to explore hollow-earth theory in *Flying Saucers* over the following years, his research found itself incorporated into Bernard's compilatory synthesis, *The Hollow Earth* (1963 or 1964). The book, kept in publication by a plurality of fringe publishers over subsequent decades, introduced new readerships to the heterodox geology of the late nineteenth and early twentieth centuries in 'a mishmash of pages-long quotes' from predecessors like Gardner and William Reed.[34] Bernard also exacerbated the confusion about the authorial intent of hollow-earth fiction, apparently accepting the framing device of Willis George Emerson's 1908 romance *The Smoky God* as truth. He informed readers that this book 'recorded' the account of the Norwegian 'Olaf Jansen' and his adventures inside the 'polar opening', as relayed 'by Jansen to Mr. Emerson before his death'.[35] In the foreword to a 1969 edition of *The Hollow Earth*, produced by the Tarot card publisher University Books, one Robert Fieldcrest stridently told readers that the 'purpose' of such books was 'to dissipate darkness and to stir the minds of the people'.[36] 'Whether you accept or reject the content of this book is your privilege', he continued, fiercely adding that '[n]o one cares'.[37] Clearly, finding new champions in pulp fiction editors and countercultural publishers, hollow-earth writing continued to thrive by sending readers down winding paths of individual enlightenment.

It is, then, little surprise that John Uri Lloyd's provocative *Etidorhpa* also enjoyed a renaissance during this rebellious period. Revived by the ubiquitous Palmer in his series of 'Inspired Novels' in 1962 (accompanied by Emerson's *The Smoky God* in 1965), it was subsequently taken up by the Santa Fe–based

---

[32] Holly Folk, 'Raymond W. Bernard, Hollow Earth, and UFOs', in *Handbook of UFO Religions*, ed. by Benjamin E. Zeller (Leiden: Brill, 2021), 312–25.
[33] Minkel, 193–97.
[34] Minkel, 212.
[35] Raymond Bernard, *The Hollow Earth; The Greatest Geographical Discovery in History* (New York: University Books, 1969), 55–56.
[36] Robert Fieldcrest, 'Foreword', in Bernard, *The Hollow Earth*, 7–9 (7).
[37] Fieldcrest, 9.

New Age firm Sun Publishing in 1974.[38] Citing the novel's allusions to the goddess Aphrodite (added at the last minute, as we saw in chapter 3, to please Lloyd's wife), the front matter of an expanded 1976 Sun edition called it a 'transcendent masterpiece' that 'approaches being the ultimate aphrodisiac as well'.[39] While readers looking for titillation in Lloyd's genre-bending romance were likely to be disappointed, an introduction by fringe science connoisseur Neal Wilgus pointed to more plausible illicit pleasures. 'Of particular interest to modern readers', Wilgus stated, 'is the graphic psychedelic experience' beginning in chapter 34.[40] He felt that Lloyd's ostensible 'sermon on temperance' in this chapter hid a 'deeper meaning', likely inspired by 'Lloyd's own experience ... with hallucinogenic substances'.[41] Wilgus speculated about 'precisely what psychedelic agent Lloyd might have used' before finding 'the *Psilocybe mexicana* and other psilocybin producing mushrooms of Mexico' the likely candidate.[42] Lloyd's modern biographer, Michael A. Flannery, has denied these speculations that the self-avowedly empirical Cincinnati pharmacist was a pioneering predecessor of the 1970s New Age movement who had submerged his drug experimentation in labyrinthine allegories. These interpretations of *Etidorhpa* extended to one claim that John Augustus Knapp's illustrations feature a coded reference to another hallucinogenic mushroom, *Stropharia cubensis*.[43]

Far from the world of alternative narcotics, the most epochal reversal in fortunes of a field that skulked on the shadiest fringes of palaeoscience was that of young-earth creationism. Price, despite herculean efforts, had struggled to make Flood Geology an obligatory ingredient in Christian fundamentalism's anti-evolutionary recipe. After decades of uncertain status, its newfound success was heralded by the publication of a dense monograph, *The Genesis Flood* (1961), co-written by theologian John C. Whitcomb and hydraulic engineer Henry M. Morris and produced by Presbyterian and Reformed Publishing. This book was central in turning Price's marginal beliefs about the age of the Earth into standard fundamentalist doctrine during the Cold War era. Although Whitcomb and Morris profitably jettisoned their predecessor's most satirical stylings in favour of a firmly academic presentation, downplaying references to his writing in general, the book's sparse allusions to the somewhat embarrassing Seventh-day Adventist disguised vast intellectual debts to his work.[44] Despite this whitewashing, Pricean

---

[38] Michael A. Flannery, *John Uri Lloyd: The Great American Eclectic* (Carbondale: Southern Illinois University Press, 1998), 118.
[39] John Uri Lloyd, *Etidorhpa* (Santa Fe, NM: Sun Publishing, 1976).
[40] Neal Wilgus, 'Introduction: The Pharmaceutical Alchemist', in John Uri Lloyd, *Etidorhpa* (1976), xv–xx (xviii).
[41] Wilgus, xviii.
[42] Wilgus, xix.
[43] Flannery, 120–22.
[44] Numbers, *Creationists*, 223–24.

antecedents shone through, not just in the book's theorization of the Deluge but also in its scientific philosophy. Both co-authors commonsensically insisted in a new 1964 preface that geologists 'leave the strict domain of *science* when they become *historical geologists*', adding, with deliberate ignorance of the history of science, that historical geology had been developed by 'non-geologists' like 'Charles Lyell (a lawyer)' and 'Georges Cuvier (a comparative anatomist)'.[45] Their implication was not only that these non-geologists were not to be trusted on geological questions, but also that modern non-geologists—the authors themselves—had much to contribute to conceptions of Earth's history.

It was not Price alone whose work Morris and Whitcomb rejuvenated in a somewhat backhanded manner. In one of the book's many lengthy footnotes, the authors contended that '[g]lacial geologists have never answered the cogent criticisms of Sir Henry Howorth' nor refuted his 'tremendous amount of evidence' that the world was once swept by a great deluge.[46] They hinted, however, that geologists' inattention to the antiquarian MP's 'cogent criticisms' of glacial theory was to be blamed upon his books' near-indigestible profuseness, a strategy that, as I argued in chapter 4, was key to his precarious scientific respectability. Morris recalled finding 'Howorth's massive [1905] work *Ice or Water*' in 'the library of the University of Minnesota's outstanding Department of Geology', and, upon borrowing it, discovered this to be 'the first time in the forty-odd years of its residence there that it had ever been checked out or (judging from the numerous page-pairs still not cut apart from each other) even opened'.[47] This inattention was a testament to the double-edged success of the Manchester politician's imposing presentation. Given the forbiddingly scholarly nature of *The Genesis Flood*'s own literary technologies, no doubt many of the 300,000 copies sold by the book's fiftieth anniversary, most of them presumably bought by evangelicals intending to educate themselves about biblical geoscience, have suffered a similarly lonely—and not entirely unintended—fate.[48]

Challenging and controversial ideas about the planet's interior, surface, and age; about the possibility of seeing through time; and about the (pre-)history of human civilization, thus continued to evolve. The borderline palaeoscience of the mid-twentieth century not only tackled the shifting intellectual, religious, and political currents of the era, but was also promulgated and popularized using genres, and literary technologies, fit for the age: cheap paperbacks, nationalist

---

[45] John C. Whitcomb and Henry M. Morris, *The Genesis Flood: The Biblical Record and Its Scientific Implications* (Phillipsburg, NJ: Presbyterian and Reformed Publishing, 1991 [1961]), xxvii.
[46] Whitcomb and Morris, 292.
[47] Whitcomb and Morris, 293.
[48] John C. Whitcomb, 'Preface to the Fiftieth Anniversary Edition', in John C. Whitcomb and Henry M. Morris, *The Genesis Flood: The Biblical Record and Its Scientific Implications: Fiftieth Anniversary Edition* (Phillipsburg, NJ: Presbyterian and Reformed, 2011), xxxiii–xxi (xxx).

textbooks, manga, scholarly occult science journals, science fiction pulp magazines, New Age reprints, countercultural quarterlies, a jargon-heavy geophysical manual under a Presbyterian imprint. These were just some of the latest textual tools with which to pursue the ongoing quest of contesting Earth's history—tools that might, in Teilhard's words, enable one '*to see* and *to make others see*'.[49]

---

[49] Teilhard, 31.

Bibliography

Archives

Add MS 44474, British Library, London
Applications GSL/F/1/10, Geological Society of London Archives
Charles Lapworth Archive Collection, Lapworth Museum of Geology, University of Birmingham
D859 – Collingwood, Dr. Cuthbert; with Research Papers of Mrs Nora McMillan, MBE, MSc, MRIA (1853–1990s), University Library, Special Collections and Archives, University of Liverpool
Dawson-Harrington Families Fonds, CA MUA MG 1022, McGill University Archives, Montreal
Denton Family Papers, Wellesley Historical Society, Wellesley, MA
DF BOT/404/1/12, Natural History Museum Archives, London
DF PAL 100/80, Natural History Museum Archives, London
George Grantham Bain Collection, Library of Congress Prints and Photographs Division, Washington, D.C.
Glynne-Gladstone Archive, Gladstone's Library, Hawarden, UK
Irwin Porges Papers, Edgar Rice Burroughs Memorial Collection, Archives and Special Collections, University of Louisville, KY
James McBride Collection of John Symmes' Hollow Earth Theory, 1819–1859, Academy of Natural Sciences of Drexel University, Library and Archives, Philadelphia, PA
John Uri Lloyd Papers, 1849–1936, Lloyd Library and Museum, Cincinnati, OH
LDGSL/776, Geological Society of London Archives
Letters and Autographs of Zoologists with Biographies and Portraits, vols 2 and 4, Ellen S. Woodward Collection, Blacker Wood Collection, McGill University Archives, Montreal
Loma Linda University Photo Archive, LLU00691, Department of Archives and Special Collections, Loma Linda University, CA
Papers of Sir John William Dawson (1820–1899), GB237 Coll-192, University of Edinburgh Library Heritage Collections
Papers of Thomas Henry Huxley, Imperial College London, as filmed by the Australian Joint Copying Project (M876–M916)

Primary Sources

'Address of Professor William Denton', *Proceedings at the Second Annual Meeting of the Free Religious Association* (Boston: Roberts Brothers, 1869), 37–42
Armstrong, William Jackson, 'The Smoky God', *Los Angeles Times*, 30 August 1908, III13
'Atlantis', *Chicago Tribune*, 25 March 1882, 9
[Bell, Eric Temple] 'John Taine', *Before the Dawn* (Baltimore, MD: Williams and Wilkins, 1934)
Bernard, Raymond, *The Hollow Earth; The Greatest Geographical Discovery in History* (New York: University Books, 1969)

Besant, Annie, *The Pedigree of Man* (Benares [Varanasi]: Theosophical Publishing Society, 1904)
Besant, Annie, and Charles Webster Leadbeater, *Man: Whence, How and Whither: A Record of Clairvoyant Investigation* (Adyar: Theosophical Publishing House, 1913)
Bible Readings for the Home Circle (Battle Creek, MI: Review and Herald Publishing Company, 1889)
Blades, Braxton, 'Dr. Bell, alias John Taine', *Los Angeles Times*, 1 June 1930, J1
Blake, Charles Carter, 'The Third Eye', *Lucifer*, 4 (1889), 341–45
Blavatsky, Helena Petrovna, *Isis Unveiled: A Master-Key to the Mysteries of Ancient and Modern Science and Theology*, 2 vols (New York: J. W. Bouton, 1877)
Blavatsky, Helena Petrovna, *The Secret Doctrine: The Synthesis of Science, Religion, and Philosophy*, 2 vols (London: Theosophical Publishing Company, 1888)
[Bogart, William Henry], *Who Goes There? Or, Men and Events* (New York: Carleton, 1866)
'Book Notices', *Green's Fruit Grower*, 30 (1910), 2
Bradshaw, William R., *The Goddess of Atvatabar; Being the History of the Discovery of the Interior World and Conquest of Atvatabar* (New York: J. F. Douthitt, 1892)
Brandt, Conrad A., 'Overthrowing Old Traditions', *Amazing Stories*, 6 (1931), 762
Brandt, Conrad A., 'In Memoriam', *Amazing Stories*, 10 (1936), 134
Buchanan, Joseph Rodes, 'Psychometry', *Buchanan's Journal of Man*, 1 (1850 [1849]), 49–62, 97–113, 145–56, 208–27
Buckland, William, *Geology and Mineralogy Considered with Reference to Natural Theology*, 2 vols (London: William Pickering, 1836)
Burgan, W. L., 'Newspaper Work in Korea', *Advent Review and Sabbath Herald*, 101 (1924), 21–22
Burroughs, Edgar Rice, 'At the Earth's Core [1/4]', *All-Story Weekly*, 30 (1914), 1–21
Burroughs, Edgar Rice, 'Pellucidar [1/5]', *All-Story Weekly*, 44 (1915), 385–411
Burroughs, Edgar Rice, 'At the Earth's Core [2/11]', *Pluck*, 1 (1923), 651–53
Burroughs, Edgar Rice, 'Tanar of Pellucidar [2/6]', *Blue Book Magazine*, 48.6 (1929), 76–97
Burroughs, Edgar Rice, 'Tarzan at the Earth's Core [6/7]', *Blue Book Magazine*, 50.4 (1930), 28–44
Burroughs, Edgar Rice, 'Seven Worlds to Conquer [1/6]', *Argosy Weekly*, 270 (1937), 4–27
Burroughs, Edgar Rice, *Back to the Stone Age* (Tarzana, CA: Edgar Rice Burroughs Incorporated, 1937)
Cater, Edwin, 'Geological Speculation and the Mosaic Account of Creation', *Southern Presbyterian Review*, 10 (1858), 534–73
Chase, Warren, *Forty Years on the Spiritual Rostrum* (Boston: Colby & Rich, 1888), 268–69
Chettiar, Satappa Ramanatha Muthiah Ramaswami, 'Tamil Language—A Phase of Its History and an Aspect of Its Modern Requirements', *The Light of Truth*, 11 (1910–11), 87–93, 134–44
Chipman, De Witt C., *Beyond the Verge: Home of Ten Lost Tribes of Israel* (Boston: James H. Earle, 1896)
Churchward, James, *The Lost Continent of Mu: The Motherland of Man* (New York: William Edwin Rudge, 1926)
Churchward, James, *Cosmic Forces of Mu* (New York: Ives Washburn, 1934)
Churchward, Jack E., ed., *Lost Gems of the Lost Continent of Mu: Reviews and Correspondence from James Churchward's Scrapbooks Volume 1* (Clearwater, FL: Churchward, 2006)
Clark, P., 'The Symmes Theory of the Earth', *Atlantic Monthly*, 31 (1873), 471–80
Coblentz, Stanton A., 'Man through the Ages', *New York Times*, 17 January 1932, BR20
Cole, Grenville Arthur James, untitled review of *Illogical Geology: The Weakest Point in the Evolution Theory*, by George McCready Price (1906), *Nature*, 74 (1906), 513
Collingwood, Cuthbert, *A Vision of Creation: A Poem* (London: Longmans, Green, 1872)
Collingwood, Cuthbert, *A Vision of Creation: A Poem*, 2nd edn (Edinburgh: William Paterson, 1875)
[Collingwood, Cuthbert] 'A Graduate of Oxford', *New Studies in Christian Theology: Being Thirty-Three Lectures on the Life and Teaching of Our Lord* (London: Elliot Stock, 1883)
Collingwood, Cuthbert, *The Bible and the Age; or, An Elucidation of the Principles of A Consistent and Verifiable Interpretation of Scripture* (New York: James Pott, 1887)
'The Comfortable Word "Evolution"', *Illustrated London News*, 20 April 1912, 582

Corns, Albert R., and Archibald Sparke, *A Bibliography of Unfinished Books in the English Language* (London: Bernard Quaritch, 1915)

'The Cosmogony of Genesis: Professor Driver's Critique of Professor Dana', *Bibliotheca Sacra*, 45 (1888), 356–65

Dana, James Dwight, *Manual of Geology: Treating of the Principles of the Science with Special Reference to American Geological History, for the Use of Colleges, Academies, and Schools of Science* (Philadelphia, PA: Theodore Bliss, 1863)

Dana, James Dwight, 'Creation; or, the Biblical Cosmogony in the Light of Modern Science', *Bibliotheca Sacra*, 42 (1885), 201–24

Darwin, Charles, *On the Origin of Species by Means of Natural Selection, or, The Preservation of Favoured Races in the Struggle for Life* (London: John Murray, 1859)

Davis, Andrew Jackson, *The Principles of Nature, Her Divine Revelations, and A Voice to Mankind* (New York: S. S. Lyon and William Fishbough, 1847)

Dawkins, William Boyd, 'Sir Henry Hoyle Howorth, K.C.I.E., D.C.L., F.R.S.', *Man*, 23 (1923), 138–39

Dawson, John William, *The Story of the Earth and Man*, 2nd rev. edn (London: Hodder and Stoughton, 1873)

Dawson, John William, *Nature and the Bible* (New York: Robert Carter and Brothers, 1875)

Dawson, John William, *The Origin of the World, According to Revelation and Science* (New York: Harper & Brothers, 1877)

Dawson, John William, *The Meeting-Place of Geology and History* (Chicago: Fleming H. Revell, 1894)

Day, Langston, and George De La Warr, *New Worlds beyond the Atom* (London: Vincent Stuart, 1956)

Denton, Sherman F., *As Nature Shows Them: Moths and Butterflies of the United States East of the Rocky Mountains*, 2 vols (Boston, MA: Bradlee Whidden, 1900)

Denton, William, *Poems for Reformers* (Dayton, OH: William and Elizabeth Denton, 1856)

Denton, William, *Our Planet, Its Past and Future; or, Lectures on Geology* (Boston: William Denton, 1868)

Denton, William, *The Soul of Things; or, Psychometric Researches and Discoveries*, vol. III (Boston: William Denton, 1874)

Denton, William, and Elizabeth M. Foote Denton, *The Soul of Things; or, Psychometric Researches and Discoveries* (Boston: Walker, Wise and Company, 1863)

Denton, William, and Elizabeth M. Foote Denton, *Nature's Secrets or Psychometric Researches*, ed. by [W. L. Thompson] (London: Houlston and Wright, 1863)

Denton, William, and Elizabeth M. F. Denton, *The Soul of Things; or, Psychometric Researches and Discoveries*, 3rd edn (Boston: Walker, Wise, 1866)

Descriptive Guide Book of the California-Pacific International Exposition at San Diego California 1935 (Chicago: American Autochrome, 1935)

Donnelly, Ignatius, *Atlantis: The Antediluvian World* (New York: Harper & Brothers, 1882)

'Donnelly's Atlantis', *American Naturalist*, 16 (1882), 729–31

Doyle, Arthur Conan, *The Lost World* (London: Hodder and Stoughton, 1912)

Doyle, Arthur Conan, *The Wanderings of a Spiritualist* (London: Hodder & Stoughton, 1921)

Driver, Samuel Rolles, 'The Cosmogony of Genesis: A Defense and a Critique', *Andover Review*, 8 (1887), 639–49

[Eberty, Felix], *The Stars and the Earth; or, Thoughts Upon Space, Time, and Eternity, Part II* (London: H. Bailliere, 1847)

Emerson, Willis George, *The Smoky God; or, A Voyage to the Inner World* (Chicago: Forbes, 1908)

'Eminent Living Geologists: William Carruthers', *Geological Magazine*, 9 (1912), 193–99

Fawcett, Brian, 'Epilogue', in Percy Harrison Fawcett, *Exploration Fawcett* (UK: Arrow, 1963)

'Fiction', *Literary World*, 16 (1885), 405

Fieldcrest, Robert, 'Foreword', in Raymond Bernard, *The Hollow Earth; The Greatest Geographical Discovery in History* (New York: University Books, 1969), 7–9

Figuier, Louis, *The World before the Deluge*, trans. by W. S. O. (London: Chapman and Hall, 1865)
'Finds All Religion Springs from Mu', *New York Times*, 3 April 1933, 8
'Franklin Titus Ives', *New York Times*, 31 January 1910, 7
Gager, Charles Stuart, 'At the Top Is Magic', *Science*, 74 (1931), 569–70
Gardner, Marshall B., *A Journey to the Earth's Interior; or, Have the Poles Really Been Discovered* (Aurora, IL: Marshall B. Gardner, 1913)
Gardner, Marshall B., *A Journey to the Earth's Interior; or Have the Poles Really Been Discovered*, rev. edn (Aurora, IL: Marshall B. Gardner, 1920)
Georg, Eugen, *The Adventure of Mankind*, trans. by Robert Bek-Gran (New York: E. P. Dutton, 1931)
Gladstone, W. E., 'Proem to Genesis: A Plea for a Fair Trial', *Nineteenth Century*, 19 (1886), 1–21
Gladstone, W. E., 'The Creation Story', *Good Words*, 31 (1890), 300–11
'Gladstone or Salisbury', *Congregationalist*, 13 (1884), 715–26
'Going to Look for a Big Hole at the Top of the World', *Marion Daily Mirror* [OH], 25 April 1908, 9
Grover, John William, *Conversations with Little Geologists on the Six Days of Creation, Illustrated with a Geological Chart* (London: Edward Stanford, 1878)
Guyot, Arnold, *Creation; or, The Biblical Cosmogony in the Light of Modern Science* (New York: Charles Scribner, 1884)
'Guyot's View of Creation', *Science*, 3 (1884), 599–601
Hallowell, Alfred Irving, untitled review of *The Phenomenon of Man*, by Pierre Teilhard de Chardin, *Isis*, 52 (1961), 439–41
Hawthorne, Julian, 'Introduction', in William R. Bradshaw, *The Goddess of Atvatabar; Being the History of the Discovery of the Interior World and Conquest of Atvatabar* (New York: J. F. Douthitt, 1892), 9–12
'Heart to Heart Talks', *All-Story Weekly*, 44 (1915), 322–25
Henderson, W. P., 'Working Among the Troops in Shanghai', *Advent Review and Sabbath Herald*, 105 (1928), 19–20
Higgins, Lisetta Neukom, 'When Mu Ruled the World', *Washington Post*, 27 November 1932, n.p.
[Hitchcock, Edward] 'Poetaster', 'The Sandstone Bird', *Knickerbocker*, 8 (1836), 750–52
Hitchcock, Edward, 'First Anniversary Address before the Association of American Geologists', *American Journal of Science*, 41 (1841), 232–75
Hitchcock, Edward, *The Religion of Geology and Its Connected Sciences* (Boston: Phillips, Sampson, 1851)
Holbrook, David L., *The Panorama of Creation as Presented in Genesis Considered in Its Relation with the Autographic Record as Deciphered by Scientists* (Philadelphia, PA: Sunday School Times, 1908)
Howorth, Henry Hoyle, *The Mammoth and the Flood: An Attempt to Confront the Theory of Uniformity with the Facts of Recent Geology* (London: Sampson Low, Marston, Searle, & Rivington, 1887)
Howorth, Henry Hoyle, *The Glacial Nightmare and the Flood: A Second Appeal to Common Sense from the Extravagance of Some Recent Geology*, 2 vols (London: Sampson Low, Marston & Company, 1893)
'Hugh Miller and Geology', *Dublin University Magazine*, 50 (1857), 596–610
Hunt, Robert, *Panthea, the Spirit of Nature* (London: Reeve, Benham, and Reeve, 1849)
Huxley, T. H., *American Addresses, with a Lecture on the Study of Biology* (New York: Appleton, 1877)
Huxley, T. H., and Henry Drummond, 'Mr. Gladstone and Genesis', *Nineteenth Century*, 19 (1886), 191–214
'Hy-Brasil', *Times of India*, 17 April 1925, 5
[Irving, Washington] 'Dietrich Knickerbocker', *A History of New York, from the Beginning of the World to the End of the Dutch Dynasty*, 2 vols (New York: Inskeep & Bradford, 1809)
'Is There a World Inside of the World?', *The Age-Herald* [Birmingham, AL], 3 August 1913, n.p.

Ives, Franklin Titus, *The Hollow Earth* (New York: Broadway Publishing, 1904)
Jinarajadasa, Curuppumullage, *First Principles of Theosophy*, 2nd edn (Adyar: Theosophical Publishing House, 1922)
Jinarajadasa, Curuppumullage, 'Introduction', in Charles Webster Leadbeater, *The Astral Plane*, rev. edn (Adyar: Theosophical Publishing House, 1973 [1933]), vii–xxi
Jones, Alonzo T., '"Evolution" and Evolution (*Concluded*)', *Signs of the Times*, 11 (1885), 404
Joyce, Thomas Athol, 'The Problem of Atlantis', *Times Literary Supplement*, 24 July 1924, 456
Judd, Neil M., 'The Rising Quality of New World Archeology', *Scientific Monthly*, 63 (1946), 391–94
Keil, Carl Friedrich, and Franz Delitzsch, *Biblical Commentary on the Old Testament*, 4 vols, trans. by James Martin (Edinburgh: T. & T. Clark, 1872)
Kinns, Samuel, *The Harmony of the Bible with Science; or, Moses and Geology*, 2nd edn (New York: Cassell, Petter, Galpin, 1882)
Kinns, Samuel, *Moses and Geology; or, The Harmony of the Bible with Science* (London: Cassell, 1883)
Kinns, Samuel, *Moses and Geology; or, The Harmony of the Bible with Science*, 8th thousand (London: Cassell, 1885)
Knapp, George [sic] Christian, *Lectures on Christian Theology*, 2 vols (New York: G. & C. Carvill, 1831–33)
Kurtz, John Henry, *The Bible and Astronomy; An Exposition of the Biblical Cosmology, and Its Relations to Natural Science*, 3rd rev. German edn, trans. by T. D. Simonton (Philadelphia, PA: Lindsay & Blakiston, 1857)
[Lankester, Edwin], untitled review of *The Testimony of the Rocks*, by Hugh Miller, *Athenæum*, 1536 (1857), 429–31
Lapworth, Charles, 'Presidential Address to the Geological Section', *Report of the Sixty-Second Meeting of the British Association for the Advancement of Science held at Edinburgh in August 1892* (London: John Murray, 1893), 695–707
Leadbeater, Charles Webster, *The Perfume of Egypt, and Other Weird Stories* (Adyar: Theosophist Office, 1911)
Leslie, J. Ben, *Submerged Atlantis Restored; or, Rĭn-Gä'-Sĕ Nud Sī-ī-Kĕl'Zē (Links and Cycles)* (Rochester, NY: Austin Publishing, 1911)
Lethbridge, Thomas Charles, *Ghost and Ghoul* (London: Routledge and Kegan Paul, 1967 [1961])
Lloyd, John Uri, *Etidorhpa; or The End of Earth* (Cincinnati, OH: John Uri Lloyd, 1895)
Lloyd, John Uri, *Etidorhpa, or The End of Earth*, 2nd edn (Cincinnati, OH: Robert Clarke Company, 1896)
Lloyd, John Uri, *Etidorhpa; or The End of Earth*, 11th rev. edn (New York: Dodd, Mead, 1901)
Lloyd, John Uri, *Etidorhpa* (Santa Fe, NM: Sun Publishing, 1976)
Lovecraft, H. P., 'The Call of Cthulhu', *Weird Tales*, 11 (1928), 159–78
Lovecraft, H. P., 'The Shadow Out of Time', *Astounding Stories*, 17 (1936), 110–54
Lovecraft, H. P., *Collected Essays Volume 3: Science* (New York: Hippocampus Press, 2004)
'The Lunar Pitris', *Light*, 15 (1895), 430
Lyell, Charles, *Travels in North America, in the Years 1841–2; with Geological Observations on the United States, Canada, and Nova Scotia*, 2 vols (New York: Wiley and Putnam, 1845)
Lyell, Charles, *A Second Visit to the United States of North America*, 2 vols (London: John Murray, 1849)
'The Mammoth and the Flood', *Quarterly Review*, 166 (1888), 112–29
'The Mammoth and the Flood', *Spectator*, 61 (1888), 275–76
Mantell, Gideon, *The Wonders of Geology; or, A Familiar Exposition of Geological Phenomena*, 2 vols (London: Relfe and Fletcher, 1838)
[McBride, James], *Symmes's Theory of Concentric Spheres; Demonstrating that the Earth Is Hollow, Habitable Within, and Widely Open about the Poles* (Cincinnati, OH: Morgan, Lodge and Fisher, 1826)

McBride, James, *Pioneer Biography: Sketches of the Lives of Some of the Early Settlers of Butler County, Ohio*, 2 vols (Cincinnati, OH: Robert Clarke, 1869–1871)
'Men Who Won the West: John C. Frémont—Explorer', *Blue Book Magazine*, 48.5 (1929), 6
'Men Worth Knowing', *Blue Book Magazine*, 50.3 (1930), 5
Miller, Hugh, *The Old Red Sandstone; or, New Walks in an Old Field, Edited with a Critical Study and Notes*, 2 vols, ed. by Ralph O'Connor and Michael A. Taylor (Edinburgh: National Museums Scotland Publishing, 2023)
Miller, Hugh, *The Testimony of the Rocks; or, Geology in Its Bearing on the Two Theologies, Natural and Revealed* (Edinburgh: Thomas Constable, 1857)
Miller, Hugh, *The Cruise of the Betsey; or, A Summer Ramble among the Fossiliferous Deposits of the Hebrides* (Edinburgh: Thomas Constable, 1858)
Miller, Hugh, *Sketch Book of Popular Geology: A Series of Lectures Read before the Philosophical Institution of Edinburgh*, ed. by Lydia Miller (Boston: Gould and Lincoln, 1859)
Milton, John, *Paradise Lost*, ed. by Stephen Orgel and Jonathan Goldberg (Oxford: Oxford University Press, 2004)
Morton, Henry, 'The Cosmogony of Genesis and Its Reconcilers', *Bibliotheca Sacra*, 54 (1897), 264–92, 436–68
'Moses and Geology', *The Times*, 23 August 1882, 4
'Mr. Gladstone on Recent Corroborations of Scripture', *Manchester Guardian*, 23 September 1890, 6.
Mulhall, Marion McMurrough, *Beginnings, or Glimpses of Vanished Civilizations* (London: Longmans, Green, 1911)
Murie, James, 'On the Systematic Position of the *Sivatherium giganteum* of Falconer and Cautley', *Geological Magazine*, 8 (1871), 438–48
Naden, Constance, 'Geological Epochs', *Agnostic*, 1 (1885), 304–8
Nelson, Byron C., *'After Its Kind': The First and Last Word on Evolution* (Minneapolis, MN: Augsburg Publishing House, 1927)
Nelson, Byron C., *The Deluge Story in Stone: A History of the Flood Theory of Geology* (Minneapolis, MN: Augsburg Publishing House, 1931)
'New Publications', *New York Times*, 18 February 1882, 5
'New Worlds for Old', *Blue Book Magazine*, 48.5 (1929), 5
Olcott, Henry Steel, *Old Diary Leaves: The Only Authentic History of the Theosophical Society, Fifth Series: January, 1893–April, 1896* (Adyar: Theosophical Publishing House, 1932)
P., L. B., untitled review of *The Panorama of Creation*, by David L. Holbrook, *Hartford Seminary Record*, 19 (1909), 311–12
Palmer, Raymond, 'Editorial', *Hidden World*, A-6 (1962), 958–63
'Paradise Found', *Atlantic Monthly*, 56 (1885), 126–32
Parkinson, James, *Organic Remains of a Former World: An Examination of the Mineralized Remains of the Vegetables and Animals of the Antediluvian World; Generally Termed Extraneous Fossils*, 3 vols (London: Sherwood, Neely, and Jones, 1804–11)
Pavgee, Narayan Bhavanrao, *The Vedic Fathers of Geology* (Poona [Pune]: Arya-Bhushan Press, 1912)
Peart, Robert, 'The Problem of Atlantis', *Observer*, 10 August 1924, 14
'The Phantom of the Poles', *The Four-Track News*, 11 (1906), 172
'The Phantom of the Poles', *Newtown Bee* [CT], 21 February 1908, 5, 12
[Plummer, George Winslow] 'Khei X', *Rosicrucian Fundamentals: An Exposition of the Rosicrucian Synthesis of Religion, Science and Philosophy in Fourteen Complete Instructions* (New York: Flame Press, 1920)
'Polar Voids', *Chicago Daily Tribune*, 22 February 1875, 3
Powell, J. H., *William Denton, the Geologist and Radical: A Biographical Sketch* (Boston, MA: J. H. Powell, 1870)
Price, George McCready, *Illogical Geology: The Weakest Point in the Evolution Theory* (Los Angeles, CA: Modern Heretic Company, 1906)

Price, George McCready, *God's Two Books; Or Plain Facts about Evolution, Geology, and the Bible* (South Bend, IN: Review and Herald Publishing, 1911)
Price, George McCready, *Q. E. D., or New Light on the Doctrine of Creation* (New York: Fleming H. Revell, 1917)
Price, George McCready, *The New Geology: A Textbook for Colleges, Normal Schools, and Training Schools; and for the General Reader* (Mountain View, CA: Pacific Press Publishing Association, 1923)
Price, George McCready, 'The Significance of Fundamentalism', *Advent Review and Sabbath Herald*, 104 (1927), 13–14
Price, George McCready, *A History of Some Scientific Blunders* (Chicago: Fleming H. Revell, 1930)
Price, George McCready, 'The Man from Mars', *Signs of the Times*, 57 (1930), 2–3, 10, 15
Price, George McCready, *The Geological-Ages Hoax: A Plea for Logic in Theoretical Geology* (New York: Fleming H. Revell, 1931)
Price, George McCready, *Back to the Bible; or, The New Protestantism*, 3rd rev. edn (Taokoma Park, Washington, D.C.: Review and Herald Publishing Association, 1932)
Price, George McCready, untitled review of *The Deluge Story in Stone*, by Byron C. Nelson, *Ministry*, 5 (1932), 25
Price, George McCready, *Modern Discoveries Which Help Us to Believe* (New York: Fleming H. Revell, 1934)
Price, George McCready, *Some Scientific Stories and Allegories* (Grand Rapids, MI: Zondervan, 1936)
Price, George McCready, *Genesis Vindicated* (Washington, D.C.: Review and Herald Publishing Association, 1941)
Price, George McCready, *The Man from Mars* (Washington, D.C.: Review and Herald Publishing Association, 1950)
Price, George McCready, *Poems of my Long Ago*, unpublished typescript, Center for Adventist Research, Andrews University (1959)
'The Problem of Atlantis', *Nature*, 114 (1924), 409–10
R., H. H., 'How Does He Do It?', *All-Story Weekly*, 30 (1914), 671
Reed, William, *The Phantom of the Poles* (New York: Walter S. Rockey, 1906)
'Reed's "Phantom of the Poles"', *American Journal of Clinical Medicine*, 14 (1907), 931
Rimmer, Harry, *Noah's Ark, Modern Science and the Deluge* (Los Angeles, CA: Research Science Bureau, 1925)
Rimmer, Harry, *Monkeyshines: Fakes, Fables, Facts Concerning Evolution* (Los Angeles, CA: Research Science Bureau, 1926)
Robinson, John Talbot, and Margaret Donnelly, 'An Investigation of Clairvoyance as Applied to the Study of a Chapter in Primate Evolution', *Science Group Journal*, 4 (1960), 9–16
Roehm, Rob, and Rusty Burke, eds., *The Collected Letters of Robert E. Howard*, 3 vols (Plano, TX: Robert E. Howard Foundation Press, 2007–2008)
Rose, William, *An Explanation of the Author's Opinions on Geology: Showing the Fallacy of the Extreme Believer in That So-Called Science, and How Contradictory Geology Is When Compared with the Bible* (Birmingham: Martin Billing, Son, and Co., 1867)
Rudmose-Brown, Robert Neal, 'The Problem of Atlantis', *Geographical Journal*, 64 (1924), 181–82
Ryan, Charles James, 'Archaeological Notes', *Theosophical Path*, 23 (1922), 61–68
S., W., untitled review of *The Principles of Nature*, by Andrew Jackson Davis, *Christian Examiner*, 43 (1847), 452–55
Saint-Hilaire, Isidore Geoffroy, 'Note on Some Bones and Eggs Found at Madagascar, in Recent Alluvia, Belonging to a Gigantic Bird', *Magazine of Natural History*, 7 (1851), 161–66
Sayers, Dorothy L., *Unpopular Opinions* (London: Victor Gollancz, 1946)
Schultz, David E., and S. T. Joshi, eds., *Dawnward Spire, Lonely, Hill: The Letters of H. P. Lovecraft and Clark Ashton Smith: 1922–1937*, 2 vols continuously paginated (New York: Hippocampus Press, 2017)

Schultz, David E., and Scott Connors, eds., *Selected Letters of Clark Ashton Smith* (Sauk City, WI: Arkham House, 2003)
'The Scientific Accuracy of the Bible', *The Times*, 4 January 1884, 10
'The Scientific Accuracy of the Bible', *The Times*, 5 January 1884, 7
'The Scientific Accuracy of the Bible', *The Times*, 8 January 1884, 7
'The Scientific Accuracy of the Bible', *The Times*, 9 January 1884, 7
'The Scientific Accuracy of the Bible', *The Times*, 19 January 1884, 8
Scott-Elliot, William, *The Story of Atlantis: A Geographical, Historical, and Ethnological Sketch* (London: Theosophical Publishing Society, 1896)
Scott-Elliot, William, *The Lost Lemuria with Two Maps Showing Distribution of Land Areas at Different Periods* (London: Theosophical Publishing Society, 1904)
Scott-Elliot, William, *The Story of Atlantis: A Geographical, Historical, and Ethnological Sketch*, 2nd rev. edn (London: Theosophical Publishing Society, 1909)
'Seaborn, Adam', *Symzonia: A Voyage of Discovery* (New York: J. Seymour, 1820)
Sedgwick, Adam, 'Proceedings of the Geological Society', *Philosophical Magazine*, 7 (1830), 309–10
Sherman, Manly L., and William F. Lyon, *The Hollow Globe; or The World's Agitator and Reconciler* (Chicago: Religio-Philosophical Publishing House, 1871)
'Sir Henry Howorth', *Manchester Guardian*, 17 July 1923, 16
'Sir Henry Howorth: A Life of Wide Interests: Politics, Science, and Art', *The Times*, 17 July 1923, 14
Sinnett, Patience, and William Scott-Elliot, 'The Lunar Pitris', *Transactions of the London Lodge of the Theosophical Society*, 26 (1895), 3–30
Smith, Charlotte, *Beachy Head with Other Poems* (London: J. Johnson, 1807)
Smith, Clark Ashton, *The Star-Treader and Other Poems* (San Francisco, CA: A. M. Robertson, 1912)
Smith, Clark Ashton, *Ebony and Crystal: Poems in Verse and Prose* (Auburn, CA: Auburn Journal, 1922)
Smith, Clark Ashton, 'An Offering to the Moon', *Weird Tales*, 45.4 (1953), 54–65
'The Soul of Things', *Athenæum*, 1871 (1863), 295–97
'The Soul of Things', *Theosophist*, 4 (1883), 239–40
Spence, Lewis, *An Encyclopædia of Occultism: A Compendium of Information on the Occult Sciences, Occult Personalities, Psychic Science, Magic, Demonology, Spiritism and Mysticism* (London: George Routledge, 1920)
Spence, Lewis, *The Problem of Atlantis* (London: William Rider & Son, 1924)
Spence, Lewis, 'The National Party of Scotland', *Edinburgh Review*, 248 (1928), 70–87
Spence, Lewis, 'The Lost Continent of Mu: A Critical Appreciation', *Occult Review*, 55 (1932), 102–04
Spence, Lewis, *The Problem of Lemuria: The Sunken Continent of the Pacific* (London: Rider, 1932)
Spence, Lewis, 'Investigating a Subject', *Library Review*, 26 (1933), 45–50
Spence, Lewis, 'Ignatius Donnelly', in Ignatius Donnelly, *Atlantis: The Antediluvian World*, rev. edn, ed. by Egerton Sykes (London: Sidgwick and Jackson, 1950), xvii–xix
Stopes, Marie, *Love's Creation: A Novel* (London: John Bale, Sons & Danielsson, 1928)
'Story of Lost Lemuria', *Chronicle* (Adelaide), 28 December 1933, 3
Symmes, Americus, *The Symmes Theory of Concentric Spheres, Demonstrating that the Earth Is Hollow, Habitable Within, and Widely Open about the Poles* (Louisville, KY: Bardley & Gilbert, 1878)
'Symmes and His Theory', *Harper's New Monthly Magazine*, 65 (1882), 740–44
Symmes, Ida Elmore, 'John Cleves Symmes, The Theorist', *Southern Bivouac*, 2 (1886–86), 555–66, 621–31, 682–93
'Tanar of Pellucidar', *Blue Book Magazine*, 48.4 (1929), 5
Teilhard de Chardin, Pierre, *The Phenomenon of Man*, trans. by Bernard Wall (London: Harper Perennial, 1959)

[Tennyson, Alfred], *In Memoriam* (London: Edward Moxon, 1850)
Thoreau, Henry David, *Walden; or, Life in the Woods* (Boston, MA: Ticknor and Fields, 1854)
'To Understand the World', *Blue Book Magazine*, 49.3 (1929), 5
Tower, Washington L., *Interior World: A Romance Illustrating a New Hypothesis of Terrestrial Organization, with an Appendix Setting Forth an Original Theory of Gravitation* (Oakland, OR: Milton H. Tower, 1885)
Tuttle, Hudson, *Scenes in the Spirit World; or, Life in the Spheres* (New York: Partridge and Brittan, 1855)
Tuttle, Hudson, *Arcana of Nature; or, the History and Laws of Creation*, 2 vols (Boston: Colby & Rich, 1859)
Tuttle, Hudson, *Arcana of Nature*, 2nd rev. edn (New York: Stillman Publishing, 1909)
'The Undiscovered World', *Helena Weekly Herald* [MT], 2 December 1875, 3
Unger, Franz, 'The Sunken Island of Atlantis', *Journal of Botany*, 25 (1865), 12–26
Unpaginated advertisement for *God's Two Books*, by George McCready Price, *Oriental Watchman*, 15 (1912)
Untitled review of *A Journey to the Earth's Interior*, by Marshall B. Gardner, *Bulletin of the American Geographical Society*, 46 (1914), 543
Untitled review of *Atlantis: The Antediluvian World*, by Ignatius Donnelly, *Nature*, 26 (1882), 341
Untitled review of *Atlantis: The Antediluvian World*, by Ignatius Donnelly, *Western Christian Advocate* (Cincinnati, OH), 22 March 1882, 95
Untitled review of *Atlantis: The Antediluvian World*, by Ignatius Donnelly, *Popular Science Monthly*, 22 (1883), 131–32
Untitled review of *Beginning, or Glimpses of Vanished Civilizations*, by Marion McMurrough Mulhall, *America*, 5 (1911), 284
Untitled review of *Etidorhpa*, by John Uri Lloyd, *Saturday Review*, 82 (1896), 271–72
Untitled review of *Is Darwin Right? Or, the Origin of Man*, by William Denton, *Scientific American*, 44 (1881), 250
Untitled review of *Moses and Geology*, by Samuel Kinns, *Dublin Review*, 9 (1883), 239–41
Untitled review of *Paradise Found*, by William Fairfield Warren, *Spectator*, 58 (1885), 886
Untitled review of *Symmes's Theory of Concentric Spheres*, *American Quarterly Review*, 1 (1827), 235–53
Untitled review of *Symzonia*, by 'Adam Seaborn', *North American Review*, 13 (1821), 134–43
Untitled review of *The Bible and the Age*, by Cuthbert Collingwood, *London Quarterly Review*, 68 (1887), 159–60
Untitled review of *The Bible and the Age*, by Cuthbert Collingwood, *Wesleyan-Methodist Magazine*, 11 (1887), 558
Untitled review of *The Hollow Earth*, by Franklin Titus Ives, *Current Literature*, 37 (1904), 476
Untitled review of *The Mammoth and the Flood*, by Henry Hoyle Howorth, *Zoologist*, 11 (1887), 438–40
Untitled review of *The Panorama of Creation*, by David L. Holbrook, *Zion's Herald* [Boston, MA], 26 May 1909, 660
Untitled review of *The Story of the Earth and Man*, by John William Dawson, *Spectator*, 46 (1873), 1314–15
Untitled review of *The Testimony of the Rocks*, by Hugh Miller, *Westminster Review*, 68 (1857), 176–85 (181, 184)
Valentine, Edward A. Uffington, 'Mu, the Lost Atlantis of the Pacific Ocean', *New York Times*, 7 June 1931, 8
Verne, Jules, *Vingt mille lieues sous les mers* (Paris: Hetzel, 1872)
Warren, William Fairfield, *President's First Baccalaureate Sermon and Tenth Annual Report* (Boston, MA: University Offices, 1884)
Warren, William Fairfield, *Paradise Found: The Cradle of the Human Race at the North Pole: A Study of the Prehistoric World* (Boston, MA: Houghton, Mifflin, 1885)
Wells, H. G., 'The Time Machine [5/5]', *New Review*, 12 (1895), 577–88

Wells, H. G., Julian Huxley, and G. P. Wells, *The Science of Life: A Summary of Contemporary Knowledge about Life and Its Possibilities*, 3 vols (London: Amalgamated Press, 1930)

'Where Is Eden', *New York Times*, 5 April 1885, 5

Whitcomb, John C., and Henry M. Morris, *The Genesis Flood: The Biblical Record and Its Scientific Implications* (Phillipsburg, NJ: Presbyterian and Reformed Publishing, 1991 [1961])

Whitcomb, John C., 'Preface to the Fiftieth Anniversary Edition', in John C. Whitcomb and Henry M. Morris, *The Genesis Flood: The Biblical Record and Its Scientific Implications: Fiftieth Anniversary Edition* (Phillipsburg, NJ: Presbyterian and Reformed, 2011), xxxiii–xxi

White, Ellen G., *Spiritual Gifts III: Important Facts of Faith, in Connection with the History of Holy Men of Old* (Battle Creek, MI: Steam Press of the Seventh-day Adventist Publishing Association, 1864)

White, Ellen G., *Counsels to Teachers, Parents and Students regarding Christian Education* (Mountain View, CA; Pacific Press Publishing Association, 1913)

Wilgus, Curtis, untitled review of *The Sacred Symbols of Mu*, by James Churchward, *Hispanic American Historical Review*, 14 (1934), 85–86

Wilgus, Neal, 'Introduction: The Pharmaceutical Alchemist', in John Uri Lloyd, *Etidorhpa* (Santa Fe, NM: Sun Publishing, 1976), xv–xx

Winchell, Alexander, 'Ancient Myth and Modern Fact', *Dial*, 2 (1882), 284–86

Winchell, Alexander, 'Ignatius Donnelly's Comet', *Forum*, 4 (1887), 105–15

Woodward, Arthur Smith, 'The Relative Age of Rocks containing Fossils', *Nature*, 117 (1926), 21–23

Wordsworth, William, *The Excursion, being a portion of The Recluse, a Poem* (London: Longman, Hurst, Rees, Orme, and Brown, 1814)

Secondary Sources

Adelman, Juliana, 'Eozoön: Debunking the Dawn Animal', *Endeavour*, 31 (2007), 94–8

Aït-Touati, Frédérique, *Fictions of the Cosmos: Science and Literature in the Seventeenth Century*, trans. by Susan Emanuel (Chicago: University of Chicago Press, 2011)

Ashley, Mike, 'Blue Book—The Slick in Pulp Clothing', *Pulp Vault*, 14 (2011), 210–53

Asprem, Egil, *The Problem of Disenchantment: Scientific Naturalism and Esoteric Discourse, 1900–1939* (Leiden: Brill, 2014)

Baldwin, Melinda, *Making* Nature: *The History of a Scientific Journal* (Chicago: University of Chicago Press, 2015)

Baldwin, Melinda, 'The Business of Being an Editor: Norman Lockyer, Macmillan and Company, and the Editorship of *Nature*, 1869–1919', *Centaurus*, 62 (2020), 1–14

Barnett, Lydia, *After the Flood: Imagining the Global Environment in Early Modern Europe* (Baltimore, MD: Johns Hopkins University Press, 2022)

Bashford, Alison, Emily M. Kern, and Adam Bobbette, eds., *New Earth Histories: Geo-Cosmologies and the Making of the Modern World* (Chicago: University of Chicago Press, 2023)

Bebbington, David, *Evangelicalism in Modern Britain: A History from the 1730s to the 1980s* (London: Unwin Hyman, 1989)

Bebbington, David, *The Mind of Gladstone: Religion, Homer, and Politics* (Oxford: Oxford University Press, 2004)

Beer, Gillian, *Darwin's Plots: Evolutionary Narrative in Darwin, George Eliot and Nineteenth-Century Fiction*, 3rd edn (Cambridge: Cambridge University Press, 2009)

Beringer, Alex, '"Some Unsuspected Author": Ignatius Donnelly and the Conspiracy Novel', *Arizona Quarterly*, 68 (2012), 35–60

Bhattacharya, Sumangala, 'The Victorian Occult Atom: Annie Besant and Clairvoyant Atomic Research', in *Strange Science: Investigating the Limits of Knowledge in the Victorian Age*, ed. by Lara Karpenko and Shalyn Claggett (Ann Arbor: University of Michigan Press, 2017), 197–214

Bleiler, Everett F., *Science-Fiction: The Early Years* (Kent, OH: Kent State University Press, 1990)
Blum, Hester, 'John Cleves Symmes and the Planetary Reach of Polar Exploration', *American Literature*, 84 (2012), 243–71
Boone, Kathleen C., *The Bible Tells Them So: The Discourse of Protestant Fundamentalism* (Albany: State University of New York Press, 1989)
Bootsman, Cornelis Siebe, 'The Nineteenth Century Engagement between Geological and Adventist Thought and Its Bearing on the Twentieth Century Flood Geology Movement', unpublished PhD thesis, Avondale College of Higher Education (2016)
Bowler, Peter J., *Life's Splendid Drama: Evolutionary Biology and the Reconstruction of Life's Ancestry, 1860–1940* (Chicago: University of Chicago Press, 1996)
Boyle, Tanner F., *The Fortean Influence on Science Fiction* (Jefferson, NC: McFarland, 2021)
Bramwell, Valerie, and Robert M. Peck, *All in the Bones: A Biography of Benjamin Waterhouse Hawkins* (Philadelphia, PA: The Academy of Natural Sciences of Philadelphia, 2008)
Breidbach, Olaf, and Michael Ghiselin, 'Lorenz Oken and *Naturphilosophie* in Jena, Paris and London', *History and Philosophy of the Life Sciences*, 24 (2002), 219–47
Broman, Thomas, 'The Habermasian Public Sphere and "Science *in* the Enlightenment"', *History of Science*, 34 (1998), 123–49
Brown, Andrew J., *The Days of Creation: A History of Christian Interpretation of Genesis 1:1—2:3* (Leiden: Brill, 2012)
Brown, C. Mackenzie, 'The Western Roots of Avataric Evolutionism in Colonial India', *Zygon*, 42 (2007), 425–49
Buckland, Adelene, *Novel Science: Fiction and the Invention of Nineteenth-Century Geology* (Chicago: University of Chicago Press, 2013)
Buckland, Adelene, '"Inhabitants of the Same World": The Colonial History of Geological Time', *Philological Quarterly*, 97 (2018), 219–40
Buckland, Adelene, 'The World beneath Our Feet', in *Time Travelers: Victorian Encounters with Time & History*, ed. by Adelene Buckland and Sadiah Qureshi (Chicago: University of Chicago Press, 2020), 42–64
Buggs, Richard J. A., 'The Origin of Darwin's "Abominable Mystery"', *American Journal of Botany*, 108 (2021), 22–36
Bull, Malcolm, and Keith Lockhart, *Seeking a Sanctuary: Seventh-day Adventism and the American Dream*, 2nd edn (Bloomington: Indiana University Press, 2007)
Burger, Phillip R., 'Afterword', in Edgar Rice Burroughs, *At the Earth's Core* (Lincoln: University of Nebraska Press, 2000), 279–90
Burke, Peter, 'The Polymath: A Cultural and Social History of an Intellectual Species', in *Explorations in Cultural History: Essays for Peter Gabriel McCaffery*, ed. by David F. Smith and Hushang Philsooph (Aberdeen: Centre for Cultural History, 2010)
Byerly, Alison, 'Effortless Art: The Sketch in Nineteenth-Century Painting and Literature', *Criticism*, 41 (1999), 349–64
Cain, Joe, 'Synthesis Period in Evolutionary Studies', in *The Cambridge Encyclopedia of Darwin and Evolutionary Thought*, ed. by Michael Ruse (Cambridge: Cambridge University Press, 2013), 282–92
Cain, Victoria E. M., '"The Direct Medium of the Vision": Visual Education, Virtual Witnessing and the Prehistoric Past at the American Museum of Natural History, 1890–1923', *Journal of Visual Culture*, 9 (2010), 284–303
Campbell, James L., Sr, 'John Taine', in *Science Fiction Writers: Critical Studies of the Major Authors from the Early Nineteenth Century to the Present Day*, ed. by E. F. Bleiler (New York: Scribner, 1981), 75–82
Card, Jeb J., *Spooky Archaeology: Myth and the Science of the Past* (Albuquerque: University of New Mexico Press, 2018)
Carhart, Michael C., *The Science of Culture in Enlightenment Germany* (Cambridge, MA: Harvard University Press, 2007)
Chakrabarti, Pratik, *Inscriptions of Nature: Geology and the Naturalization of Antiquity* (Baltimore, MD: Johns Hopkins University Press, 2020)

Chang, Elizabeth Hope, 'Hollow Earth Fiction and Environmental Form in the Late Nineteenth Century', *Nineteenth-Century Contexts*, 38 (2016), 387–97

Churchward, Jack, 'Resources', *my-mu.com*, my-mu.com/resources.html

Clark, Constance Areson, '"You Are Here": Missing Links, Chains of Being, and the Language of Cartoons', *Isis*, 100 (2009), 571–89

Clark, Harold W., *Crusader for Creation: The Life and Writings of George McCready Price* (Mountain View, CA: Pacific Press Publishing Association, 1966)

Connor, Steven, 'The Birth of Humility: Frazer and Victorian Mythography', in *Sir James Frazer and the Literary Imagination: Essays in Affinity and Influence* (Basingstoke: Macmillan, 1990), 61–80

Corsi, Pietro, *Science and Religion: Baden Powell and the Anglican Debate, 1800–1860* (Cambridge: Cambridge University Press, 1988)

Cregan-Reid, Vybarr, *Discovering Gilgamesh: Geology, Narrative and the Historical Sublime in Victorian Culture* (Manchester: Manchester University Press, 2013)

Csiszar, Alex, *The Scientific Journal: Authorship and the Politics of Knowledge in the Nineteenth Century* (Chicago: University of Chicago Press, 2018)

Daston, Lorraine, and Peter Galison, *Objectivity* (New York: Zone Books, 2007; repr. 2010)

Davis, Winston, *Dojo: Magic and Exorcism in Modern Japan* (Stanford, CA: Stanford University Press, 1980)

Dawson, Gowan, *Show Me the Bone: Reconstructing Prehistoric Monsters in Nineteenth-Century Britain and America* (Chicago: University of Chicago Press, 2016)

Dawson, Gowan, '"An Independent Publication for Geologists": The Geological Society, Commercial Journals, and the Remaking of Nineteenth-Century Geology', in *Science Periodicals in Nineteenth-Century Britain: Constructing Scientific Communities*, ed. by Gowan Dawson, Bernard Lightman, Sally Shuttleworth, and Jonathan R. Topham (Chicago: University of Chicago Press, 2020), 137–71

De Camp, L. Sprague, *Lost Continents: The Atlantis Theme in History, Science, and Literature*, rev. edn (New York: Dover Publications, 1970)

Dean, Dennis R., 'Tennyson and Creation', *Tennyson Research Bulletin*, 9 (2007), 22–41

Desmond, Adrian, *Huxley: Evolution's High Priest* (London: Michael Joseph, 1997)

Eames, Rachel, 'Geological *Katabasis*: Geology and the Christian Underworld in Kingsley's *The Water-Babies*', *Victoriographies*, 7 (2017), 195–209

Edelstein, Dan, 'Hyperborean Atlantis: Jean-Sylvain Bailly, Madame Blavatsky, and the Nazi Myth', *Studies in Eighteenth-Century Culture*, 25 (2006), 267–91

England, Richard, 'Aubrey Moore and the Anglo-Catholic Assimilation of Science in Oxford', unpublished PhD thesis, University of Toronto (1997)

Fallon, Richard, *Reimagining Dinosaurs in Late Victorian and Edwardian Literature: How the 'Terrible Lizard' Became a Transatlantic Cultural Icon* (Cambridge: Cambridge University Press, 2021)

Fenton, Elizabeth, *Old Canaan in a New World: Native Americans and the Lost Tribes of Israel* (New York: New York University Press, 2020)

Ferguson, Christine, 'The Luciferian Public Sphere: Theosophy and Editorial Seekership in the 1880s', *Victorian Periodicals Review*, 53 (2020), 76–101

Ferguson, Christine, 'Beyond Belief: Literature, Esotericism Studies, and the Challenges of Biographical Reading in Arthur Conan Doyle's *The Land of Mist*', *Aries*, 22 (2022), 205–30

Ferguson, Christine, and Efram Sera-Shriar, 'Spiritualism and Science Studies for the Twenty-First Century', *Aries*, 22 (2022), 1–11

Finlay, Richard J., *Independent and Free: Scottish Politics and the Origins of the Scottish National Party, 1918–1945* (Edinburgh: John Donald, 1994)

Finlayson, Geoffrey B. A. M., *The Seventh Earl of Shaftesbury 1801–1885* (Vancouver: Regent College Publishing, 1981)

Finnegan, Diarmid A., *The Voice of Science: British Scientists on the Lecture Circuit in Gilded Age America* (Pittsburgh, PA: University of Pittsburgh Press, 2021)

Fitting, Peter, *Subterranean Worlds: A Critical Anthology* (Middletown, CT: Wesleyan University Press, 2004)

Flaherty, Clare, 'A Recently Rediscovered Unpublished Manuscript: The Influence of Sir Humphry Davy on Anne Brontë', *Brontë Studies*, 38 (2013), 30–41

Flannery, Michael A., *John Uri Lloyd: The Great American Eclectic* (Carbondale: Southern Illinois University Press, 1998)

Flipse, Abraham C., 'The Origins of Creationism in the Netherlands: The Evolution Debate among Twentieth-Century Dutch Neo-Calvinists', *Church History*, 81 (2012), 104–47

Folk, Holly, 'Raymond W. Bernard, Hollow Earth, and UFOs', in *Handbook of UFO Religions*, ed. by Benjamin E. Zeller (Leiden: Brill, 2021), 312–25

Frank, Lawrence, *Victorian Detective Fiction and the Nature of Evidence: The Scientific Investigations of Poe, Dickens, and Doyle* (Basingstoke: Palgrave Macmillan, 2003)

Fraser, Robert, ed., 'Introduction', in James George Frazer, *The Golden Bough* (Oxford: Oxford University Press, 1994), ix–xliii

Freedgood, Elaine, *The Ideas in Things: Fugitive Meaning in the Victorian Novel* (Chicago: University of Chicago Press, 2006)

Fresonke, Kris, and Mark Spence, eds., *Lewis & Clark: Legacies, Memories, and New Perspectives* (Berkeley: University of California Press, 2004)

Frow, John, *Genre* (London: Routledge, 2005)

Frye, Northrop, *Anatomy of Criticism: Four Essays* (Princeton, NJ: Princeton University Press, 1957)

Fyfe, Aileen, *Science and Salvation: Evangelical Popular Science Publishing in Victorian Britain* (Chicago: University of Chicago Press, 2004)

Geric, Michelle, *Tennyson and Geology: Poetry and Poetics* (Cham: Palgrave Macmillan, 2017)

Gess, Nicola, *Primitive Thinking: Figuring Alterity in German Modernity*, trans. by Erik Butler and Susan L. Solomon (Berlin: De Gruyter, 2022)

Greene, Mott T., *Geology in the Nineteenth Century: Changing Views of a Changing World* (Ithaca, NY: Cornell University Press, 1982)

Godwin, Joscelyn, *The Theosophical Enlightenment* (Albany: State University of New York Press, 1994)

Godwin, Joscelyn, *Atlantis and the Cycles of Time: Prophecies, Traditions, and Occult Revelations* (Rochester, VT: Inner Traditions, 2011)

Goldhill, Simon, 'Ad Fontes', in *Time Travelers: Victorian Encounters with Time and History*, ed. by Adelene Buckland and Sadiah Qureshi (Chicago: University of Chicago Press, 2020), 67–85

Gordin, Michael D., *The Pseudoscience Wars: Immanuel Velikovsky and the Birth of the Modern Fringe* (Chicago: University of Chicago Press, 2012)

Gordin, Michael D., *On the Fringe: Where Science Meets Pseudoscience* (Oxford: Oxford University Press, 2021)

Hammer, Olav, *Claiming Knowledge: Strategies of Epistemology from Theosophy to the New Age* (Leiden: Brill, 2004)

Hanegraaff, Wouter J., 'Romanticism and the Esoteric Connection', in *Gnosis and Hermeticism from Antiquity to Modern Times*, ed. by Roelof van den Broek and Wouter J. Hanegraaff (Albany: State University of New York Press, 1998), 237–68

Hanegraaf, Wouter J., 'The Theosophical Imagination', *Correspondences*, 5 (2017), 3–39

Harris, Mark, 'Natural Selection at New College: The Evolution of Science and Theology at a Scottish Presbyterian Seminar', *Zygon*, 57 (2022), 525–44

Heringman, Noah, *Deep Time: A Literary History* (Princeton, NJ: Princeton University Press, 2023)

Hester, Greg L., 'Into the Celestial Spheres of Divine Wisdom: Joseph Rodes Buchanan and Nineteenth-Century Esotericism', unpublished MA thesis, University of Amsterdam (2015)

Hughes, Thomas McKenny, 'The Causes of Glacial Phenomena', *Nature*, 48 (1893), 242–44

Huhtamo, Erkki, *Illusions in Motion: Media Archaeology of the Moving Panorama and Related Spectacles* (Cambridge, MA: MIT Press, 2013)

Hunter, Howard Eugene, 'William Fairfield Warren: Methodist Theologian', unpublished PhD thesis, Boston University (1957)
Huskinson, Benjamin L., *American Creationism, Creation Science, and Intelligent Design in the Evangelical Market* (Basingstoke: Palgrave Macmillan, 2020)
Introvigne, Massimo, 'Paintings the Masters in Britain: From Schmiechen to Scott', in *The Occult Imagination in Britain, 1875–1947*, ed. by Christine Ferguson and Andrew Radford (London: Routledge, 2018), 206–26
Kaalund, Nanna Katrine Lüders, 'Of Rocks and "Men": The Cosmogony of John William Dawson', in *Historicizing Humans: Deep Time, Evolution, and Race in Nineteenth-Century British Science*, ed. by Efram Sera-Shriar (Pittsburgh, PA: University of Pittsburgh Press, 2018), 44–67
Kafton-Minkel, Walter, *Subterranean Worlds: 100,000 Years of Dragons, Dwarfs, the Dead, Lost Races & UFOs from Inside the Earth* (Port Townsend, WA: Loompanics Unlimited, 1989)
Karpenko, Lara, and Shalyn Claggett, eds., *Strange Science: Investigating the Limits of Knowledge in the Victorian Age* (Ann Arbor: University of Michigan Press, 2017)
Keane, Patrick J., *Emerson, Romanticism, and Intuitive Reason: The Transatlantic 'Light of All Our Day'* (Columbia: University of Missouri Press, 2005)
Keene, Melanie, 'Object Lessons: Sensory Science Education, 1830–1870', unpublished PhD thesis, University of Cambridge (2008)
Keene, Melanie, 'Familiar Science in Nineteenth-Century Britain', *History of Science*, 52 (2014), 53–71
Keene, Melanie, *Science in Wonderland: The Scientific Fairy Tales of Victorian Britain* (Oxford: Oxford University Press, 2015)
Keep, Christopher, 'Life on Mars?: Hélène Smith, Clairvoyance, and Occult Media', *Journal of Victorian Culture*, 25 (2020), 537–52
Kidd, Colin, *The World of Mr Casaubon: Britain's Wars of Mythography, 1700–1870* (Cambridge: Cambridge University Press, 2016)
Killian, Crawford, 'The Cheerful Inferno of James De Mille', *Journal of Canadian Fiction*, 1 (1972), 61–67
Klaver, J. M. I., *Geology and Religious Sentiment: The Effect of Geological Discoveries on English Society and Literature between 1829 and 1859* (Leiden: Brill, 1997)
Kloes, Andrew, *The German Awakening: Protestant Renewal After the Enlightenment, 1815–1848* (Oxford: Oxford University Press, 2019)
Knell, Simon J., *The Culture of English Geology, 1815–1851: A Science Revealed through Its Collecting* (Aldershot: Ashgate, 2000)
Kuljian, Christa, *Darwin's Hunch: Science, Race and the Search for Human Origins* (Johannesburg: Jacana, 2016)
Lachman, Gary, *Dreaming Ahead of Time: Experiences with Precognitive Dreams, Synchronicity and Coincidence* (Edinburgh: Floris Books, 2022)
Land, Gary, *Historical Dictionary of Seventh-day Adventists* (Lanham, MD: Scarecrow Press, 2005)
Lanset, Andy, 'WNYC and the Land of Mu', *WNYC*, https://www.wnyc.org/story/179746-wnyc-and-land-mu/
Larson, Edward J., *Summer for the Gods: The Scopes Trial and America's Continuing Debate over Science and Religion*, rev. edn (New York: Basic Books, 2006)
Lawrence, James F., '*Correspondentia*: A Neologism by Aquinas Attains Its Zenith in Swedenborg', *Correspondences*, 5 (2017), 41–63
Lightman, Bernard, *Victorian Popularizers of Science: Designing Nature for New Audiences* (Chicago: University of Chicago Press, 2009)
Lightman, Bernard, 'Science at the Metaphysical Society: Defining Knowledge in the 1870s', in *The Age of Scientific Naturalism: Tyndall and His Contemporaries*, ed. by Bernard Lightman and Michael S. Reidy (Pittsburgh, PA: University of Pittsburgh Press, 2016), 188–206
Lightman, Bernard, and Bennett Zon, eds., *Evolution and Victorian Culture* (Cambridge: Cambridge University Press, 2014)

Livingstone, David N., *Dealing with Darwin: Place, Politics, and Rhetoric in Religious Engagements with Evolution* (Baltimore, MD: Johns Hopkins University Press, 2014)
Looney, Dennis, 'Dante Alighieri and the Divine Comedy in Nineteenth-Century America', in *The Routledge History of Italian Americans* (New York: Routledge, 2018), 91–104
Loxton, Daniel, and Donald R. Prothero, *Abominable Science! Origins of the Yeti, Nessie, and Other Famous Cryptids* (New York: Columbia University Press, 2013)
Luciano, Dana, 'Sacred Theories of Earth: Matters of Spirit in The Soul of Things', *American Literature*, 86 (2014), 713–36
Lupoff, Richard A., *Master of Adventure: The Worlds of Edgar Rice Burroughs*, new edn (Lincoln: University of Nebraska Press, 2005)
Lyons, Sherrie Lynne, *Species, Serpents, Spirits, and Skulls: Science at the Margins in the Victorian Age* (New York: State University of New York Press, 2009)
Mahady, Christine, 'No World of Difference: Examining the Significance of Women's Relationships to Nature in Mary Bradley Lane's *Mizora*', *Utopian Studies*, 15 (2004), 93–115
Marsden, George M., *Fundamentalism and American Culture*, 2nd edn (Oxford: Oxford University Press, 2006)
Mathieson, Stuart, *Evangelicals and the Philosophy of Science: The Victoria Institute, 1865–1939* (London: Routledge, 2021)
Mayer, Anna-K., 'Reluctant Technocrats: Science Promotion in the Neglect-of-Science Debate of 1916–1918', *History of Science*, 43 (2005), 139–59
McDougall-Waters, Julie, and Aileen Fyfe, 'The Rise of the *Proceedings*, 1890–1920s', in *A History of Scientific Journals: Publishing at the Royal Society, 1665–2015*, ed. by Aileen Fyfe, Noah Moxham, Julie McDougall-Waters, and Camilla Mørk Røstvik (London: UCL Press, 2022), 363–402
McIver, Tom, 'Formless and Void: Gap Theory Creationism', *Creation/Evolution*, 8 (1988), 1–24
McMillan, Nora Fisher, 'Picture Quiz: Cuthbert Collingwood (1826–1908)', *Linnean*, 17 (2001), 9–20
Miller, Brook, *America and the British Imaginary in Turn-of-the-Twentieth-Century Literature* (Basingstoke: Palgrave Macmillan, 2010)
Moore, James R., 'Geologists and Interpreters of Genesis in the Nineteenth Century', in *God and Nature: Historical Essays on the Encounter between Christianity and Science*, ed. by David C. Lindberg and Ronald L. Numbers (Berkeley: University of California Press, 1986), 322–50
Morgan, David, *Protestants & Pictures: Religion, Visual Culture, and the Age of American Mass Production* (Oxford: Oxford University Press, 1999)
Morris, David, *The Masks of Lucifer: Technology and the Occult in Twentieth-Century Popular Literature* (London: B. T. Batsford, 1992)
Morrison, Mark S., *Modern Alchemy: Occultism and the Emergence of Atomic Theory* (Oxford: Oxford University Press, 2007)
Morrison, Mark S., 'The Periodical Culture of the Occult Revival: Esoteric Wisdom, Modernity and Counter-Public Spheres', *Journal of Modern Literature*, 31 (2008), 1–22
Mortenson, Terry, *The Great Turning Point: The Church's Catastrophic Mistake on Geology— Before Darwin* (Green Forest, AR: Master Books, 2004)
Moss, Sarah, *Scott's Last Biscuit* (Oxford: Signal Books, 2006)
Murphy, Gretchen, '*Symzonia, Typee*, and the Dream of U.S. Global Isolation', *ESQ: A Journal of the American Renaissance*, 49 (2003), 249–84
Musgrave, David, *Grotesque Anatomies: Menippean Satire since the Renaissance* (Newcastle upon Tyne: Cambridge Scholars Publishing, 2014)
Nelson, Paul A., 'Introduction', in *The Creationist Writings of Byron C. Nelson*, vol. 5 of Ronald L. Numbers, ed., *Creationism in Twentieth-Century America: A Ten-Volume Anthology of Documents, 1903–1961* (New York: Garland, 1995), ix–xxi
Nickerson, Sylvia, 'Darwin's Publisher: John Murray III at the Intersection of Science and Religion', in *Rethinking History, Science, and Religion: An Exploration of Conflict and the Complexity Principle*, ed. by Bernard Lightman (Pittsburgh, PA: University of Pittsburgh Press, 2019), 110–28

Numbers, Ronald L., *The Creationists: From Scientific Creationism to Intelligent Design*, rev. edn (Cambridge, MA: Harvard University Press, 2006)
Numbers, Ronald L., *Science and Christianity in Pulpit and Pew* (Oxford: Oxford University Press, 2007)
Numbers, Ronald L., 'Science, Secularization, and Privatization', in *Eminent Lives in Twentieth-Century Science and Religion*, ed. by Nicolaas A. Rupke, 2nd edn (Frankfurt: Peter Lang, 2009), 349–62
Numbers, Ronald L., and Rennie B. Schoepflin, 'Science and Medicine', in *Ellen Harmon White: American Prophet*, ed. by Terrie Dopp Aamodt, Gary Land, and Ronald L. Numbers (Oxford: Oxford University Press, 2014), 196–23
Numbers, Ronald L., and T. Joe Willey, 'Baptizing Dinosaurs: How Once-Suspect Evidence of Evolution Came to Support the Biblical Narrative', *Spectrum*, 43 (2015), 57–68
O'Connor, Jean G., and A. J. Meadows, 'Specialization and Professionalization in British Geology', *Social Studies of Science*, 6 (1976), 77–89
O'Connor, Ralph, *The Earth on Show: Fossils and the Poetics of Popular Science, 1802–1856* (Chicago: University of Chicago Press, 2007)
O'Connor, Ralph, 'Young-Earth Creationists in Early Nineteenth-Century Britain: Towards a Reassessment of "Scriptural Geology"', *History of Science*, 45 (2007), 357–403
O'Connor, Ralph, 'From the Epic of Earth History to the Evolutionary Epic in Nineteenth-Century Britain', *Journal of Victorian Culture*, 14 (2009), 207–23
O'Connor, Ralph, 'Introduction: Varieties of Romance in Victorian Science', in *Science as Romance, vol. VII of Victorian Science and Literature*, gen. eds Gowan Dawson and Bernard Lightman (London: Pickering & Chatto, 2012), xi–xxxvi
O'Connor, Ralph, 'The Meanings of "Literature" and the Place of Modern Scientific Nonfiction in Literature and Science', *Journal of Literature and Science*, 10 (2017), 37–45
Orvell, Miles, *American Photography* (Oxford: Oxford University Press, 2003)
Owen, Alex, *The Darkened Room: Women, Power and Spiritualism in Late Victorian England* (London: Virago, 1989)
Owen, Alex, *The Place of Enchantment: British Occultism and the Culture of the Modern* (Chicago: University of Chicago Press, 2004)
Patrick, Arthur, 'Author', in *Ellen Harmon White: American Prophet*, ed. by Terrie Dopp Aamodt, Gary Land, and Ronald L. Numbers (Oxford: Oxford University Press, 2014), 91–107
Peterson, Stephen J., *Gladstone's Influence in America: Reactions in the Press to Modern Religion and Politics* (Cham: Palgrave Macmillan, 2018)
Pettit, Clare, 'At Sea', in *Time Travelers: Victorian Encounters with Time and History*, ed. by Adelene Buckland and Sadiah Qureshi (Chicago: University of Chicago Press, 2020), 196–220
Podgorny, Irina, 'Fossil Dealers, The Practices of Comparative Anatomy, and British Diplomacy in Latin America, 1820–1840', *British Journal for the History of Science* 46 (2013), 647–74
Porges, Irwin, *Edgar Rice Burroughs: The Man Who Created Tarzan* (Provo, UT: Brigham Young University Press, 1975)
Price, Robert M., 'HPL and HPB: Lovecraft's Use of Theosophy', *Crypt of Cthulhu*, 5 (1982), 3–9
Rainger, Ronald, *An Agenda for Antiquity: Henry Fairfield Osborn & Vertebrate Paleontology at the American Museum of Natural History, 1890–1935* (Tuscaloosa: University of Alabama Press, 1991)
Ramaswamy, Sumathi, *The Lost Land of Lemuria: Fabulous Geographies, Catastrophic Histories* (Berkeley: University of California Press, 2004)
Rappaport, Rhoda, *When Geologists Were Historians 1665–1750* (Ithaca, NY: Cornell University Press, 1997)
Reid, Constance, *The Search for E. T. Bell, Also Known as John Taine* (Washington, D.C.: Mathematical Association of America, 1993)
Reid, Julia, 'Archaeology and Anthropology', in *The Routledge Research Companion to Nineteenth-Century British Literature and Science*, ed. by John Holmes and Sharon Ruston (Abingdon: Routledge, 2017), 357–71

Richmond, Jesse, 'Design and Dissent: Religion, Authority, and the Scientific Spirit of Robert Broom', *Isis*, 100 (2009), 485–504

Ridge, Martin, *Ignatius Donnelly: The Portrait of a Politician* (Chicago: University of Chicago Press, 1962)

Rieppel, Lukas, *Assembling the Dinosaur: Fossil Hunters, Tycoons, and the Making of a Spectacle* (Cambridge, MA: Harvard University Press, 2019)

Rinker, Harry L., ed., *Warman's Americana & Collectibles*, 5th edn (Radnor, PA: Wallace-Homestead Company, 1991)

Roggenkamp, Karen, *Narrating the News: New Journalism and Literary Genre in Late Nineteenth-Century American Newspapers and Fiction* (Kent, OH: Kent State University, 2005)

Roukema, Aren, 'The Esoteric Roots of Science Fiction: Edward Bulwer-Lytton, H. G. Wells, and the Occlusion of Magic', *Science Fiction Studies*, 48 (2021), 218–42

Rubin, Joan Shelley, *The Making of Middlebrow Culture* (Chapel Hill: University of North Carolina Press, 1992)

Rudbøg, Tim, and Erik Reenberg Sand, eds., *Imagining the East: The Early Theosophical Society* (Oxford: Oxford University Press, 2020)

Rudwick, Martin J. S., 'The Emergence of a Visual Language for Geological Science 1760–1840', *History of Science*, 14 (1976), 149–95

Rudwick, Martin J. S., 'Charles Darwin in London: The Integration of Public and Private Science', *Isis*, 73 (1982), 186–206

Rudwick, Martin J. S., *Scenes from Deep Time: Early Pictorial Representations of the Prehistoric World* (Chicago: University of Chicago Press, 1992)

Rudwick, Martin J. S., *Bursting the Limits of Time: The Reconstruction of Geohistory in the Age of Revolution* (Chicago: University of Chicago Press, 2005)

Rudwick, Martin J. S., *Worlds Before Adam: The Reconstruction of Geohistory in the Age of Reform* (Chicago: University of Chicago Press, 2008)

Rudwick, Martin J. S., *Earth's Deep History: How It Was Discovered and Why It Matters* (Chicago: University of Chicago Press, 2014)

Rupke, Nicolaas A., 'Down to Earth: Untangling the Secular from the Sacred in Late-Modern Geology', in *Science without God: Rethinking the History of Scientific Naturalism*, ed. by Peter Harrison and Jon H. Roberts (Oxford: Oxford University Press, 2019), 182–96

Ruse, Michael, *The Evolution-Creation Struggle* (Cambridge, MA: Harvard University Press, 2005)

Ryan, Robert M., *Charles Darwin and the Church of Wordsworth* (Oxford: Oxford University Press, 2016)

Sack, Susan Kassman, *America's Teilhard: Christ and Hope in the 1960s* (Washington, D.C.: The Catholic University of America Press, 2019)

Saler, Michael, *As If: Modern Enchantment and the Literary Prehistory of Virtual Reality* (Oxford: Oxford University Press, 2012)

Scafi, Alessandro, *Mapping Paradise: A History of Heaven on Earth* (Chicago: University of Chicago Press, 2006)

Schultz, Roger Daniel, 'All Things Made New: The Evolving Fundamentalism of Harry Rimmer, 1890–1952', unpublished PhD thesis, University of Arkansas (1989)

Secord, James A., *Victorian Sensation: The Extraordinary Publication, Reception, and Secret Authorship of* Vestiges of the Natural History of Creation (Chicago: University of Chicago Press, 2000)

Secord, James A., 'Knowledge in Transit', *Isis*, 95 (2004), 654–72

Secord, James A., 'Monsters at the Crystal Palace', in *Models: The Third Dimension of Science*, ed. by Soraya de Chadarevian and Nick Hopwood (Stanford, CA: Stanford University Press, 2004), 138–69

Secord, James A., 'Science, Technology and Mathematics', in *The Cambridge History of the Book in Britain: Vol. VI 1830–1914*, ed. by David McKitterick (Cambridge: Cambridge University Press, 2009), 443–74

Secord, James A., *Visions of Science: Books and Readers at the Dawn of the Victorian Age* (Oxford: Oxford University Press, 2014)

Secord, James A., 'Global Geology and the Tectonics of Empire', in *Worlds of Natural History*, ed. by H. A. Curry, N. Jardine, J. A. Secord, and E. C. Spary (Cambridge: Cambridge University Press, 2018), 401–17

Seed, David, ed., 'Breaking the Bounds: The Rhetoric of Limits in the Works of Edgar Allan Poe, his Contemporaries and Adaptors', in *Anticipations: Essays on Early Science Fiction and Its Precursors* (Liverpool: Liverpool University Press, 1995), 75–97

Sellers, Ian, 'The Swedenborgian Church in England', in *Reinventing Christianity: Nineteenth-Century Contexts*, ed. by Linda Woodhead (Ashgate: Aldershot, 2001), 97–104

Sera-Shriar, Efram, *Psychic Investigators: Anthropology, Modern Spiritualism, and Credible Witnessing in the Late Victorian Age* (Pittsburgh, PA: Pittsburgh University Press, 2022)

Shapin, Steven, and Simon Schaffer, *Leviathan and the Air-Pump: Hobbes, Boyle, and the Experimental Life*, new edn (Princeton, NJ: Princeton University Press, 2011)

Sheets-Pyenson, Susan, *John William Dawson: Faith, Hope, and Science* (Montreal: McGill-Queen's University Press, 1996)

Shteir, Ann B., 'Botany in the Breakfast Room: Women and Early Nineteenth-Century British Plant Study', in *Uneasy Careers and Intimate Lives: Women in Science 1789–1979*, ed. by Pnina G. Abir-Am and Dorinda Outram (New Brunswick, NJ: Rutgers University Press, 1987), 31–43

Simpson, R. S., 'Kinns, Samuel (1826–1903)', in *Oxford Dictionary of National Biography*, online edn, September 2004 (Oxford: Oxford University Press, 2004), https://doi.org/10.1093/ref:odnb/34333

Sinnema, Peter W., '10 April 1818: John Cleves Symmes's "No. 1 Circular"', in *BRANCH: Britain, Representation and Nineteenth-Century History*, ed. by Dino Franco Felluga, Extension of *Romanticism and Victorianism on the Net* (2012)

Sinnema, Peter W., '"We Have Adventured To Make the Earth Hollow": Edmond Halley's Extravagant Hypothesis', *Perspectives in Science*, 22 (2014), 423–38

Sinnema, Peter W., 'Gender Trouble in the Hollow Earth: *Pantaletta*, *Mizora*, and the American Antifeminist Romance', *ESQ: A Journal of Nineteenth-Century American Literature and Culture*, 69 (2023), 201–34

Smajić, Srdjan, *Ghost-Seers, Detectives, and Spiritualists: Theories of Vision in Victorian Literature and Science* (Cambridge: Cambridge University Press, 2010)

Smail, Daniel Lord, *On Deep History and the Brain* (Berkeley: University of California Press, 2008)

Smith, Robert W., 'The "Great Plan of the Visible Universe": William Huggins, Evolutionary Naturalism and the Nature of the Nebulae', in *The Age of Scientific Naturalism: Tyndall and His Contemporaries*, ed. by Bernard Lightman and Michael S. Reidy (London: Pickering and Chatto, 2014), 113–36

Sommer, Marianne, 'Seriality in the Making: The Osborn-Knight Restorations of Evolutionary History', *History of Science*, 48 (2010), 461–82

Stiling, Rodney L., 'Scriptural Geology in America', in *Evangelicals and Science in Historical Perspective*, ed. by David N. Livingstone, D. G. Hart, and Mark A. Noll (Oxford: Oxford University Press, 1999), 177–92

Stocking, George W., Jr, *After Tylor: British Social Anthropology 1888–1951* (London: Athlone Press, 1996)

Strang, Cameron B., 'Measuring Souls: Psychometry, Female Instruments, and Subjective Science, 1840–1910', *History of Science*, 58 (2020), 76–100

Taliaferro, John, *Tarzan Forever: The Life of Edgar Rice Burroughs, Creator of Tarzan* (New York: Scribner, 2002)

Tattersdill, Will, *Science, Fiction, and the Fin-de-Siècle Periodical Press* (Cambridge: Cambridge University Press, 2016)

Taves, Ann, *Fits, Trances, and Visions: Experiencing Religion and Explaining Experience from Wesley to James* (Princeton, NJ: Princeton University Press, 1999)

Taves, Ann, 'Visions', in *Ellen Harmon White: American Prophet*, ed. by Terrie Dopp Aamodt, Gary Land, and Ronald L. Numbers (Oxford: Oxford University Press, 2014), 30–48

Tebbel, John, *A History of Book Publishing in the United States: Volume II: The Expansion of an Industry, 1865–1919* (New York: R. R. Bowker, 1972)

Tebbel, John, *A History of Book Publishing in the United States: Volume III: The Golden Age between Two World Wars 1920–1940* (New York: R. R. Bowker, 1978)

Tebbel, John, *Between Covers: The Rise and Transformation of Book Publishing in America* (New York: Oxford University Press, 1987)

Thurs, Daniel Patrick, *Science Talk: Changing Notions of Science in American Popular Culture* (New Brunswick, NJ: Rutgers University Press, 2007)

Tillett, Gregory, *The Elder Brother: A Biography of Charles Webster Leadbeater* (London: Routledge and Kegan Paul, 1982)

Tison, Richard Perry, II, 'Lords of Creation: American Scriptural Geology and the Lord Brothers' Assault on "Intellectual Atheism"', unpublished PhD Thesis, University of Oklahoma (2008)

Topham, Jonathan R., 'John Limbird, Thomas Byerley, and the Production of Cheap Periodicals in the 1820s', *Book History*, 8 (2005), 75–106

Topham, Jonathan R., 'Rethinking the History of Science Popularization/Popular Science', in *Popularizing Science and Technology in the European Periphery 1800–2000*, ed. by Faidra Papanelopoulou et al. (Basingstoke: Ashgate, 2009), 1–20

Topham, Jonathan R., *Reading the Book of Nature: How Eight Best Sellers Reconnected Christianity and the Sciences on the Eve of the Victorian Age* (Chicago: University of Chicago Press, 2022)

Tucker, Herbert F., *Epic: Britain's Heroic Muse 1790–1910* (Oxford: Oxford University Press, 2008)

Ungureanu, James C., *Science, Religion, and the Protestant Tradition: Retracing the Origins of Conflict* (Pittsburgh, PA: University of Pittsburgh Press, 2019)

Vaninskaya, Anna, 'The Late Victorian Romance Revival: A Generic Excursus', *English Literature in Transition, 1880–1920*, 51 (2008), 57–79

Vieira, Fátima, 'The Concept of Utopia', in *The Cambridge Companion to Utopian Literature*, ed. by Gregory Claeys (Cambridge: Cambridge University Press, 2010), 3–27

Weinberg, Carl R., '"Ye Shall Know Them by Their Fruits": Evolution, Eschatology, and the Anticommunist Politics of George McCready Price', *Church History*, 83 (2014), 684–722

Weinbrot, Howard D., *Menippean Satire Reconsidered: From Antiquity to the Eighteenth Century* (Baltimore, MD: Johns Hopkins University Press, 2005)

Williams, Nathaniel, *Gears and God: Technocratic Fiction, Faith, and Empire in Mark Twain's America* (Tuscaloosa: University of Alabama Press, 2018)

Williamson, Jamie, *The Evolution of Modern Fantasy: From Antiquarianism to the Ballantine Adult Fantasy Series* (Basingstoke: Palgrave Macmillan, 2015)

Wise, Kurt P., 'Contributions to Creationism by George McCready Price', *Proceedings of the International Conference on Creationism*, 8 (2018), 683–94

Yeo, Richard, 'Science and Intellectual Authority in Mid-Nineteenth Century Britain: Robert Chambers and *Vestiges of the Natural History of Creation*', *Victorian Studies*, 28 (1984), 5–31

Yeo, Richard, 'An Idol of the Market-Place: Baconianism in Nineteenth-Century Britain', *History of Science*, 23 (1985), 251–98

Yost, Michelle Kathryn, 'American Hollow Earth Narratives from the 1820s to 1920', unpublished PhD thesis, University of Liverpool (2014)

Zimmerman, Virginia, *Excavating Victorians* (Albany: State University of New York Press, 2008)

Zirkle, Conway, 'The Theory of Concentric Spheres: Edmond Halley, Cotton Mather, & John Cleves Symmes', *Isis*, 37 (1947), 155–59

Index

For the benefit of digital users, indexed terms that span two pages (e.g., 52–53) may, on occasion, appear on only one of those pages.

Acland, Henry, 51–52, 56–57
Agassiz, Louis, 35, 67–68, 76
akashic records, 89–90, 93, 95, 99–102, 169–171.
   see also Theosophical Society
American antiquity, 130, 151, 167–168, 180
American Museum of Natural History (New York), 10, 22–23, 207, 215–216
Ashley-Cooper, Anthony, 7th Earl of Shaftesbury, 50, 52, 54
Atlantis, the lost continent of, 1, 146–147, 182, 231–232, 235–236. *see also* Donnelly, Ignatius; Scott-Elliot, William; Spence, Lewis
   natural history, 64, 67–68
   object of elite scientific interest, 147, 150, 167
   object of textual research, 147–148, 167–168
   syncretic interpretations, 16, 68–69, 89–90, 94–95, 130, 235–236

Baconian induction. *see* inductive presentation in science
Bell, E. T. (John Taine), 98–99
   Before the Dawn, 99–100
Bernard, Raymond (occultist), 235–236
Besant, Annie, 88–89, 231, 233–234
   Man: Whence, How and Whither, 95
   The Pedigree of Man, 68
Blavatsky, Helena Petrovna, 88–89, 91, 166–167
   Isis Unveiled, 91–92
   The Secret Doctrine, 90–91
Bonney, Thomas George, 50, 138
borderline palaeoscience
   critical role of literary language, 4, 17–21. *see also* literary technologies
   defined, 12–17, 227–228
   inspiring authors of fiction, 21, 99–100, 101–104, 137–138, 181, 182–183, 231–232, 234
   later incarnations, 228–239
   as logical, 157, 178, 183–184, 199–200, 201–202, 213–215
   nondisciplinary nature, 29–30, 147–148, 150–151, 157

   participatory attitudes, 104, 112–115, 118, 130–131, 146, 171–172, 184, 202, 205–207
   proponents' limited access to resources, 18, 66
   unsettling conventional genres, 21, 109–110
   vision, role of, 21–26, 145, 169, 172–173, 182, 203, 213–214, 228–229. *see also* day-age theory; clairvoyance; psychometry; White, Ellen G.
borderline palaeoscience, publishing
   in Christian media, 42, 46, 58–59, 197–199, 205, 207–208
   in elite periodicals, 56–58
   in occult media, 75–77, 93, 167–168, 233–234
   in specialist scientific journals, 159–162
   with prestigious publishers, 155–156, 158, 165, 175–177
   pulp fiction, 231–232, 235–237
borderline palaeoscience, reading
   active reading encouraged, 60, 112, 135, 207
   among Lovecraft circle, 101, 179–183
   reviews in Christian periodicals, 41, 49–50, 62
   reviews in generalist periodicals, 15, 47–48, 120, 124–125, 127–128, 136–137, 152–156, 158, 165, 171–172, 177–179
   reviews in specialist periodicals, 120–124, 136–137, 152–154, 165, 171–172, 177, 198–199
   sympathetic but noncommittal attitude, 15, 154–155, 158, 165, 171–172, 177–178
   unread books, 163, 238
borderline palaeoscience, textual research in, 9, 167–168. *see also* human testimony of prehistory
   as encyclopaedic, 165–166
   as inductive method, 162–164
   insufficient to gain expertise, 147–148
   legitimacy defended, 158–159, 163–164, 165–166, 169–171, 174–175, 183–184
Bradshaw, William Richard, *The Goddess of Atvatabar*, 129
British Association for the Advancement of Science, the, 35–36, 46, 106, 107–108, 165–166

Buchanan, Joseph Rodes, 73–74, 77–78, 131
Buckland, Adelene, 18–19, 189
Buckland, William, 67–68, 90–91, 216–217
　Geology and Mineralogy, Considered with Reference to Natural Theology, 209
Buffon, Georges-Louis Leclerc, Comte de, 8–9, 13, 23
Bulwer-Lytton, Edward, 78, 116–117
Burroughs, Edgar Rice, 126, 137–144, 231
　and Atlantis, 146–147
　Pellucidar series, 138, 139–140, 141–145, 231

Carruthers, William, 46–47, 50–55
catastrophists, 156, 159–160, 174. *see also* Churchward, James; Flood, the; Howorth, Henry Hoyle; Price, George McCready
Cater, Edwin (pastor), 191, 209
Chambers, Robert, *Vestiges of the Natural History of Creation*, 9–10, 15, 71–73
Chipman, De Witt C., *Beyond the Verge*, 130–131, 235
Christian fundamentalism, 11, 14, 188–190, 197, 223–224, 231
Churchward, James, 174–180, 181–182
　The Lost Continent of Mu, 174–178
clairvoyance, 64–66, 69–71, 104, 149. *see also* Denton family; psychometry; Spiritualism
　and E. T. Bell, 98–100
　and H. P. Lovecraft, 100–104
　influenced by general reading, 78, 232–233
　influenced by palaeoscientific texts, 67–68, 71–72, 78–80, 105
　influenced by palaeoscientific visual language, 72–73, 85–86, 96
　and Lewis Spence, 168, 169–171, 172–173
　mocked by young-earth creationists, 192, 213–214
　overlap with day-age theory, 40–41, 68–69
　role of gender, 75, 92–93, 104
　Theosophical versions, 88–96
Collingwood, Cuthbert, 35–36, 48–49, 56
　The Bible and the Age, 40–41
　New Studies in Christian Theology, 39–40
　A Vision of Creation, 35–40
common sense in science, 115–116, 156, 164, 188–190, 193, 196–197, 201–202
continents, movements of, 6, 150, 169–171, 230
Culmer, Frederick, *The Inner World*, 119, 129–130
Cuvier, Georges, 57, 66–68, 90–91, 112, 201, 237–238

Dana, James Dwight, 31–32, 55–56, 58
　Manual of Geology, 34–35, 59

Darwin, Charles, 6–7, 10–11, 23, 53, 81, 96, 117, 214
Davis, Andrew Jackson, *The Principles of Nature*, 71–72
Dawkins, William Boyd, 152–154, 160–162, 165–166
Dawson, John William, 4–5, 12, 19, 31–32, 34–35, 39–40, 41–46, 54–55, 59, 81, 90, 189–190
　The Meeting-Place of Geology and History, 59
　Nature and the Bible, 44
　The Origin of the World, According to Revelation, 42–44
　The Story of the Earth and Man, 42–44
Day, Langston, and George de la Warr, *New Worlds beyond the Atom*, 232–233
day-age theory, 30–35. *see also* Collingwood, Cuthbert; Dawson, John William; Gladstone, W. E.; Holbrook, David L.; Kinns, Samuel; Miller, Hugh
　debated in the public sphere, 29–30, 55–59
　divisive among Protestants, 51–55, 59, 63, 190
　literary in nature, 29–30, 33, 36–39, 42–44, 47–50, 62–63
　mixed metaphors, 58–62
　overlap with occultism, 40–41, 68–69
　relationship with scenes from deep time, 27, 38–39, 44–46, 48–49
　visionary in nature, 30, 32–33, 36–39, 44, 48–49, 60–62, 68–69, 191
deep time, 8–9, 12–14, 66–67, 89–90, 104, 194, 209
deluges. *see* Flood, the
Denton family, 74–88
　Denton, Elizabeth M. Foote, 2, 75–77, 78–81, 86–88, 92–93, 100
　Denton, Sherman F., 81–89
　Denton, William, 24, 65–66, 74–76, 77–78, 80–81, 85–89
　Our Planet, Its Past and Future, 80
　The Soul of Things, 75–86
Donnelly, Ignatius, 148–149, 150–151
　Atlantis, 150–155, 169, 173–174, 179–182
　The Great Cryptogram, 151–152
　Ragnarok, 155–156
Doyle, Arthur Conan, 125–126, 181–182
　and Atlantis, 146
　and Sherlock Holmes, 125–126, 201
　The Lost World, 10–11, 125–128
Driver, Samuel Rolles, 58

Eclectic medicine, 73, 131
Eden, the Garden of, 34–35, 216–217. *see also* Warren, William Fairfield

Emerson, Willis George, *The Smoky God*, 127–128, 166–167, 236–237
Epic of Gilgamesh, 13, 150
epic poetry. *see* Collingwood, Cuthbert
evolutionary epic, 2, 71–73, 214
evolutionary theory
    contested mechanisms of, 6–7, 229–230
    occult interpretations of, 89–90, 91–92, 96
    opposition to, 27, 35–36, 47–48, 53, 174, 188–190, 195–198, 205–207, 215–216, 220, 222
extinct animals, 90–92. *see also* mammoths
    destroyed by deluge, 146, 160, 193, 195–196, 200, 216–217
    examined clairvoyantly, 64, 71–73, 78–79, 80–86, 93–95, 100, 102, 232–234
    in Genesis creation story, 27, 36–37, 44, 48–49, 57, 216–217
    inside hollow earth, 124, 125–126, 137–138, 145, 235–236

Falconer, Hugh, 34–35, 91–92
Figuier, Louis, *The World before the Deluge*, 38–39, 48–49, 51, 85–86
Flood, the, 146–148, 224. *see also* Howorth, Henry Hoyle; Price, George McCready; young-earth creationism
    object of elite scientific interest, 13, 150, 207–208
    photographic evidence, 203, 207–208
    as true identity of ice ages, 159–160, 174, 195–196

gap theory creationism, 31, 190, 205
Gardner, Marshall B., *A Journey to the Earth's Interior*, 125–126, 231
Geikie, Archibald, 79–80, 227–228
geohistory
    as contingent, 12–13, 110–111
    establishment of conventional narrative, 1–2, 5–8, 23–24, 189–190, 215–216
    insufficiently verifiable, 21–23, 185–186, 213–214, 227–228, 237–238
    resembling a plot, 18–19, 189–190
    role of imagination in reconstructing, 66–67, 105, 107–108, 189
Geological Society of London, the, 8–9, 46–47, 52, 189
geology, visual language of, 22–23, 174–175, 186–187, 208–213
Georg, Eugen, *The Adventure of Mankind*, 15
geotheory, 8–9, 10–13. *see also* hollow-earth theory

Genesis, the Book of. *see* day-age theory; young-earth creationism
Gladstone, W. E., 27–29, 39–40, 56–59, 154, 223–224
Gordin, Michael D., 17–18
Gosse, Philip Henry, 93–94, 201–202
Gould, Charles, *Mythical Monsters*, 1–3
Grover, John William, *Conversations with Little Geologists on the Six Days of Creation*, 27–29, 38, 209
Guyot, Arnold, 31–32, 44–46

Haeckel, Ernst, 7–8, 94
Haggard, H. Rider, 126–127, 140
Hawkins, Benjamin Waterhouse, 44–46, 59–60, 76–77, 85–86, 98–99
higher criticism, biblical, 31–32, 33–34, 36
Hitchcock, Charles Henry, 76–78
Hitchcock, Edward, 69–71, 76, 191
    The Religion of Geology, 69–71
Hodson, Geoffrey (clairvoyant), 233–234
Holbrook, David L., *The Panorama of Creation*, 59–62
hollow-earth theory, 110–117, 119. *see also* Burroughs, Edgar Rice; Symmes, John Cleves Jr
    authorship, 115–116, 117–118, 119–120, 130–131, 133–135, 139–141
    blurring fact and fiction, 108–110, 124–128, 130–131, 134–135, 138–141
    and colonial exploration, 115, 119–120, 124–126, 129, 137–138, 140–141
    participatory nature, 112–115, 118, 130–131, 134–136, 145, 236
    postwar revival, 231–232, 235–237
    and religious belief, 128–137, 141–144
    tonal ambiguity, 110, 115–116, 124–125, 127–128, 133–134, 137, 144
Howorth, Henry Hoyle, 12, 14, 159–166, 196–197, 207–208, 238
    The Glacial Nightmare, 163–164, 196–197
    Ice or Water, 163, 238
    The Mammoth and the Flood, 162–164
human testimony of prehistory
    as legitimate source of knowledge, 1, 3–4, 13–16, 18, 90–92, 147–148, 158–159, 169–171, 174–175
    subjected to literary analysis, 29–30, 58, 60–62, 168
Hunt, Robert, *Panthea*, 68–69
Huxley, T. H., 9–10, 27–29, 33–34, 56–58, 160, 202, 223–224

inductive presentation in science, 151–152, 157, 162–164, 188–190, 193, 196–197, 201–202, 223

Ives, Franklin Titus, *The Hollow Earth*, 119

Jinarajadasa, Curuppumullage, 24, 88–89, 95–96
  First Principles of Theosophy, 96

Kinns, Samuel, 46–55, 56–57
  Moses and Geology, 47–51, 54–55
Knapp, Georg Christian, 32, 36, 38
Kurtz, Johann Heinrich, 32, 36

Lane, Mary E. Bradley, *Mizora*, 126–127, 128–129
Lapworth, Charles, 106–108
Leadbeater, Charles Webster, 88–89, 93–95, 102, 231, 233–235
  Man: Whence, How and Whither, 95
  'A Test of Courage', 96–98
Lemuria, the lost continent of, 6, 7–8, 89–90, 93–95, 101–102, 169–171, 182, 234–236
Leslie, J. Ben, and Carrie C. Van Duzee, *Submerged Atlantis Restored*, 64
Lethbridge, T. C., *Ghost and Ghoul*, 232–233
literary technologies, 22–23, 29–30, 50–51, 62–63, 64–65, 92–93, 137, 158–159, 214, 215–216, 238–239
  criticised by young-earth creationists, 185–186, 208–213
Lloyd, John Uri, 21, 131, 134–136
  Etidorhpa, 131–137, 236–237
lost world romance, 124–127, 130–131, 137–144, 231–232
Lovecraft, H. P., 21, 100–104, 120, 179–181
  The Call of Cthulhu, 101
  The Shadow Out of Time, 101–104
Lyell, Charles, 6–7, 18–19, 189–190, 193, 198–199, 237–238
  and George McCready Price, 185, 195–197, 198–199, 223–224
  imaginative power, 23, 66–67
  and John William Dawson, 31–32, 34–35, 43–44

Macaulay, James (*Leisure Hour* editor), 42, 54
mammoths, 25, 124–125, 135–136, 146, 160, 162–163, 175, 236
Mantell, Gideon, *The Wonders of Geology*, 190–192, 223
Mahatma Letters, the, 88–89, 174–175
McBride, James, *Symmes's Theory of Concentric Spheres*, 111–112, 115, 117–118

Menippean satire, 110, 222
Miller, Hugh, 9–10
  influence on successors, 36–37, 43–44, 49–50, 73, 76, 78–80
  The Old Red Sandstone, 74–75, 216–217
  The Testimony of the Rocks, 32–33
Milton, John, *Paradise Lost*, 33, 37, 56, 90–91
Mu, the lost continent of, 174, 182–183, 234. see also Churchward, James
Mulhall, Marion McMurrough, *Beginnings, or Glimpses of Vanished Civilizations*, 16, 18
mythology as source of facts, 1, 13–14, 156–157, 168

Naden, Constance, 55–56
Naturphilosophie. see Romantic conception of nature
Natural History Museum, London, 10, 46–47, 51–53, 62, 197–198. see also Carruthers, William
natural history galleries of the British Museum, 32–33, 47–48
Nelson, Bryon C., 207–208
  'After Its Kind', 207–209
  The Deluge Story in Stone, 207–208

O'Connor, Ralph, 18–20, 23–24, 37, 186–187, 190–191
Olcott, Henry Steel, 88–89
Osborn, Henry Fairfield, 7–8, 10–11, 22–23, 168, 215–216
Owen, Richard, 46–47, 56–57, 67–68, 71–72, 91–92

Palaeoart. see scenes from deep time
palaeobotany, 34–35, 42–44, 53
palaeoscience, demarcation of legitimate, 10, 18, 33–34, 41–42, 178–179, 181, 197, 229. see also borderline palaeoscience
  commercial needs overriding demarcation, 155–156, 160–162
  democratic use of sources criticised, 152–154, 158, 171–172, 177
  forensic methods criticised, 152–154, 155–156, 158, 165, 180, 183–184, 230–231
  hierarchies of expertise, 8–10, 198–199, 202
  identity of practitioner deemed suspect, 18, 158–159, 237–238
  literary analysis as way of understanding, 17–21
palaeoscientific research
  and imperialism, 5–6, 23, 115, 129
  and judicial metaphors, 57–58, 60–62, 154, 198–200

and Orientalism, 68, 79–80, 90–92, 105, 192
  nondisciplinary aspects of, 10–11, 230
  participatory aspects of, 11, 74–75
  writing, role of, 18–21
Palmer, Raymond, 235–237
Pavgee, Narayan Bhavanrao, *The Vedic Fathers of Geology*, 15–16, 18
Payne Smith, Richard (Dean of Canterbury), 50–52
Pellucidar. *see* Burroughs, Edgar Rice
Plummer, George Winslow (Rosicrucian), 67–69
popularization of science, 16, 19, 42, 98–99, 190–191, 215–216, 220
Powell, Baden, 31–32
prehistoric humans, 1, 7–8, 207
  Cro-Magnons, 167–168—169, 181–182, 215
  more advanced than scientists believed, 102, 151–152, 174, 179
  Theosophical interpretation of, 89–90, 93–95, 233–234
Price, George McCready, 4–5, 17, 24, 185–188, 194–195, 229–231, 237–238. *see also* Seventh-day Adventist Church
  criminological language, 198–202
  literary background, 194–195, 217
  use of imagination, 217–220, 222–225
  Back to the Bible, 217–220
  The Geological-Ages Hoax, 197
  God's Two Books, 196–197, 200, 216–217
  A History of Some Scientific Blunders, 214
  Illogical Geology, 196–197, 198–199
  Modern Discoveries Which Help Us To Believe, 201–202
  The New Geology, 20–21, 185, 197–200, 203–205, 207–209, 214–216
  Poems of My Long Ago, 220–221, 225–226
  Some Scientific Stories and Allegories, 220, 222–225
Pritchard, Charles (clergyman and astronomer), 51–52
professionalization of science, resistance to, 11, 165–166
pseudoscience, 17–18. *see also* palaeoscience, demarcation of legitimate
psychometry, 73–88, 99–100, 232–234. *see also* Buchanan, Joseph Rodes; Denton family
public sphere, scientific knowledge in the, 9, 15–17, 29–30, 50, 55–59, 112, 144, 158–159, 163–164, 183–184

racism and racial hierarchies, 93–94, 115–116, 168–169, 174, 179
Reed, William, *The Phantom of the Poles*, 119, 124
religious belief, relationship with science, 9–11, 14, 20–21, 181–182
Robinson, John Talbot (palaeontologist), 233–234
Romantic conception of nature, 68, 75–76, 105
Rimmer, Harry, 205
  Monkeyshines, 205–207
Riou, Édouard, 38–39, 48–49, 85–86
Rose, William, *An Explanation of the Author's Opinions on Geology*, 192
Rudwick, Martin J. S., 5–6, 8–9, 18, 22–24, 38, 44–46, 66–67, 186–187, 210–213

Saler, Michael, 21, 100–101, 109, 128, 133
'Seaborn, Adam', *Symzonia*, 115–116
scenes from deep time, 22–23, 27, 38–39, 44–46, 59–60, 72–73, 81–86, 175
Seventh-day Adventist Church, 193–195, 197–200, 209, 220–221, 227–228. *see also* Price, George McCready; White, Ellen G.
Scott-Elliot, William, 93–95, 101–102, 146, 234–235
science fiction, 96–104, 222–223, 231–232, 235
scientific language, technical, 10–11, 17–20, 105, 163–164, 237–238
scientific naturalism, 9–10, 34–35, 147–148, 157–158, 159–160, 195–196
  open-ended naturalism, 14, 71, 195–196
Sclater, Philip, 6–8
Shaver, Richard S., 235–236
Sherman, M. L., *The Hollow Globe*, 116–118
Sinnett, Alfred Percy, *Esoteric Buddhism*, 88–89
Sinnett, Patience, 93–94
Smith, Clark Ashton, 100–101, 179–183, 231–232
  'An Offering to the Moon', 182–183
Spence, Lewis, 167–168, 177, 179–180
  Encyclopædia of Occultism, 168, 172–173
  The Problem of Atlantis, 168–169, 171–173
  The Problem of Lemuria, 169–171, 172–173
Spencer, Herbert, 96, 189, 196–197
Spiritualism, 12, 64, 71–73, 74–75, 146, 181–182
Stopes, Marie, *Love's Creation*, 62–63
Suess, Eduard, *The Face of the Earth*, 6, 13, 23, 68, 150
suppositional synthesis, 148–149, 166–167, 178–180, 183–184, 198, 230–231. *see also* Donnelly, Ignatius; Howorth, Henry Hoyle; Spence, Lewis; Warren, William Fairfield
Swedenborgianism, 39–41
Symmes, Americus Vespucius, 117
  The Symmes Theory of Concentric Spheres, 117–118
Symmes, Ida Elmore, 118–119

Symmes, John Cleves Jr, 106, 111–119, 157

Teilhard de Chardin, Pierre, 228–230
Tennyson, Alfred, Lord, *In Memoriam*, 35–37
Theosophical Society, the, 4–5, 11, 14, 15–16, 88–96, 181–182, 231, 235–236. *see also* akashic records; Besant, Annie; Blavatsky, Helena Petrovna; Jinarajadasa, Curuppumullage; Leadbeater, Charles Webster
   and George McCready Price, 213–214
   and James Churchward, 174–175
   and hollow-earth fiction, 129–130
   and Lewis Spence, 168–171
   literary style of members, 92–95, 101–104
   museum of the Adept Brotherhood, 95–96, 102
   periodicals, 91, 93, 95, 233–234
   Root Races, 89–90, 93–95, 101, 234
Tower, Washington L., *Interior World*, 124–125
Tuttle, Hudson, *The Arcana of Nature*, 72–73

uniformitarianism, 6–7, 160, 162–164, 195–197, 200, 202, 222
utopian romance, 109, 115–116, 126–127

Verne, Jules, 155–156
   Journey to the Centre of the Earth, 49–50, 116–117, 125
   Twenty Thousand Leagues under the Seas, 152

Vestiges of the Natural History of Creation, 9–10, 15, 71–73
Victoria Institute, the, 46, 189–190
Voysey, Charles (heretical theist), 54

Wallace, Alfred Russel, 12, 39–40, 104
Warren, William Fairfield, *Paradise Found*, 156–159
Wegener, Alfred, 6, 230
weird fiction, 101–104, 179–183. *see also* Lovecraft, H. P.; Smith, Clark Ashton
Wells, H. G., 95, 197–198, 222–223
Woodward, Henry, 51, 160–162
Whitcomb, John C., and Henry M. Morris, *The Genesis Flood*, 237–238
White, Ellen G., 21–22, 193–196, 220–221, 222–223
   Spiritual Gifts, 193
Winchell, Alexander, 154, 158

young-earth creationism, 185–188, 190–192. *see also* Price, George McCready; Seventh-day Adventist Church; White, Ellen G.
   and detective fiction, 201–202
   face-value interpretations of geological phenomena, 200–202, 208, 223
   and Henry Hoyle Howorth, 164, 196–198, 238
   opposition to narrative and imagination, 190–192, 194, 199–201, 207, 209–214, 223
   photographic aesthetic, 187–188, 203–213
   recent incarnations, 225, 237–238